To,

The Quantum Revolution in Philosophy

The Quantum Revolution in Philosophy

Richard Healey

UNIVERSITY PRESS

Great Clarendon Street, Oxford, OX2 6DP,
United Kingdom

Oxford University Press is a department of the University of Oxford.
It furthers the University's objective of excellence in research, scholarship,
and education by publishing worldwide. Oxford is a registered trade mark of
Oxford University Press in the UK and in certain other countries

© Richard Healey 2017

The moral rights of the author have been asserted

First Edition published in 2017

Impression: 1

All rights reserved. No part of this publication may be reproduced, stored in
a retrieval system, or transmitted, in any form or by any means, without the
prior permission in writing of Oxford University Press, or as expressly permitted
by law, by licence or under terms agreed with the appropriate reprographics
rights organization. Enquiries concerning reproduction outside the scope of the
above should be sent to the Rights Department, Oxford University Press, at the
address above

You must not circulate this work in any other form
and you must impose this same condition on any acquirer

Published in the United States of America by Oxford University Press
198 Madison Avenue, New York, NY 10016, United States of America

British Library Cataloguing in Publication Data

Data available

Library of Congress Control Number: 2017931027

ISBN 978–0–19–871405–7

Printed and bound by
CPI Group (UK) Ltd, Croydon, CR0 4YY

Links to third party websites are provided by Oxford in good faith and
for information only. Oxford disclaims any responsibility for the materials
contained in any third party website referenced in this work.

Contents

Preface — ix
Acknowledgments — xiii
List of Figures — xv
List of Boxes — xvii

1. Overview: A New Kind of Science — 1

Part I. Quantum Theory

2. Superposition — 15
   2.1 Beams — 15
   2.2 Some Interference Phenomena — 18
   2.3 Polarization — 21
   2.4 More Interference Phenomena — 24
   2.5 Modeling Interference — 28
   2.6 Measurement — 34
   2.7 What's Really Happening? — 36
   2.8 Summary — 38

3. Entanglement — 40
   3.1 What is Entanglement? — 40
   3.2 Entanglement and Incompleteness — 45
   3.3 Entanglement Exhibited — 47

4. "Non-locality" — 53
   4.1 Spooky Action at a Distance? — 53
   4.2 Mermin's Demonstration — 56
   4.3 A Simplified Bell Argument — 59
   4.4 Two Views of a Quantum State — 65
   4.5 Quantum States Offer Prescriptions — 68
   4.6 Exorcising the Spook — 72

5. Assigning Values and States — 75
   5.1 Making Quantum State Assignments — 75
   5.2 Measurement-free Quantum Probabilities — 79
   5.3 The Grounds of Quantum State Assignments — 86

6. Measurement — 90
   6.1 The Measurement Problem — 90
   6.2 The EPR Argument Revisited — 100

7. Interlude: Some Alternative Interpretations — 102
　7.1 Bohmian Mechanics — 103
　7.2 Non-linear Theories — 109
　7.3 Many-outcomes Theories — 110

Part II. Philosophical Revelations

8. Theories, Models, and Representation — 121
　8.1 Representation and Theories — 121
　8.2 Applying Quantum Models — 126
　8.3 Truth, Content, and Objectivity — 132

9. Probability and Explanation — 138
　9.1 Probability — 143
　9.2 Explanation — 148

10. Causation and Locality — 165
　10.1 Introduction — 165
　10.2 Relativistic Spatiotemporal Relations — 166
　10.3 Relativistic Chance — 168
　10.4 Chances from Quantum Probabilities — 170
　10.5 Chance, Causation, and Intervention — 172
　10.6 Locality and Local Causality — 179

11. Observation and Objectivity — 184
　11.1 Introduction — 184
　11.2 Objectivity — 186
　11.3 Wigner's Friend — 188
　11.4 Paradox Resolved — 190
　11.5 Objective Content Secured — 193
　11.6 Independent Verifiability — 196
　11.7 Conclusion: A Limit to Transcendental Objectivity — 198

12. Meaning — 202
　12.1 Introduction — 202
　12.2 Some Novel Quantum Concepts — 204
　12.3 The Content of Magnitude Claims — 210
　12.4 Some Conceptual Mutations — 218
　12.5 The Content of Denoting Terms — 223

13. Fundamentality — 228
　13.1 Philosophy and Fundamental Physics — 229
　13.2 Entities — 230
　13.3 Properties — 234
　13.4 Laws — 238
　13.5 How Quantum Theory is Fundamental — 242

14. Conclusion	246
Appendix A. Operators and the Born Rule	259
A.1 Vectors and Inner Products	259
A.2 Operators and the Simple Born Rule	261
A.3 More About Operators	265
A.4 Commutation, Compatibility, and the Compound Born Rule	266
A.5 Density Operators and Mixed States	266
A.6 Interacting Systems	267
Appendix B. Two Arguments Against Naive Realism	270
B.1 First Argument	270
B.2 Second Argument	272
Appendix C. A Simple Model of Decoherence	275
Bibliography	277
Index	285

Preface

This book is about the philosophical significance of the quantum revolution. How can we understand the world in which we find ourselves? What is the nature of reality? Stephen Hawking was right to say such questions are traditionally for philosophy. However, he went on:

but philosophy is dead. Philosophers have not kept up with modern developments in science. Particularly physics. [Hawking, 2011]

Such reports of the death of philosophy are greatly exaggerated. But I believe it is not currently enjoying the best of health. My diagnosis is not insufficient intake of (still speculative) cutting-edge physics but failure to digest the rich sustenance supplied by the quantum revolution of the past century. The symptoms are most pronounced in contemporary metaphysics, but they can be found even where philosophers of science examine the foundations of physics.

Philosophers need to face questions like "How does quantum theory help us to understand the world in which we find ourselves?" and "What does quantum theory tell us about the nature of reality?" Almost a century after Heisenberg's and Schrödinger's seminal advances there is still no consensus on how to answer these questions. Perhaps because of their resistance to experimental investigation or mathematical resolution many practicing physicists have dismissed them as "merely" philosophical.

Philosophers should regard such dismissal as a call to action. It is the philosopher's job to reformulate vaguely worded questions about matters of general intellectual importance so that they admit of clear and illuminating answers. This requires not ill-informed speculation but the close analysis, rigorous argumentation, and questioning of fundamentals that are the professional tools of the practicing philosopher.

These tools are freely available. Albert Einstein, Erwin Schrödinger, John Bell, and other physicists used them to great effect in improving our understanding of quantum theory—though mostly by showing clearly how it *cannot* be understood! A lively interdisciplinary community of physicists, mathematicians, and philosophers continues actively to develop a variety of approaches to quantum theory. Detailed analysis has now posed sharp questions that have been answered by mathematical theorems and experimental results. But, beside this progress, heated debate continues between people with radically different views on how to understand quantum theory and its relation to our world.

Despite their radical differences, I think each of many opposed views gets something importantly right. I cannot improve on an ancient but overworked metaphor.[1] Their

---

[1] In Rumi's version, from "Tales from the Masnavi".

proponents are like men who give conflicting accounts of an animal they all scrutinize in the dark. One says it is a throne, another a pillar, a third a fan, a fourth a waterspout. While none of them is wholly wrong, none of them is completely correct. For the animal is an elephant, whose back is like a throne, whose leg is like a pillar, whose ear is like a fan, and whose trunk is like a waterspout. Each has incorrectly generalized to the nature of the whole animal from how it appears from one perspective. Quantum theory is an elephant, whose complete understanding requires the integration of many partially correct perspectives. A philosopher must shed light on the whole beast.

But why should a philosopher make this effort when excuses are available? Many philosophers regard the subject as too technical and esoteric to be relevant to their concerns: "There are no elephants in my back yard". Some have been put off by what they consider the naive or confused views expressed on this subject even by great physicists: "I'll pay attention only when the experts come up with a clear account of the elephant". There are metaphysicians concerned with the most general categories of being who think these may be applied without change in a merely schematic understanding of quantum theory: "All I need to know is that it's some kind of animal".

Philosophers of science have no such excuse. Their subject covers the philosophical foundations of individual sciences, including physics. There are philosophers now who know a lot of physics and relevant mathematics and some have made significant technical contributions to these fields. But for many, quantum theory has become the elephant in the room—what everyone is thinking about but no one mentions.

I see two reasons for this. Over the years philosophers have tried to make sense of quantum theory in different ways but without success.² Stymied by lack of progress toward solving long-standing conceptual problems with measurement and non-locality, some have moved on to investigate new problems posed by quantum information theory and the more exotic mathematics of algebraic quantum field theory. Sharing John Bell's exasperation that we still lack an exact formulation even of non-relativistic 'particle' quantum mechanics, others have diverted attention to what they consider closely related but better-defined alternatives to that theory.

The elephant is still here: quantum theory remains fundamental to contemporary science. We need to remove the barriers to our understanding of it—to shed light on the elephant by opening the window shades. Here is where the philosopher's penchant for questioning fundamentals can help.

The main barrier to understanding quantum theory is not our inability to imagine the world it describes but the presumption that it be understood as describing the world we experience. Rather than itself representing reality, quantum theory should be understood as a source of reliable advice on how to represent and what to expect from the physical world. Quantum theory helps us explain physical phenomena, and not only the results of observation and experiment. But the phenomena do not arise as consequences of fundamental quantum laws, and while their explanation may be

---

² I include my own [1989] work in this category.

understood to involve causal relations these have some novel features (though no instantaneous action at a distance).

Understanding quantum theory both requires and motivates rethinking generally accepted accounts of representation, content, probability, causation, explanation, laws, and ontology. Philosophers have developed explicit accounts of these concepts, while physicists' views are usually only implicit in their practice. But to understand quantum theory a lot of common ground must be turned over and replanted with hardier varieties.

The failure of many brilliant minds to reach agreement on these issues over decades makes it seem naive to hope for consensus. Nevertheless, I remain optimistic. Physicists have given us this wonderful theory, and surely philosophers can help us understand it. By rising to the challenge they may restore their discipline to a more vibrant state of health.

I've been trying to understand quantum theory and its philosophical significance ever since first encountering it as an undergraduate studying both physics and philosophy. I have Patrick Sandars and David Brink to thank for supervising this encounter; Tony Leggett for deepening my knowledge and interest while I was a graduate student in theoretical physics at Sussex University and after; and Hilary Putnam and Abner Shimony for what they taught me as a philosophy Ph.D. student.

My present understanding has been developing since a sabbatical in 2009–10 when my research was supported by the National Science Foundation under Grant No. SES-0848022. That year I was lucky enough to visit the Institute for Quantum Optics and Quantum Information of the Austrian Academy of Sciences under the Templeton Research Fellows Program 'Philosophers–Physicists Cooperation Project on the Nature of Quantum Reality'. Special thanks go to my hosts in Vienna, Prof. A. Zeilinger and Prof. M. Aspelmeyer, who made this visit possible.

I wish to acknowledge the support of the Perimeter Institute for Theoretical Physics under their sabbatical program, and to thank its members (particularly Chris Fuchs) and visitors for the intellectual stimulation they provided during my visit in fall 2009. Research at the Perimeter Institute is supported by the Government of Canada through Industry Canada and by the Province of Ontario through the Ministry of Research and Innovation. I also wish to thank the University of Sydney and Huw Price, then Director of the Centre for the Study of Time, for inviting me to give a series of lectures early in 2010 in conjunction with the award of an International Visiting Research Fellowship, thereby providing the stimulus and opportunity to share my early ideas.

Grant No. 21112 from the John Templeton Foundation supported publication of several papers on whose contents I have drawn in this book: the opinions expressed in these works are those of the author and do not necessarily reflect the views of the John Templeton Foundation. I thank the Yetadel Foundation, and J. P. Jones, Dean of the School of Social and Behavioral Sciences at the University of Arizona, who between them provided the financial support and relief from teaching that gave me the time to write this book.

Thanks also to the valuable contributions by the students to seminars I taught on this material, two at the University of Arizona and one at Princeton University whom I thank for inviting me to visit in fall 2012 as Class of 1932 Fellow in Philosophy and the Humanities Council. Bernard d'Espagnat, David Glick, and especially Paul Teller all provided helpful written comments on earlier versions of the manuscript, as did two anonymous readers for Oxford University Press. As always, I gratefully acknowledge the stimulation provided by conversations with my colleagues in the Philosophy Department at the University of Arizona, and especially with Terry Horgan and Jenann Ismael.

My pleasure in thanking all these people and institutions is tinged with the disappointment I feel at the inadequacy of the ideas expressed in this book they have all made possible. Moreover I am aware that many colleagues and friends who disagree with its conclusions will doubtless share that disappointment. Abner, Bernard, and Hilary, you deserved better. You are sorely missed. I dedicate this book to your memory.

Acknowledgments

Part I makes use of examples and diagrams drawn from two introductory texts: *Quantum Physics: a First Encounter*, by V. Scarani (Oxford: Oxford University Press, 2006); and *Sneaking a Look at God's Cards*, by G.-C. Ghirardi (Princeton: Princeton University Press, 2005). The arguments of Appendix B closely follow Mermin [1993].

This book contains material I have adapted from several previous publications, including (especially in Chapters 8 and 9) two of my papers in *The British Journal for the Philosophy of Science*: "Quantum Theory: A Pragmatist Approach", Volume 63 (2012), 729–71; and "How Quantum Theory Helps us Explain", Volume 66 (2015), 1–43.

Chapter 11 contains material from "Observation and Quantum Objectivity", ©2013 by the Philosophy of Science Association.

I thank Springer Science+Business Media, LLC for permission to reprint passages from "Quantum Decoherence in a Pragmatist View: Dispelling Feynman's Mystery", *Foundations of Physics* 42 (2012), 1534–55; "Quantum States as Objective Informational Bridges", *Foundations of Physics*, DOI 10.1007/s10701-015-9949-7, published online 09 September 2015; and "A Pragmatist View of the Metaphysics of Entanglement", *Synthese*, published online 21 September 2016, DOI 10.007/s-11229-016-1204-2.

Chapter 12 contains material from my "Quantum Meaning", *The Harvard Review of Philosophy*, XX Spring 2014, 45–61.

I thank Cambridge University Press for permission to reprint material appearing in Chapter 10 from "Local Causality, Probability and Explanation", in *Quantum Non-locality and Reality: 50 years of Bell's Theorem*, edited by Mary Bell and Shan Gao (Cambridge: Cambridge University Press, 2016), 172–94.

List of Figures

2.1	Neutron interferometer	18
2.2	Beam splitter	19
2.3	Three beam splitters	19
2.4	Balanced Mach–Zehnder interferometer	19
2.5	Polarization analysis	22
2.6	Analysis with a third polarizer	23
2.7	Action of calcite crystal	24
2.8	Half-wave plate in balanced interferometer	25
2.9	Two-slit interference	26
2.10	Superposition of states	29
2.11	Representation of complex numbers	30
2.12	Calculating intensities	30
3.1	Franson interferometer, balanced	48
3.2	Unbalanced Franson interferometer	49
4.1	A possible experimental arrangement	56
6.1	Stern–Gerlach apparatus	95
6.2	After a second apparatus	95
10.1	The light cone at O	168
10.2	The chances of e	170
10.3	Alice's and Bob's space-time diagram	171
10.4	The chances of v_A	172
10.5	Local causality	180
10.6	A late explanatory chance	182
A.1	The vector space $V(E^3)$	259
A.2	Some subspaces of $V(E^3)$	260
A.3	The inner product in $V(E^3)$	261
A.4	Projection operators on $V(E^3)$	263
A.5	Probabilities for non-degenerate eigenvalues	264
B.1	Some related three-photon polarization observables	272

List of Boxes

3.1	Some Non-separable State Vectors	42
4.1	Proof of a Bell Inequality, and its Experimental Violation	59
8.1	Kepler's Laws Approached Syntactically and Semantically	124
9.1	Kepler's Laws Explained by Incorporation into Newtonian Models	149

1
Overview: A New Kind of Science

Physicists welcomed (some more enthusiastically than others) two revolutionary theories during the early part of the twentieth century: relativity and quantum theory. In the twenty-first century these theories continue to play a foundational role in physical thought as well as underlying much of current and likely future technology. That role had previously been played by the classical physics of Isaac Newton, James Clerk Maxwell, and essentially all other eighteenth- and nineteenth-century physicists. Relativity and quantum theory have so successfully usurped this foundational role that perhaps, by now, they deserve the title of "classical", and indeed some have come to apply that term to relativity—but never to quantum theory.

Why not? Almost everyone familiar with these theories—physicist and non-physicist—agrees that quantum theory breaks more radically than relativity with the physics of Newton, Maxwell, and their ilk. But this consensus dissolves if one asks what makes quantum theory such a revolutionary break with classical physics. Here are a few ways some prominent physicists have answered that question:[1]

There is no quantum world. There is only an abstract quantum description. It is wrong to think that the task of physics is to find out how nature is. Physics concerns what we can say about nature.   (Niels Bohr, quoted by Petersen [1963, p. 12])

When the province of physical theory was extended to encompass microscopic phenomena through the creation of quantum mechanics, the concept of consciousness came to the fore again. It was not possible to formulate the laws of quantum mechanics in a fully consistent way without reference to the consciousness.   (Eugene Wigner [1967, p. 172])

The doctrine that the world is made up of objects whose existence is independent of human consciousness turns out to be in conflict with quantum mechanics and with facts established by experiment.   (Bernard d'Espagnat [1979])

Although the next three quotes do not explicitly mention quantum theory, in each case the context makes it clear that was what prompted the remark:

Observations not only disturb what is to be measured, they produce it.
    (Pascual Jordan [1934] paraphrased from the German by Max Jammer [1974, p. 161])

---

[1] Heisenberg and Jordan each made major contributions as founders of modern quantum theory. As well as Heisenberg, Bohr and Wigner both won Nobel prizes for physics: Bohr was a central figure in attempts to interpret the theory's significance, while Wigner made major contributions to its applications in atomic physics and elsewhere: d'Espagnat and Mermin have more recently made profound studies of the conceptual foundations of quantum theory.

[T]he atoms or elementary particles themselves are not real; they form a world of potentialities or possibilities rather than one of things or facts.   (Werner Heisenberg [1958])

We now know that the moon is demonstrably not there when nobody looks.
(N. David Mermin [1981])

A common thread runs through all these quotes. The alleged consequences of quantum theory concern not just the weird behavior of the world, at least at a microscopic level with which we, in any case, have no familiarity in our daily lives. In each of the quotes, quantum theory is claimed to challenge some more basic assumption—about the world and our relation to it as observers, conscious beings, or scientists.

To be sure, popularizations of quantum theory tend to highlight the alleged weirdness of quantum behavior at the microscopic scale. They say such things as: a quantum particle can be in two places at once; some pairs of particles can instantaneously coordinate their behavior through a kind of "spooky" action no matter how far apart they are; "virtual" particles randomly but briefly pop in and out of existence in empty space; or the mass of every massive particle originates from the resistance it experiences while moving through the mud/glue/molasses of the Higgs field. I'll "deconstruct" some of these popular myths in the course of this book.

But such claims may be taken with the same pinch of metaphorical salt required to render palatable such strange claimed consequences of relativity as: moving objects contract along their direction of motion; moving clocks run slow (and time stops altogether for a clock moving at the speed of light); gravity is not a force but the curvature of space; and the universe began in a "big bang" which occurred both everywhere and nowhere but is now expanding at an accelerating rate even though it is likely infinite. Surely, the universe is queerer than we supposed—maybe even queerer than we can suppose (as J. B. S. Haldane famously surmised). But we can trust physicists to cash out the confusing metaphors that arise when trying to use familiar language to describe the weird, even unimaginable, behavior displayed by objects like photons, quarks, and the Higgs field through skillful use of mathematics to represent such behavior.

I believe such trust is well placed in the case of relativity. The layperson may have trouble getting his or her mind around the theory, but physicists have no trouble mathematically representing the relativistic behavior of objects, and (most) philosophers see no outstanding barriers to understanding this activity as a natural extension of the practice of physics since Galileo—the practice of using mathematical representation as an extremely effective way of conveying briefly but precisely the content of a description of what may happen in the physical world that cannot, for practical reasons, be given in any other way.

But many a physicist, deeply familiar with quantum theory, has delivered a considered judgment that's not easily reconciled with the view that quantum theory offers mathematical representations of what may happen in the physical world. These judgments may, of course, be incorrect, and they are far from unanimous. Moreover, the above quotes already suggest a variety of incompatible alternatives to this view. So far I have merely suggested that quantum theory makes a radical break with previous physics not because of the weirdness of the physical behavior it represents, but for some other reason. And I have raised the suspicion that the reason is that, unlike classical physics, quantum theory is not simply in the business of representing what happens in the physical world.

Physicists are not the only ones to base sweeping conclusions on their understanding of quantum theory. Often following up a physicist's tantalizing suggestion, one philosopher after another has argued that quantum theory implies some radical change in our view of the world or the conceptual scheme we use to make sense of it. Putnam (at one time [1968]) argued that quantum theory shows the world obeys a non-classical logic. Pitowsky [2006] and others have reasoned that probability in quantum theory conforms to principles different from the familiar laws that apply to dice and other games of chance. Many continue to believe that quantum theory's commitment to objective chance has freed us from the iron determinism of Newton's laws: Anscombe [1971] is one philosopher to take seriously the purported implication that it thereby makes room for genuine freedom in the physical world.

Quantum theory has often been taken to have radical implications for our view of the mind. Though regrettable, it is hardly surprising that some have taken remarks like those of Jordan and Mermin out of context as claims that quantum theory shows reality to be purely mental. Such idealism has even been seriously defended as the best response to a number of so-called paradoxes of quantum physics. The suggestion has often been made that while conscious mind is not everything it does have a special role to play in quantum theory. Wigner took this to be active, so that only mind is able to affect the behavior of physical objects in a peculiar way. Lockwood [1989] and others have allotted mind the merely passive role of experiencing what appear as multiple, contrary yet simultaneous, outcomes of a physical process. Such "many-minds" views thereby seek to reconcile our experience as of chancy quantum events with an underlying physical determinism.

The philosopher Reichenbach once said that the philosophy of physics is as technical and intricate as is physics itself.[2] Indeed, since his day, it has become harder and harder to distinguish books and papers on quantum theory by philosophers from those written by physicists or mathematicians. This is a good thing insofar as it manifests the ability and willingness of philosophers of physics to base their analyses on a thorough understanding of physics: moreover, it makes it easier for philosophers and physicists to work together or at least reach a mutual understanding. But it comes at the cost of an increased specialization that can distance philosophers of physics from the concerns and insights of their philosophical colleagues. This can encourage a narrowness of philosophical vision and prompt philosophers to focus their energies on

---

[2] "The philosophical significance of the theory of relativity", in [Schilpp, ed. 1949], p. 292.

just a few approaches—especially those that connect to some currently active (though not necessarily mainstream) research program in physics.

Approaches to what? A philosopher is likely to respond: "Approaches to the interpretation of quantum theory". Asked what that might amount to, she may well repeat a formula that owes much to her fellow philosopher, Bas van Fraassen [1991]: An interpretation of quantum theory is an answer to the question: "How could the world possibly be how quantum theory says it is?" I'll call someone trying to answer that question a quantum Interpreter.

Already we see a narrowing of options. Perhaps quantum theory is radical precisely because it does not say how the world is. This is an option at least worthy of serious consideration. It will receive such consideration in this book. After reading the book you may even decide this is the right option. Meanwhile, set it aside while we pursue the quest of the quantum Interpreter.

Embarking on the task of saying how the world could possibly be how quantum theory says it is, one naturally looks to quantum theory to supply descriptions or representations of what is going on in the world—the physical world, since quantum theory is a physical theory. Classical physics represents features of a system of things like particles of matter or electric fields by specifying its state at one time, and describes the behavior of these things by saying how this state evolves in various circumstances. Quantum theory also assigns states to systems and says how these evolve—according to the Schrödinger equation.[3] One naturally expects this to be how the theory represents the features of a system and describes the behavior of its constituents. Surprisingly, it turns out to be hard to square this expectation with results of experiments on such systems.[4]

This presents our quantum Interpreter with a dilemma, pithily expressed by the physicist John Stewart Bell [2004]: either the Schrödinger equation is not everything, or it is not right. What the Schrödinger equation cannot represent is a definite, unique outcome of a typical experiment on a quantum system, as recorded in the experimenter's notebook.

Grasping one horn of the dilemma, some quantum Interpreters have appealed to an additional representational aid such as the actual positions of particles making up a quantum system. These are not specified by the quantum state, and the Schrödinger equation does not describe how particles move. But it is reasonable to suppose that the outcome of an experiment is always eventually recorded in the positions of some collection of particles (making up a pointer on a dial, ink on paper, etc.). It has long been known that for a system of a fixed number of particles involved in an experiment there is a way consistently to supplement the Schrödinger equation by an equation

---

[3] I'll introduce this equation in the next chapter and state it in Chapter 8. It says how states evolve in ordinary quantum mechanics. Evolution equations in other forms of quantum theory all share its important mathematical properties.

[4] This is one way of putting the notorious quantum measurement problem that I'll discuss in Chapter 6.

representing the motions of the particles so that some of their positions evolve into a configuration capable of representing the experimental outcome.[5]

But there are reasons why few physicists take the resulting theory seriously, either as a way of understanding quantum theory or as a potential rival to it. Moreover, quantum theory is now widely and successfully applied to systems of various kinds that are not composed of a fixed number of particles. It remains an open question whether there is a similarly consistent way of supplementing the quantum state of all these systems. This commits the Interpreter who grasps this horn of the dilemma to an open-ended research program in physics which few physicists consider worthy of pursuit, especially if it promises merely to duplicate the predictions of unadorned quantum theory.

By grasping the other horn of the dilemma, the Interpreter becomes committed to a different research program, energetically pursued by a few physicists and mathematicians but which has yet to receive any experimental support. As explained in the next chapter, the Schrödinger equation has the key property of linearity. This is the property that prevents the evolving quantum state from representing a definite, unique outcome of a typical experiment. Perhaps this barrier could be overcome by some suitable non-linear variant or alternative evolution equation? Indeed physicists have investigated this possibility, and the properties of some non-linear equations have been studied.

If the Schrödinger equation is not right, then a theory incorporating some non-linear equation will imply that in some situations quantum theory makes incorrect predictions. It turns out to be very hard to compare the predictions of proposed alternatives to those of quantum theory, either because all the details of the rival theory have not been fully spelled out or because the theory's predictions differ so little in presently experimentally accessible circumstances. But this does not affect the main point: by pinning hopes on a rival theory, the Interpreter has moved the goalposts. We began by asking what is so radical about quantum theory. If some incompatible theory incorporating a non-linear evolution equation should turn out to be right where quantum theory is wrong that would not help answer this question, though it might diminish its importance in some eyes. The second horn of the dilemma provides an even clearer illustration of the power of the phrase "How could the world possibly be...?" to distract attention away from the primary task of understanding quantum theory and what makes *it* so radical.

There may be a way for an Interpreter of quantum theory to reject Bell's dilemma. The idea is to argue that the Schrödinger equation can correctly represent a definite outcome of a typical experiment on a quantum system as recorded in the experimenter's notebook, but that outcome is not in fact unique. This goes back to the proposal by Hugh Everett III [1957], now popularly known as the many worlds Interpretation. The proposal has been ingeniously defended recently by some philosophers,

---

[5] I'll give details of this de Broglie–Bohm pilot wave theory in the Interlude (Chapter 7).

and has long been a favorite of a minority of physicists, especially cosmologists. To the obvious objection that we never experience more than one outcome of a quantum measurement, a many worlds Interpreter responds that while there are indeed multiple outcomes (each uniquely observed by a different "version" of the experimenter), one outcome and the observer experiencing it is then confined to (what is represented by) one "branch" of the evolving quantum state, which subsequently evolves independently of all the other "branches". In this way quantum theory itself conspires to conceal from each observer any evidence of all the equally real outcomes but the one he has experienced.

The many worlds Interpretation has the distinct advantage of calling for no new physics. Controversy continues as to whether it can make sense of the quantum rule that assigns one outcome of an experiment a greater probability than another, given that every outcome always occurs. But even its devotees admit it sounds crazy. They have a response to the objection "What could be a more egregious flouting of the principle of Ockham's Razor than the proposal that indefinitely many worlds arise out of every quantum measurement?" Ockham (supposedly) enjoined us only not to multiply entities beyond necessity: here we are justified in multiplying worlds by dire need. This response will convince only someone who agrees the need to answer the question "How could the world possibly be how quantum theory says it is?" is sufficiently dire, and can find no other way to meet it.

There is a view of understanding that ties in nicely with the Interpreter's aim of answering that question. The view is that to understand a theory is to grasp its content, where the content of a theory is how things would have to be for it to be true—the set of possible worlds at which it is true, as some philosophers would have it. Suppose one thinks of a theory as supplying a set of mathematical models, each able to represent its own "world". The content of the theory would then consist of all these "worlds", and the theory would be true just if the real world is one of them. Understanding the theory would consist in being able to tell whether something is in the set of "worlds" that determine the theory's content. A scientist learns the theory by acquiring this ability, starting with the very simple "worlds" first encountered as examples in lectures and texts.

Such a view of the content of a scientific theory—of what it is for a theory to be true and what is involved in understanding it—is quite popular among philosophers. It adapts to scientific theories a widespread view of the content of any statement: that a statement's content is given by the conditions under which it would be true ("the possible worlds at which it is true"), and that understanding a statement is a matter of grasping these truth conditions. But it takes for granted that the function of a mathematical model in a scientific theory is always to represent a "possible world". Moreover, not all philosophers have accepted the underlying representational view of content—after all what is it to "grasp" a truth condition? The difficulties faced by an Interpreter of quantum theory warrant a closer look at philosophical doctrine here.

We don't have to look far to find an alternative non-representational tradition in philosophical thought—the pragmatist tradition of William James and John Dewey, especially as developed in the work of more recent thinkers. For a pragmatist, what counts is how something is used, not what it stands for. One understands a theory or a statement if one knows how to use it. A statement is used in communication and reasoning, while a theory is used by being applied to the world.

According to a pragmatist, understanding quantum theory is a matter of knowing how it is applied, and how the theory's various elements function in these applications. While the theory can be said to acquire content by being used, it may not be helpful to ask what this content consists of. Perhaps understanding a theory is not a matter of grasping something called its content but is simply having the ability to use the theory. This is an ability acquired and exercised by many physicists and other scientists. But it is not ineffable. The tacit knowledge of the scientist can be made explicit by investigating how, and for what purposes, scientists use the various elements of the mathematical models made available by quantum theory.

That investigation should be undertaken with no prior commitment to the assumption that it is the function of each element of a mathematical model of quantum theory to represent something physical. Indeed, a parallel assumption would be false even for a theory of classical physics, where a mathematical model often includes elements (such as the absolute value of the gravitational potential at some location, or the multidimensional phase space physicists find it convenient to use when modeling the position and velocity of every molecule in a gas) that are not used to represent some magnitude or entity in the physical world. Bell introduced the term 'beable' to refer to those elements in a theory which might correspond to elements of (physical) reality—and would, if the theory should turn out to be true. In Newton's theory, the acceleration of a falling body is a beable, but the gravitational potential where it is at any moment is not.

But despite the pragmatist's suspicion of representation, it would be equally wrong to assume from the outset that it is the function of no element of a mathematical model of quantum theory to represent something physical. It is surely a broad function of physical theory to predict natural phenomena. One of the universally acknowledged triumphs of quantum theory is that by using it scientists have been able to make a large number of correct predictions, some of incredible accuracy, while encountering no predictions whose falsity could be blamed on quantum theory. It is also generally believed that quantum theory explains a wide variety of otherwise inexplicable phenomena, from the stability of ordinary matter to the extraordinary properties of different isotopes of helium at very low temperatures. These varied predictive and explanatory achievements have won quantum theory its foundational place in twenty-first-century physics. But to make a prediction, a physicist must describe or represent what is supposed to happen: and to use a theory to explain a phenomenon, a physicist must base an account of what is happening on what that theory says.

Prediction and explanation require representation. There is a prima facie case that quantum theory can only have earned its predictive and explanatory credentials because some elements of its mathematical models are capable of representing something in the physical world—that is, because quantum theory has at least some beables. There is a stronger way to put the point. The evidence for any scientific theory comes from successful experimental and observational tests of its predictions, including those that confirm the correctness of its explanations. If quantum theory had no beables, it could make no predictions and explain nothing. So quantum theory must contain at least some beables if we are to have any reason to accept or believe it. Some elements of its mathematical models must function to represent something in the physical world in order for us to have any reason to take it seriously, let alone to accept it as a fundamental theory within twenty-first-century physics.

I hope you found the arguments of that last paragraph convincing. If you did, you are ready to appreciate what I believe to be truly radical about quantum theory. My main aim in this book is to try to convince you that quantum theory is a new kind of science because it warrants our acceptance of its fundamental status within twenty-first-century physics despite lacking any beables of its own. Once we see how the key elements of its mathematical models function in furthering the predictive and explanatory aims of physics we shall understand how quantum theory has led to such significant advances in physics, even though none of these elements serve to represent novel elements of physical (or mental!) reality.

Of course we need to know what these key elements are and how they function. I begin to explain this in Part I of the book. There I give an elementary but opinionated presentation of some of the concepts and mathematics of quantum theory and their uses, emphasizing those aspects that have prompted popular stories about objects in two places at once, "spooky" action at a distance, and cats suspended between life and death, to show how these stories depend on questionable assumptions about the representational function of a quantum state. Most of this material will already be familiar to many scientific readers as well as my fellow philosophers of physics. But they might want to read it first because of the way it treats the quantum state (no collapses), and then to refer back to Chapters 4 and 5 later while reading Part II.[6]

Armed with some acquaintance with quantum theory, Part II details what I take to be radical about that theory, and traces consequences of this radicalism for physics, for the philosophy of science, for metaphysics, and for philosophy more generally. Here is where quantum theory has a lot to teach us—not about the weirdness of our world

---

[6] I originally intended this presentation of quantum theory to remain neutral on disputed issues. But views on central topics like locality, measurement, and the nature of quantum states diverge so widely that this proved impossible. So while Chapters 4–6 present quantum theory as I now see it, this view is not universally accepted. It incorporates insights of physicists, including some sympathetic to ideas of Niels Bohr (Brukner and Zeilinger [2003], Brukner [2017]), and others self-styled QBists (Fuchs [2016], Mermin [2014], Fuchs, Mermin, and Schack [2014]). I believe philosophers have often overlooked these insights while raising justified objections to their frequently instrumentalist setting.

but about how we can come to grips with the world only by dropping preconceptions about what that involves.

It can be hard to shed deeply held preconceptions, especially when these are treated as common knowledge among those engaged in heated debate on other disputed issues. Most contemporary philosophers trying to understand quantum theory simply assume this involves saying how the world could possibly be the way quantum theory says it is. They then go on to champion rival Interpretations, each proposed as an account of our world according to quantum theory. Anyone favoring such an account is unlikely to be receptive to the suggestion that we can understand quantum theory (and what makes it radical) only by rejecting not just that account but the project of finding *any* account of our world according to quantum theory.

So before embarking on Part II I have included an Interlude (Chapter 7) in which I briefly say why I find all existing "Interpretations" of quantum theory inadequate—not as a prelude to yet another Interpretation, but to motivate a different way of thinking about quantum theory that rejects the Interpretative project itself as based on a misunderstanding of how that theory works, and so what makes it radical. Given its purpose of weaning would-be Interpreters of quantum theory away from the mother's milk of their shared commitment, the Interlude cannot be as self-contained as the rest of the book. Debates between rival Interpretations have become quite technical, and so must be objections to each Interpretation to prompt serious reconsideration of the Interpretative project. Readers unencumbered by prior Interpretative commitments are encouraged to skip the Interlude and go straight to Part II.

The quantum revolution makes a radical break with classical physical theorizing, and this has implications for philosophy that should prompt a reassessment of a variety of views in the philosophy of science, metaphysics, and philosophy more generally. Whether this amounts to a revolution in philosophy depends on the extent to which these views are radically modified as a result. Lacking anything resembling Kuhn's [1962] normal science, philosophy always manifests a wide variety of divergent opinions on basic issues, undermining the metaphor of a revolution in philosophy. Some philosophers of a broadly pragmatist persuasion may consider their philosophical views well able to accommodate the quantum revolution without modification. But the views of many analytic metaphysicians, scientific realists, and philosophers of science and language cannot so easily be reconciled with the lessons of the quantum revolution. The quantum revolution is at least a revelation in philosophy. It reveals inadequacies in many currently popular philosophical views while highlighting the virtues of some of their less fashionable rivals.

Chapter 8 argues for a non-representational view of quantum models by contrasting their function with that of models of physics and other sciences since Newton. Briefly, a quantum model advises an agent about how to represent aspects of the physical world without itself representing novel physical entities or magnitudes: and it goes on to advise that agent on what to believe about these aspects, so represented. This does not mean quantum theory represents agents—it is no more about agents (or what they know) than it is about the world. Recall the pragmatist perspective: a model of a scientific theory is always created by agents for the use of agents. The agent is the wielder of the model, not its target. But since agents are themselves in the physical world, in principle one agent may apply a model to another agent regarded as part of the physical world. This second agent then becomes the target of the first, who wields the model. I say "in principle" since human agents are so complex and so "mixed up" with their physical environment that to effectively model a human (or a cat!) in quantum theory is totally impracticable.

This account of how an agent uses a quantum model raises serious questions about several topics dear to the heart of philosophers of science. In Chapter 9 I introduce a pragmatist account of quantum probabilities while showing how these underlie the theory's explanatory triumphs. The paradox is resolved as to how quantum models can be of explanatory value despite their inherent inability to represent novel physical entities or magnitudes. But the resulting account of quantum explanation sits uneasily with views that stress the importance of causation in scientific explanation. What are often called "non-local" phenomena bring this issue to a head. In Chapter 10 I argue that quantum theory helps us locally to explain non-localized correlations that violate Bell inequalities, and that there is no question of instantaneous action at a distance.

Quantum theory has presented a persistent threat to objectivity. Chapter 11 surveys the scope of the threat before setting out to neutralize it. The paradox of "Wigner's friend" presents the threat in stark terms. It has been used to argue that quantum probabilities must be about conscious experiences, not what happens to physical objects including experimental apparatus. I refute this argument by appealing to the relational nature of quantum state assignments and the role of decoherence in giving objective content to statements about properties of physical objects. This makes it clear that quantum probabilities are neither about conscious experiences nor solely about the outcomes of measurements. Then I assess the suggestion that objective properties emerge from subjective quantum states through quantum Darwinism. Quantum states are not subjective, even though one cannot reliably determine the wholly unknown quantum state of an individual system. Quantum Darwinism is helpful in accounting for the fact that many agents can come to agree on the properties they attribute to a quantum system in its environment. But the agreement would be illusory if these were not objective properties of the system—which we can indeed take them to be, but only when environmental decoherence of a quantum state permits.

Chapter 12 touts the virtues of the pragmatist approach to content that underlies the present view of quantum theory. I show how their uses in applications of the theory endow novel quantum concepts with content and at the same time modify the contents of other physical concepts. The concepts of quantum state, observable, and probability may be said to acquire objective representational content in this way even though it is not their function to represent any novel physical magnitude or entity. The main focus of this chapter is then the content of physical magnitude attributions to a system. Such statements figure in inferences, sometimes as premise and sometimes as conclusion. This suggests an inferentialist account of content like that of Robert Brandom [1994]. The account has the flexibility to allow the content of a statement about a physical system to depend on details of interaction with the environment. The more robustly interactions decohere the quantum state in a model, the greater the content of magnitude attributions specified by its decoherence basis—as indicated by such statements' increased inferential power. I illustrate this by contrasting the content of statements about traffic signals and memory elements of a contemporary digital computer with analogous statements about properties of a possible element of a quantum computer.

The inferentialist account of content helps clarify the status of fields and particles in a quantum field theory. Physicists successfully apply such theories to electrons and the Higgs field. Yet there are strong reasons why the theories cannot themselves be understood to describe either particles or fields (nor should they be taken to *misdescribe* them, of course). We can resolve this tension by showing that each application of a quantum field theory is in an environmental context that gives content to a statement about particles or about fields. *We* describe either particles or fields in a legitimate application of quantum field theory, and we have strong evidence that some of our descriptions are both significant and true, even when they concern phenomena we can detect only with the aid of complex and expensive instruments. But neither particles nor fields are beables in a relativistic quantum field theory. This inferentialist account of content has an interesting consequence even for everyday statements about ordinary things like the settings of switches and knobs on experimental equipment and the readings of instruments. Accepting quantum theory means taking the content of these statements subtly to depend on the environment in which such things are located.

Quantum theory is often claimed to be a fundamental theory. But what does this mean? Chapter 13 distinguishes different senses of 'fundamental' and goes on to evaluate the claim in each sense. Fundamental physics is often described as a search for the ultimate laws of nature: maybe physics has found some in quantum theory? I raise several problems with this idea. First, even in classical physics it is not clear what the relation is between the mathematical models supplied by a theory and laws of nature. I offer a clarification and try to apply it to quantum theory. While Newton's law of universal gravitation then counts as (at least an attempt to state) a law of nature, the Schrödinger equation does not, since it does not describe the behavior of a physical system. Next, what could be meant by calling a law of nature ultimate (or basic, or fundamental)? One suggestion is that the laws of nature all reduce to the ultimate laws. Some claim that Newton's second law of motion reduces to the Schrödinger equation. But even waiving doubts about the precise content of the claim and whether the Schrödinger equation expresses a law of nature, there are strong reasons to deny the claimed reduction. Quantum theory is not fundamental in the sense of supplying fundamental laws to which all other laws reduce.

Philosophers often talk about something they call physicalism. Without entering the thicket of attempts precisely to state this thesis, I offer this simple formulation:

everything is composed of physical things, and the facts about the physical world determine every worldly fact (e.g. that you are now thinking about the meaning of physicalism—or not, if your attention has wandered!). But what are these physical things and physical facts? Perhaps what makes quantum theory fundamental is its answer to this question—that (ultimately) the physical things are just what quantum theory says they are, and the physical facts are just what it says about them. The suggestion, in fancier language, is that quantum theory supplies a fundamental ontology for the world as well as a supervenience basis for all facts about it. Chapter 13 disposes of this suggestion by noting that, lacking beables of its own, quantum theory has no physical ontology and states no facts about physical objects or events. There are objective facts about quantum states and probabilities, but they function not as descriptions of physical objects or events, but as authoritative advice to agents on the content and credibility of statements about the physical world, stated in "classical" terms. The chapter concludes by acknowledging an important sense in which quantum theory is (currently) fundamental: by using quantum theory we are able to predict and/or explain a huge variety of physical phenomena that we cannot predict and/or explain using any other theory, and none of the predictions of quantum theory have been shown to be false. That sense of fundamentality satisfies almost all physicists (for now): it should be enough for philosophers, too.

Quantum theory may be a new kind of science, but it is after all a successful scientific theory and not a speculation, a part of psychology or philosophy, a contribution to science fiction, or a healing aid. In the conclusion I reflect on how quantum theory has been able so spectacularly to advance the aims of science, despite its radicalism. Prediction, control, and explanation are three generally acknowledged aims of science. Quantum theory has contributed enormously to our successful pursuit of these aims by extending the predictive and explanatory power of physics as well as fostering a wide variety of technologies indispensable to modern life: many believe the control of individual quantum systems puts us in the early twenty-first century on the threshold of a second technological revolution. Quantum theory has made huge contributions to quite traditional scientific tasks: in this respect it is not at all radical. What makes it radical is how it has contributed to these tasks, as Part II will explain.

Quantum theory is conventional in other respects. Accepting quantum theory does not commit one to believing there is a multiverse of countless other physical worlds, that there is instantaneous action at a distance, or that consciousness has peculiar effects on matter. Quantum theory has helped us to find out how nature is, but also what we can significantly say about it. Atoms and elementary particles are (sometimes) real, and quantum theory helps us to say when: and it shows why the moon demonstrably is there when nobody looks. Quantum theory is very metaphysically conservative. Measurement or observation plays no special physical role: it has the traditional epistemological function of better informing the one who makes it.

PART I
Quantum Theory

2

Superposition

Paul Dirac began his classic *Principles of Quantum Mechanics* [1930] with a principle of superposition of states. The principle is mathematical but he introduced it by first describing some physical phenomena, and I'll follow his example. Some more or less familiar observations can be used to motivate a simple mathematical model that makes sense of them. These are all observations of interference of beams of matter. What does this mean?

2.1 Beams

It is easy to see how the Old English word 'beam' for a living tree came to refer to a timber post. But it is not so clear why we use the same word when advising drivers to turn off their high beams when they see an oncoming vehicle. The term 'sunbeam' may have entered the English language as a translation of the seventh-century Latin '*columna lucis*' Bede used to refer to a column of fire like that which (according to Ecclesiastes) guided the Israelites by night on their flight from Egypt. After the scientific revolution people didn't think of a beam of light as a rigid rod but as something emitted by a source that seemed to instantly illuminate distant objects. Whatever it was, this "something" counted as material since it pushed distant objects around and heated them up, however slightly. Newton thought light consisted of particles because it traveled in straight lines without bending around corners. Despite the subsequent successes of the rival wave theory, Feynman [1985] could still maintain that Newton was right (though for the wrong reasons).

Atoms and molecules certainly count as matter, and we can produce a beam of them by heating a substance in an oven and letting some of it escape through a hole and pass through another hole. If atoms are material then presumably so also are their constituent parts, including electrons and neutrons. An old-fashioned television tube produced a picture through the impact of a guided beam of electrons on a phosphorescent screen: a nuclear reactor acts as an "oven" from which a beam of neutrons may be extracted.

When a beam of matter encounters an object, the beam as well as the object may be affected. Newton studied many ways of modifying a sunbeam, including shining it through a prism to produce a spectrum of light and passing it through a glass lens resting on a flat piece of glass. In each case this produces colors—not of an object but apparently of the light itself. Newton concluded that sunlight is a mixture of different types of "pure" light, where each type has some physical property other than color that makes it bend more or less when passing through a prism: we see light as colored only because when each different mixture of types of light enters our eye and affects our brain it produces a distinctive color sensation. But Newton had trouble accounting for details of how light is modified when passed through a glass lens resting on a glass plate. These are most clearly revealed when the light is of just one type, forming what we persist in calling a monochromatic beam.

Sunlight and most forms of artificial light are not monochromatic. I live in Tucson, a center for astronomy because of its clear night skies and nearby mountain-top sites for telescopes. Many street lamps here do produce very nearly monochromatic light by passing electricity through sodium vapor at low pressure. The astronomers like this because it makes it easy to remove stray street light from what they collect from distant stars. The local police are not so pleased because it makes it hard to see the natural colors of clothing, cars, hair, and skin in the streets at night, since everything appears some shade of the same yellow!

A concentric series of colored rings appears when a beam of light is shone from above onto a convex lens resting on a flat glass surface. These are called Newton's rings after their famous but puzzled discoverer. They exhibit in a more controlled environment the same phenomenon that may be observed in the natural iridescence of some insect wings, and the colors that can be seen in soap bubbles, thin films of oil on water, and the peacock's tail. All these phenomena are examples of interference of light.

Interference occurs when different beams or parts of the same beam take different paths before uniting in a single beam. When a beam of light encounters a thin layer of a different transparent medium (a soap bubble, or the air gap between a glass lens and the glass plate on which it rests) part is reflected and part is transmitted at its first surface. Some of the part that was transmitted is then reflected by the second surface, returns through the medium, and is then (mostly) emitted parallel to the other part. These two parts of the beam then merge to form a single beam which has been modified in the process. Importantly, how it has been modified depends on the difference between the lengths of the paths taken by the two parts. Newton's rings appear because the difference in path lengths varies with the distance from the center of the lens, since this governs the size of the air gap between the lens surface and the glass plate.

The dependence on path difference is clearest when the original beam is monochromatic. The rings do not appear multicolored in this case, but they exhibit a notable variation in intensity: bright rings regularly alternate with dark rings where hardly any light can be seen. This is a good example of interference: one part of the beam interferes with another part. Where a bright ring appears the interference is constructive: together the partial beams form an emitted beam whose intensity is greater than the sum of their individual intensities. Where a dark ring appears the interference is destructive: the intensity of the total beam emitted here is less than that of either individual intensity—the partial beams have canceled each other out, at least to some extent.

Changing the color of the incident beam changes the diameters of the rings, showing that they arise from some kind of interaction between the type of light and the difference in path lengths of the two parts of the beam. The multicolored rings seen when the beam is a mixture of types are a natural consequence of overlapping differently colored bright rings of different diameters.

Newton had trouble accounting for interference of light because he thought of a beam as composed of particles. Young later developed a wave model that could easily explain the interference of light by analogy to interference of waves in water— in one place a peak of a wave from some source may cancel a trough of a wave from a different source, while in another place their peaks and troughs coincide, as in this image of water waves made by the synchronized flapping of a floating bee's wings.[1] The wave theory of light became a cornerstone of classical physics after its mathematical development by Fresnel and Maxwell's theoretical identification of light as an electromagnetic field radiating from its source at an enormous speed of almost 300 million meters per second. But as we shall see, light behaves in ways that can't be encompassed by a classical physics of waves or of particles. So does atomic matter. But fortunately light and atomic matter behave in very similar ways!

If you stand in a garden in bright sunlight and view the window of a house from an angle, you will see not only inside the house but also a reflected image of flowers and trees. This is because besides letting light through from inside, the window also acts as a mirror by reflecting some light from outside. It is not difficult to construct a half-silvered mirror that reflects about as much of a beam of light as it transmits. Since the 1970s it has become possible to make something like a mirror for a beam of neutrons emerging from a nuclear reactor.

Such "mirrors" are the basic components of a neutron interferometer, depicted in Figure 2.1.

Cut from a single crystal of silicon, this looks a bit like an electric toaster for very thick slices of bread. I'll have more to say about the uses of this device later, but for now just consider what happens when the neutrons reach the first slab of silicon. A beam of neutrons all with the same speed is analogous to a monochromatic beam of light. If this beam hits the slab at just the right angle, part of it goes straight through while another part also goes through but is bent (as shown in the figure). By splitting the beam, the slab has essentially the same effect on the neutron beam that a half-silvered mirror has on a beam of light.

For beams of many other kinds of matter (atoms, molecules, electrons, etc.) analogs of mirrors have also been constructed. For each kind of matter there are detectors capable of measuring the intensity of the beam. Since passage from a source through

---

[1] See the image at http://upload.wikimedia.org/wikipedia/commons/c/c5/Ripples_waves_bee.jpg.

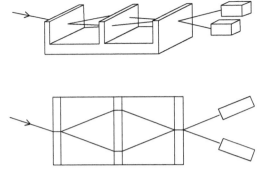

Figure 2.1 Neutron interferometer

a system of mirrors affects the beam intensity in basically the same way in each case, we can describe this without giving details of the kind of matter, its source, the mirrors, or the detectors. A generic analog of a half-silvered mirror is called a *beam splitter*.

We have now assembled all the necessary tools to exhibit the kind of peculiar observations that prompt the formulation of a principle of superposition of states. In sections 2.2 and 2.4 I'll deploy these tools by describing what is observed to happen in two series of schematic experiments, in each case building up to some clear but puzzling examples of interference phenomena.

2.2 Some Interference Phenomena

Suppose a source produces a beam of matter of intensity I, as measured by a suitable detector. For a beam of neutrons or other particles, I may indicate how many neutrons are detected every second: for a beam of light, I may indicate the power of the beam in watts (or fractions of a watt). If a beam of measured intensity I is directed at a 50–50 beam splitter that absorbs none of it, detectors placed to collect the reflected and transmitted beams as in Figure 2.2 will each measure intensity $1/2\ I$. On average, half of the particles of a particle beam are detected at D_1, the other half at D_2.

Now suppose that these detectors are removed and two more 50–50 beam splitters are placed, one to intercept the transmitted beam, the other to intercept the reflected beam (see Figure 2.3). What would you expect to be the intensity measured by detectors placed to intercept beams transmitted and reflected by these additional beam splitters?

If a beam splitter acts as some kind of filter, transmitting one sort of particle or light but reflecting another sort, then one might expect detectors marked TT and RR each to register intensity $1/2\ I$ while those marked TR and RT register nothing. (A prism does act much like a filter for light, as Newton showed. It affects a sunbeam passing through it—a continuous spectrum of different monochromatic beams emerge, each

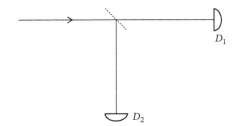

Figure 2.2 Beam splitter

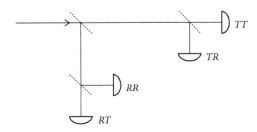

Figure 2.3 Three beam splitters

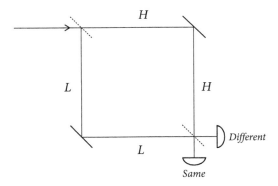

Figure 2.4 Balanced Mach–Zehnder interferometer

at a slightly different angle: these angles don't change when each monochromatic beam is bent by a second prism.) But if a beam splitter just divides up a beam without filtering it, then one would expect each detector (*TT*, *RR*, *TR*, *RT*) to register the same intensity, $1/4\, I$: and indeed that is what happens.

Now consider the arrangement in Figure 2.4, where a dotted line represents a 50–50 beam splitter and a bold line represents a fully reflecting mirror for this type of beam. When paths *H*,*L* have the same length between the two beam splitters this is said to constitute a *balanced* interferometer.

I've labeled one detector *Same* to remind you that it would be expected to detect that part of the beam either transmitted at both beam splitters (*TT*) or reflected at both beam splitters (*RR*): the detector called *Different* would be expected to detect that part of the beam transmitted at one beam splitter but reflected at the other. What happened in the arrangement of Figure 2.3 seemed to show that a beam splitter does not act as a filter. So one would expect that each partial beam following path *H,L* would be split equally by the second beam splitter, and that each detector would therefore measure intensity $1/2\,I$. But that is *not* what happens. Instead, detector *Different* measures intensity *I* while detector *Same* measures intensity 0. Partial beams following paths *H,L* appear to have interfered with each other so their intensities completely cancel each other at *Same* while wholly reinforcing each other at *Different*.

This is particularly difficult to understand if the source emits atoms or other particles. Consider a beam of neutrons emerging from a nuclear reactor. You can control their speed and intensity so they are typically separated from each other by 300 meters in the beam. The neutron interferometer pictured in Figure 2.1 effectively realizes the scheme of Figure 2.4. It is much less than a meter long. So it is very unlikely that more than one neutron is present in the interferometer at once. It is natural to assume that each neutron detected travels through the interferometer either along path *H* or along path *L*. If you have any doubts, you can check by blocking either path: this cuts the total detected intensity by half, while blocking both paths reduces the total detected intensity to 0. Moreover, it seems that no neutron splits in two, with half following path *H* and the other half following path *L*. For the two detectors almost never respond simultaneously, and nor would detectors placed along both paths inside the interferometer.

Suppose a particular particle travels along path *H*. If path *L* is blocked, then it is just as likely to be detected at *Same* as at *Different*, since each detector measures intensity $1/2\,I$ when path *L* is blocked (as the observations of Figure 2.3 already lead one to expect). But if path *L* is now reopened, it is never detected at *Same* but always at *Different*! It appears that altering a path along which a particle never travels somehow affects its behavior. Similarly, altering path *H* seems to affect the behavior of a particle assumed to travel along path *L*.

Instead of blocking a path, you can change its length. Make path *L* very slightly longer (or shorter) and you will not detect the full intensity *I* at *Different*—you will detect some of the original beam intensity at *Same*. Gradually changing the difference in path lengths by changing either length produces a continuous variation in the relative intensities measured at the two detectors, until for a particular path difference it is detector *Same* that measures the full intensity *I* while nothing is detected at *Different*. If a particle does take either path *H* or path *L* through the interferometer, then very slight variations of the length of the path it does not take are reliably correlated with variations in the relative intensities detected.

It is not always easy to gradually vary the length of a path. But there are other ways of altering a path through an interferometer to exhibit the continuous variation in relative intensity associated with interference between paths. Inserting a thin sample of the right kind of material in a path through a neutron interferometer will not significantly reduce the intensity of neutrons detected following that path: you can check this after first blocking any other path through the device. But it can temporarily slow down any neutrons that go through it, much as light slows down while it goes through glass or water. If the sample is wedge-shaped, then you can vary the thickness neutrons pass through by gradually inserting or withdrawing the wedge. So you can continuously vary how long neutrons taking that path are delayed by their passage. This is another way of continuously varying the interference between the two paths through the interferometer, as you can observe by comparing how much intensity each detector records.

Yet another way of altering a path through a neutron interferometer reveals new and important features of single-particle interference. But since this requires devices that are not exactly familiar household objects, it is best to introduce the idea by describing its analog for beams of light.

2.3 Polarization

Most people today are familiar with polarized sunglasses and most photographers are familiar with clear polarizing filters. These both reduce glare from reflected light by absorbing light that has been polarized by that reflection. A beam of direct sunlight is not polarized. But if it is passed through a clear polarizing filter its intensity is reduced while the rest of the beam goes straight through with no color change. The emerging beam is polarized along an axis at right angles to its direction of travel determined by what is called the polarization axis of the polarizer. So far 'polarized' is just a word. Its cash value is revealed by what happens when the beam that emerges from one polarizer encounters a second, similar polarizer.

If their polarization axes are perfectly aligned, the second polarizer reduces the beam intensity hardly at all. But if they are at right angles to one another (and also to the direction of travel) then the second polarizer blocks essentially all of the beam. By rotating the second polarizer in its plane one can vary the intensity of the beam it transmits continuously between these two extremes. Newton studied this phenomenon and speculated that it might occur if the particles have "sides". Fresnel developed a rival wave theory that waves of light are rapidly propagating periodic disturbances of some otherwise stationary transparent medium. He adapted this theory to the phenomena of polarization by taking these disturbances to occur in a direction at right angles to the direction in which they propagate—so-called transverse waves. But we can now observe phenomena associated with the polarization of light for which neither Newton's particle theory nor Fresnel's wave theory could account.

If you search the phrase "single photon detector" on the Web, you will find links to many companies offering to sell you such a device: you could be forgiven for assuming that what it detects are the single photons we are now convinced compose a beam of light. There are observations that support that conviction, including some I'll describe in the next chapter. Einstein revealed the power of this idea in a paper he published in 1905. In two other papers published that year he first formulated his theory of relativity and showed the equivalence of mass and energy. But the one paper Einstein published in 1905 whose ideas he himself described as "very revolutionary" and which were later mentioned by the Nobel committee when awarding him his prize, adopted the viewpoint that:

> the energy of a light ray spreading out from a point source is not continuously distributed over an increasing space but consists of a finite number of energy quanta which are localized at points in space, which move without dividing, and which can only be produced and absorbed as complete units. [1905, 1965]

Although someone else introduced the term later (to refer to something else!) these quanta came to be known as photons. I defer until later (Chapters 12 and 13) a careful attempt to answer the questions of what it means to say a light beam is composed of photons and whether we should believe it is. But for now let's just bypass the first question and assume a positive answer to the second.

Suppose that a single photon detector is placed behind a polarizer. A severely attenuated beam of light is directed toward this polarizer, but first encounters another similar polarizer. The beam is composed of so few photons that each separately encounters this first polarizer (see Figure 2.5).

While absorbing some photons from the beam, the first polarizer may pass others: these then encounter the second polarizer. Experience with more intense beams of light leads one to expect that, on average, the fraction passed by the second polarizer will depend on the relative orientations of the two polarizers' polarization axes. If these are aligned, essentially the same number detected passing the first polarizer with the second polarizer removed will still be detected when it is replaced: but few if any will be detected if the polarizers' axes are then crossed, that is, at right angles to each other. Observations confirm this expectation, in line with the assumption that photons in a beam behave independently in this setup. You may wonder whether photons that pass a polarizer differ somehow from those it absorbs: maybe a polarizer works like a grating of bars parallel to its axis, so photons with "sides" closely aligned with these bars slip through more often than those whose "sides" are badly misaligned with its axis? Observations in the setup of Figure 2.5 don't address this suggestion, but insertion

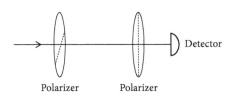

Figure 2.5 Polarization analysis

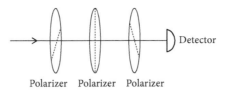

Figure 2.6 Analysis with a third polarizer

of a third similar polarizer between crossed polarizers (as in Figure 2.6) at least casts doubt on it.

If the intermediate polarizer's axis is aligned either with that of the initial polarizer or with that of the final polarizer, no photons are detected, in accord with the suggested model of photons and their interaction with polarizers. But if the axis of this intermediate polarizer is rotated around the direction of the beam, some photons will be detected. The average rate at which they are detected depends smoothly on how far it is rotated. It is greatest when the axis of the intermediate polarizer is arranged to make equal 45 degree angles with the other two polarizers' axes: in this arrangement a quarter as many photons are detected on average as when the intermediate polarizer is removed and the other polarizers' axes are aligned.

This last observation strongly suggests that polarizers don't simply act as filters, letting one type of light through unchanged while blocking light of another type. A polarizer can apparently also modify some photons while letting them pass: modification by an intermediate polarizer is sometimes enough to permit a photon to pass through each of two crossed polarizers. If a photon did have "sides" or other properties whose relation to a polarizer determined whether or not it would pass that polarizer, then its passage would have to completely "reset" these properties—for how likely it is to be detected with a third polarizer depends only on the angle between the axes of the second and third polarizers. Chapter 4 will present a famous argument against even this possibility. If cogent, this shows that photons have no properties that make some pass a polarizer while others are absorbed: qualitatively identical photons behave differently in just the same circumstances, individually at random but with statistical regularity.

Newton was not the first to observe and comment on polarization phenomena: some have speculated that the "sunstone" Vikings used to navigate the Atlantic ocean centuries earlier was a polarizing crystal of calcium carbonate still known as Iceland spar. A beam of light shone on such a crystal at the correct angle is split into two beams that separate inside the crystal to emerge parallel: one (called the ordinary ray) goes straight through, but the other (the extraordinary ray) is displaced, as shown in Figure 2.7.

The important thing is that each of these beams is polarized, with their axes of polarization at right angles to one another and to their common direction of travel on exiting the crystal. This is easily shown by placing a polarizer to intercept each

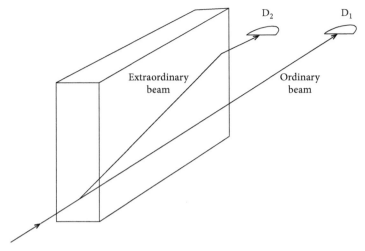

Figure 2.7 Action of calcite crystal

beam separately and noting how its intensity is observed to vary (using your eye or other detector) as that polarizer is rotated around the beam axis.

Iceland spar is a two-channel polarizer because (unlike a polarizing "filter") it yields two oppositely polarized beams: there are others. They all act as polarizing beam splitters for light. Using mirrors if necessary, it is easy to arrange for the two beams to leave a polarizing beam splitter in perpendicular directions. There are also clear "filters" capable of rotating the polarization axis of polarized monochromatic light they pass by any desired angle: for historical reasons these are called half-wave plates.[2]

2.4 More Interference Phenomena

After this brief introduction to polarization of light beams it is time to turn to a more abstract level of description of a series of phenomena that can also be displayed by observations on beams of other types of matter such as electrons, atoms, and neutrons. I already noted that a beam of neutrons all with the same speed displays interference analogous to that displayed by a monochromatic beam of light. So does a beam of electrons, atoms, molecules, etc. Some beams of each of these kinds of matter also manifest behavior closely analogous to that displayed by polarized light: indeed, physicists often talk about polarized beams of neutrons, and so will I.

The analog for neutrons of photon polarization is called spin. The behavior of electrons, atoms, molecules, etc. can be modeled by assigning them a spin state, just as the behavior of photons can be modeled by assigning them a polarization state. I will explain later how this modeling works. But now I must issue a warning. I mentioned

---

[2] See the demonstration at https://www.youtube.com/watch?v=_sUVXHfUVsY.

the problems created by the assumption that a photon has some definite polarization properties that determine what will happen when it encounters a polarizer. The same kind of problems recur if one thinks of a particle's spin as a property that determines what will happen when it encounters a device that is sensitive to its spin state. I will later address the question of how the modeling techniques introduced in this chapter are able to resolve these problems, beginning in Chapter 4.

Recall the generic balanced Mach–Zehnder interferometer depicted in Figure 2.4. Polarizing a beam of matter before it enters the interferometer reduces its intensity. But it has no other effect on what is observed: the remaining intensity is still all detected at *Different*, none at *Same*. But if you also place a half-wave plate along one of the paths H or L inside the interferometer then you will observe something new (see Figure 2.8).

Rotating the half-wave plate around the direction of that path varies the relative amounts of the beam that are detected at *Different* and *Same*. At one angle *Different* detects all the intensity, at another angle *Same* detects the same intensity as *Different* (even if one then varies the path lengths), while rotating through intermediate angles continuously varies the relative detected intensities between these two extremes. So rotating its polarization axis is another way of altering a beam of matter that affects how it interferes. It can do so without changing the length of the path this beam travels.

Suppose you fix the polarization state of the incoming beam as P by passing it through a polarizer. By adjusting the angle of the half-wave plate you can make *Different* and *Same* detect equal intensities. At this angle the plate rotates the polarization of its beam so its polarization state is opposite to P. This means the two beams are oppositely polarized when they encounter the second beam splitter. The detectors then respond as if each beam behaved just the way it would if the other beam were not there at all: they respond as if there were no interference. One could then plausibly maintain that each particle either took path H or took path L between the beam splitters. Indeed one could claim to tell which of these paths a particle took by replacing the second

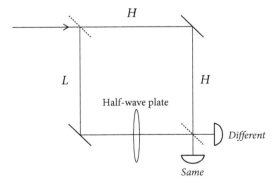

Figure 2.8 Half-wave plate in balanced interferometer

beam splitter with a polarizing beam splitter (PBS)—a device that acts as a polarizing filter as well as a beam splitter. If its axis is opposite to *P*, it seems that any particle detected must have followed a path taking it through the half-wave plate: if its axis is *P*, then any particle detected must presumably have followed the other path through the interferometer.

Instead, suppose you set the axis of the PBS half way between *P* and its opposite. A particle encountering the PBS is now just as likely to pass through whether its polarization state is *P* or its opposite. Even if such a particle had followed path *H* or path *L* through the detector, detecting it would not entitle you to say which path it had taken. You might still maintain that it must have taken one path or the other, even though you can no longer tell which. But the interesting thing is that with the axis of the PBS set like this *Same* no longer detects anything—only *Different* detects any particles! Set this way, the PBS has restored interference between the paths. If you continue to maintain that each detected particle either took path *H* or path *L* then it seems you must admit that you can detect interference only when you have removed any possibility of determining which path a detected particle took.

I have not yet mentioned a famous example of interference that Feynman [1963] once claimed contains the only mystery of quantum behavior—the two-slit experiment with electrons, as depicted in Figure 2.9.

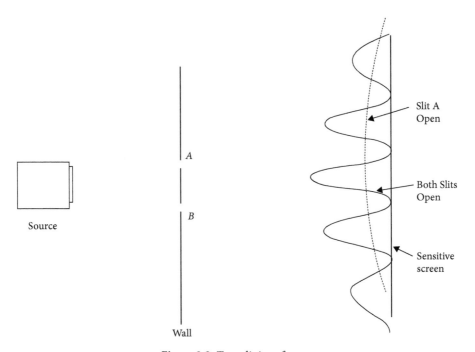

Figure 2.9 Two-slit interference

I left it till now for two reasons. First, when Feynman described it before 1963, it had not actually been done—it was merely a thought experiment: the first real experiment with individual electrons was performed (with the expected results) much later, and no experiment successfully followed Feynman's precise methodology until 2012.[3] Second, if you think in terms of direct paths from source through a slit to a point on the screen, there is only a tiny spatial separation (less than a thousandth of a millimeter) between a path through one slit and a path through the other: Figure 2.9 is certainly not drawn to scale! Analogous paths through a Mach–Zehnder interferometer can be much more widely separated: paths through a neutron interferometer are separated by centimeters, making it particularly hard to imagine how a tiny subatomic particle could follow both at once.

Popular discussions of the two-slit experiment with electrons often heighten the sense of mystery by claiming that any attempt to observe through which slit an electron went will inevitably destroy the interference pattern shown in 2.9, replacing it by a smooth pattern—a simple sum of the pattern you get when only slit A is open and the pattern you get when only slit B is open. Those single-slit patterns are smooth and show none of the rapid variations of intensity we saw characterize interference fringes (remember Newton's rings). One way to observe is to shine light on the electrons and see which slit they go through by collecting the reflected light in a microscope. Heisenberg analyzed essentially this proposal in his influential [1930] book and used it to illustrate his so-called uncertainty principle. His idea was that observing the path of an electron by shining light on it would inevitably disturb its motion: the disturbance produced by an observation accurate enough to determine through which slit it went would so change its subsequent motion as to destroy the interference pattern. Heisenberg took this as an instance of the general principle that there is a strict limit on how precisely one can observe both the position and the momentum of an object.

But as we've seen, interference between two beams may be affected, or destroyed entirely, without in any way affecting the motion of a particle in either beam—it suffices to alter the polarization state of particles in one beam. Moreover, merely altering the beam polarization to destroy interference is not a way of observing which path a particle takes through an interferometer. Instead, what this does is to make possible a subsequent observational procedure that some may interpret as an indirect determination of that path. A different subsequent procedure may on the other hand remove even that possibility, thereby restoring detectable interference. Whether interference is detectable between beams of matter is not simply a function of the inevitable clumsiness involved in our attempts to observe which path a particle takes. Quantum theory itself will help us to say when such interference will be detectable, as we shall see in Chapter 5.

---

[3] See Frabboni *et al.* [2008], Bach *et al.* [2013].

2.5 Modeling Interference

It is time to explain how these peculiar observations can be modeled. Consider a beam of matter of intensity I incident on a Mach–Zehnder interferometer with two 50–50 beam splitters. Assume none of the beam is absorbed while passing through the interferometer, and that the detectors *Different, Same* are perfectly efficient, so together they detect the full intensity I of the beam. We wish to develop a model we can use to represent how the intensity detected by each of *Different, Same* individually can be expected to depend on what the beam encounters inside the interferometer.

For that purpose, we represent the state of the beam as it passes through the device by a *vector* we write as $|\psi\rangle$. The velocity of an automobile and the force the road exerts on it when accelerating or braking are both vectors: each has both a magnitude represented by its length, and a direction in which it points. The length of the velocity vector represents the car's speed, while its direction indicates where the vehicle is headed. If a force **F** of magnitude F moves a body through a distance r in a direction making an angle θ with its line of action, then the work done by the force is $Fr\cos\theta$: this may be written as a scalar product of two vectors **F.r**. Only the component $F\cos\theta$ of **F** in the direction of **r** contributes to this work: $F\cos\theta$ is the *projection* of **F** along that direction. The magnitude of a vector is easily written in terms of this scalar product as, for example, $F^2 = $ **F.F**. The vector $|\psi\rangle$ also has a length and a direction: you can think of it as like an arrow—not in physical space but in a mathematical vector space. There is a generalization of the scalar product of two vectors $|\psi\rangle, |\varphi\rangle$ to this space called the inner product, written as $\langle\psi|\varphi\rangle$. The length or *norm* $\|\psi\|$ of $|\psi\rangle$ is defined by $\|\psi\|^2 = \langle\psi|\psi\rangle$. I give further details of vectors and inner products at the beginning of Appendix A (see § A.1).

The norm, $\|\psi\|$ of $|\psi\rangle$, determines the intensity of the beam $|\psi\rangle$ represents. The intensity I to be expected if a beam is detected is represented by the square of the norm of the vector that represents it—$I = \|\psi\|^2$. Since the state of the beam changes as it passes through the interferometer, the vector representing it is actually a function of time: $|\psi(t)\rangle$. We assumed none of the beam is absorbed, so $I(t) = \|\psi(t)\|^2$ is constant. How $|\psi\rangle$ changes from the initial state $|\psi(t_i)\rangle$ to the final state $|\psi(t_f)\rangle$ at the detectors depends on what the beam encounters inside the interferometer.

Write the state inside the interferometer in situation H when the first beam splitter is removed as $|high(t)\rangle$, and the state inside the interferometer in situation L when the first beam splitter is replaced by a perfect reflector as $|low(t)\rangle$. For reasons that will soon become clear we model the fact that a detector placed in the low(high) path never detects anything in the high(low) path by a condition that says the state vectors $|high\rangle, |low\rangle$ are always *orthogonal* to each other: $\langle high|low\rangle = 0$, indicating that the length of the projection of $|high\rangle$ onto $|low\rangle$ is always zero (see Figure 2.10). On the other hand, the length of the projection of $|high\rangle$ onto itself gives the norm of $|high\rangle$: $\langle high|high\rangle = \|high\|^2 = I = \|low\|^2$. It is convenient to rescale the units of intensity so that $I = 1$, and then $|high\rangle, |low\rangle$ and $|\psi\rangle$ all have norm 1—they are *normalized*.

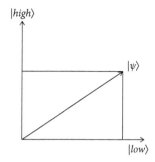

Figure 2.10 Superposition of states

To model the patterns of intensity detected by *Different, Same* in various circumstances, we postulate that the final state is just a *superposition* of the final states corresponding to the high and low paths:

$$|\psi(t_f)\rangle = h|high(t_f)\rangle + l|low(t_f)\rangle \quad \text{(Superposition)}$$

What this says is that the final state of the beam can be represented by a vector formed by combining *component* vectors representing its final state in situations H and L: these vectors are simply added, each in due proportion. The coefficients h and l specify their respective contributions to the resulting vector. Each of h, l is a *complex* number of the form $r(\cos\theta + i\sin\theta)$, where the *modulus* r and *argument* θ are real numbers and $i = \sqrt{-1}$. These numbers are not wholly independent, because we required $|high\rangle, |low\rangle$, and $|\psi\rangle$ all to be normalized. This implies that the moduli $|h|, |l|$ of h, l satisfy:

$$|h|^2 + |l|^2 = 1 \quad (2.1)$$

while imposing no constraint on the arguments of h, l (these notions are illustrated in Figure 2.11).

Indeed the intensities one would expect to detect in high and low beams inside the interferometer with the 50–50 beam splitter in place are equal, so $|h|^2 = |l|^2 = 1/2$.

We further assume that the state evolves deterministically as follows:

$$|\psi(t_i)\rangle \implies |\psi(t_f)\rangle$$

and that this evolution is *linear*. Each component $|high\rangle, |low\rangle$ of the superposition $|\psi\rangle$ evolves independently, so if nothing is done to the beam inside the interferometer the coefficients h, l are constant, and for *every* time t between t_i and t_f:

$$|\psi(t)\rangle = h|high(t)\rangle + l|low(t)\rangle. \quad \text{(Linearity)}$$

We can now model a wide variety of different observed intensity patterns: both replacements H, L of the first beam splitter; the balanced interferometer (with the

30 SUPERPOSITION

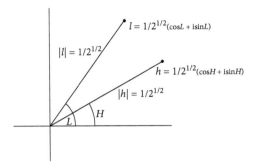

Figure 2.11 Representation of complex numbers

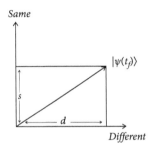

Figure 2.12 Calculating intensities

beam splitter in place); and the unbalanced interferometer with arbitrary insertions into either or both paths! This is how it works. Draw two Cartesian axes and label them *Different*, *Same* as in Figure 2.12.

Now the intensity $I_{Different}$ at the *Different* detector is represented by the squared length d of the projection of $|\psi(t_f)\rangle$ on the *Different* axis, and the intensity I_{Same} at the *Same* detector is represented by the squared length s of the projection of $|\psi(t_f)\rangle$ on the *Same* axis. Pythagoras's theorem then guarantees that

$$I_{Different} + I_{Same} = 1. \tag{2.2}$$

When the interferometer is balanced (high and low paths have the same length), *Different* detects 100% of the intensity while *Same* detects 0%, corresponding to the fact that in this case the final state

$$|D\rangle = h \left| high(t_f) \right\rangle + l \left| low(t_f) \right\rangle \tag{2.3}$$

points along the *Different* axis. Unbalancing the interferometer by adding a small amount to the length of one or other path through it alters the state inside the interferometer in a way that can be modeled by changing the arguments but not the moduli of the coefficients h, l when the new state is incident on the second beam splitter. So suppose the state in the balanced interferometer at time t_1 just before the second beam splitter is

$$|\psi(t_1)\rangle = h\,|high(t_1)\rangle + l\,|low(t_1)\rangle. \tag{2.4}$$

Changing one path length results in a different outgoing state at t_f that no longer points along the *Different* axis—it has a small component in the *Same* direction, so *Same* now detects some of the intensity while *Different* detects less. Further increasing the difference in path lengths yields an outgoing beam $|S\rangle$ that is detected only at *Same*.

Suppose, for example, the length of the lower path is slightly increased. This is modeled by specifying a corresponding alteration in the vector associated with that path, resulting in a different total state vector just before the second beam splitter. When the increase is just enough, the state of the lower path at t_1 is no longer $|low(t_1)\rangle$ but $-|low(t_1)\rangle$. So the total state vector just before the second beam splitter is then

$$|\psi^\perp(t_1)\rangle = h\,|high(t_1)\rangle - l\,|low(t_1)\rangle \tag{2.5}$$

where the notation indicates the fact that $|\psi(t_1)\rangle$, $|\psi^\perp(t_1)\rangle$ are othogonal: $\langle\psi|\psi^\perp\rangle = 0$—a simple consequence of the orthogonality of $|high\rangle$, $|low\rangle$. Since the evolution is not only deterministic and linear but also *unitary*, that is, orthogonal states evolve into orthogonal states,[4]

$$\langle D|S\rangle = 0 \tag{2.6}$$

which is indicated in Figure 2.12 by the fact that $|S\rangle$ points along the *Same* axis, at right angles to the *Different* axis. Indeed

$$|S\rangle = h\,|high(t_f)\rangle - l\,|low(t_f)\rangle. \tag{2.7}$$

So

$$2h\,|high(t_f)\rangle = |D\rangle + |S\rangle \tag{2.8}$$
$$2l\,|low(t_f)\rangle = |D\rangle - |S\rangle.$$

Equation 2.7 models the observed detection for this particular increase in the length of the lower path. An intermediate case gives rise to an outgoing superposition of the form

$$|\psi'(t_f)\rangle = c_1|D\rangle + c_2|S\rangle \quad (|c_1|^2 + |c_2|^2 = 1). \tag{2.9}$$

Now suppose we are dealing with a polarized beam in a (modified) Mach–Zehnder interferometer. We need to extend the vector representation of states to incorporate a representation of polarization. Write the state of the incoming beam as

$$|\Psi\rangle = |\psi\rangle \otimes |\pi\rangle \tag{2.10}$$

---

[4] Given linearity, this follows from its preservation of norms, i.e. the requirement that $\|\psi(t)\|^2$ be constant as t varies.

where $|\psi\rangle$ now represents everything about the current state except the polarization (what Dirac called the translational state) and $|\pi\rangle$ represents the polarization. The symbol '$\otimes$' indicates an operation of taking the tensor product of the two vectors flanking it: I'll have more to say about this in Chapter 3. For now, just focus on the fact that it makes a vector in a bigger space out of a pair of vectors in smaller spaces. Suppose the incoming beam has polarization state $|\pi\rangle = |V\rangle$: this means that a perfectly efficient polarizer would pass everything in the beam if its axis were set vertically, but a reduced intensity if it were set at any other angle. So we have

$$|\Psi(t_i)\rangle = |\psi(t_i)\rangle \otimes |V\rangle. \tag{2.11}$$

In the interferometer this becomes

$$|\Psi(t)\rangle = \left(h\,|high(t)\rangle + l\,|low(t)\rangle\right) \otimes |V\rangle \tag{2.12}$$

so the state at the detectors is

$$\left|\Psi(t_f)\right\rangle = |D\rangle \otimes |V\rangle. \tag{2.13}$$

Therefore all the intensity is detected at just one of the two detectors (each assumed to respond to an incident beam whatever its polarization).

Now insert a half-wave plate along the low path. This has the effect of rotating the polarization state of anything on that path, while leaving the polarization state of anything on the high path unchanged. For a particular setting of the half-wave plate, the effect on the low path is to change the polarization axis from vertical to horizontal

$$|low\rangle \otimes |V\rangle \implies |low\rangle \otimes |H\rangle. \tag{2.14}$$

In accordance with the superposition principle, the effect on the total state inside the interferometer after the half-wave plate is therefore

$$\left|\Psi'(t_1)\right\rangle = h\,|high(t_1)\rangle \otimes |V\rangle + l\,|low(t_1)\rangle \otimes |H\rangle. \tag{2.15}$$

After recombining the beams at the second beam splitter, this produces an outgoing beam represented by

$$\left|\Psi'(t_f)\right\rangle = h\,|high(t_f)\rangle \otimes |V\rangle + l\,|low(t_f)\rangle \otimes |H\rangle \tag{2.16}$$

$$= \frac{1}{2}\{[(|D\rangle + |S\rangle) \otimes |V\rangle] + [(|D\rangle - |S\rangle) \otimes |H\rangle]\} \tag{2.17}$$

$$= \frac{1}{\sqrt{2}}\left\{|D\rangle \otimes \frac{1}{\sqrt{2}}(|V\rangle + |H\rangle) + |S\rangle \otimes \frac{1}{\sqrt{2}}(|V\rangle - |H\rangle)\right\}.$$

But these superpositions of $|V\rangle, |H\rangle$ turn out to correspond to linear polarization along the 45° and 135° axes!

$$\left|45°\right\rangle = \frac{1}{\sqrt{2}}\,(|V\rangle + |H\rangle) \tag{2.18}$$

$$\left|135°\right\rangle = \frac{1}{\sqrt{2}}\,(|V\rangle - |H\rangle).$$

So

$$|\Psi'(t_f)\rangle = \frac{1}{\sqrt{2}}\left\{\left(|D\rangle \otimes |45°\rangle\right) + \left(|S\rangle \otimes |135°\rangle\right)\right\}. \tag{2.19}$$

So equation 2.16 (correctly) predicts intensity $1/2I$ at each detector. Because $|45°\rangle$ and $|135°\rangle$ are orthogonal vectors, making the polarization axis of the lower beam horizontal has removed all interference between the high and low beams. This can be verified experimentally by checking that each detects intensity $1/2I$ even when the path length of one or other beam is varied.

You can restore interference by rendering the high and low paths indistinguishable. Simply place a polarizer with polarization axis oriented at 45° to the vertical to intercept anything emerging from the interferometer. Whatever passes this polarizer might be supposed to have come by either the high or the low path—but if so there is now no way to tell which. (Of course this will reduce the total detected intensity by half.) Why does this work?

If the beam emerging from the interferometer is passed through a polarizer with polarization axis oriented at 45° to the vertical, the second term in equation 2.19 makes no contribution to the predicted intensity. Only detector *Different* is (correctly) predicted to fire. As the path length is now varied, the translational state in the first term of 2.19 acquires a corresponding $|S\rangle$ component, correctly predicting the relative amounts of the total reduced intensity $1/2I$ detected by *Different* and *Same*. The roles of the first and second terms are reversed in correctly predicting interference with the polarization axis at 135°.

These examples should serve to illustrate the fertility of a modeling technique that represents a beam of matter by a state vector $|\psi\rangle$ (or $|\Psi\rangle$), and partial beams into which it may be split and recombined by component state vectors whose (linear) superposition equals $|\psi\rangle$ (or $|\Psi\rangle$). But what exactly does this technique enable us to do, and how much does its success tell us about the properties and behavior of whatever it is that composes the beams themselves?

By applying the technique one is able to predict the intensities detected by different devices, each placed to intercept a different part of the beam emerging from an interferometer. Since any real device will fail to detect some of a beam it encounters, the technique is usually applied to predict the *relative* intensities detected, assuming a detector's efficiency is the same no matter which partial beam it is placed to detect (an assumption that may be independently checked). Notice that while this technique required one to assign a state vector to a beam one can do so without saying anything about what the beam is made of. It is unimportant to its success whether we think of this as a beam of particles, of waves, or of something else entirely. The physical nature of the system to which we ascribe a quantum state may play a role in the decision to assign state $|\Psi\rangle$ rather than $|\psi\rangle$ to model a *polarized* beam. But making that decision still does not commit one to a claim about its physical composition. To avoid any such commitment one should simply say that a state vector is assigned to a *quantum system*.

Nevertheless, as I shall explain in Chapter 5, it is now possible to conduct interference experiments on beams of quite large molecules in which individual molecules stick to a detection screen and are then separately imaged. Whatever attitude one adopts to the claim that a single photon detector detects the individual photons that compose a beam of light, it is hard to deny that the large molecules composing these beams are detected singly. Consider any experiment in which a beam composed of single particles passes through a balanced Mach–Zehnder interferometer and these particles are individually detected by single particle detectors *Different, Same*. Suppose we take the intensity registered by each detector to be the number of particles detected by it per second. If the beam is sufficiently attenuated, this number will vary significantly from one second to the next, as individual particles are detected at random intervals of time. What the modeling technique predicts is the *average* rate of detection, not the exact number that will be detected in any particular second.

More significantly, for a typical state $|\psi\rangle$ (or $|\Psi\rangle$) a model does not permit one to predict whether an individual particle p going through the interferometer will be detected by *Different* rather than *Same*: only for a special state such as $|S\rangle$ can one predict (with probability 1) for each particle p that it will be detected at *Same* rather than *Different*. One might hold out the hope that some further feature will distinguish a particle p_1 detected at *Different* from a particle p_2 in the same beam detected at *Same*. But no extension of the present modeling technique by analogy with the introduction of the vector $|\pi\rangle$ to represent some additional feature of the beam (its polarization, in that case) can alter its fundamentally probabilistic character: the fact that all it predicts are *expected average* detected intensities, or (equivalently) the *probability* that an individual particle in a beam will be detected by a given detector. This does not preclude the possibility of some quite different way of representing the nature, properties, and behavior of beam elements that would enable us to predict for each element that it would be detected by *Different* or that it would be detected by *Same*. But in subsequent chapters we shall see why any such scheme would at least have some very peculiar features.

2.6 Measurement

Most people are familiar with a device called a Geiger counter. When placed near a suitable radioactive object, this device emits a click every time it registers something produced by the decay of a radioactive atom in the object. Modern detectors rarely emit a sound when registering a particle in a beam emerging from an interferometer, but it is still common to say the device clicks when registering the presence of a particle. Notice that what is directly measured here is the *position* of the particle—the measurement locates it where the detector is. In some cases this position is measured very precisely to within a few millionths of a millimeter on the surface of a detection screen. What the modeling technique enables one to predict is then the probability that a particle emerging from the beam will be detected at one such location rather than another: the detector "click" then indicates a measurement of its position. But a detector "click" does not always simply indicate a measurement of the position of the particle it registers: it may, for example, indicate a measurement of the particle's polarization or energy.

Consider the example recently discussed in which an input beam represented by state $|\Psi(t_i)\rangle = |\psi(t_i)\rangle \otimes |V\rangle$ passes through a Mach–Zehnder interferometer with a half-wave plate set to intercept the low partial beam. The state of the beam at the detectors is represented by $|\Psi'(t_f)\rangle$ in equation 2.19. Any particle detected by *Different* will then be measured as polarized at 45° while any particle detected by *Same* will be measured as polarized at 135°, so a detector "click" can now be taken to indicate measurement of a particle's polarization as well as its position.

We saw that by representing the state of a beam by a vector in an abstract space it is possible to predict the probability that a detector will register the presence of a particle in that beam after it is subjected to various transformations. We now see that a detector can also be taken to measure the position or polarization of the detected particle. This generalizes. For each system (like a particle in a beam) to which quantum theory may be applied there is a set of magnitudes, often called dynamical variables because their values may be expected to take different values at different moments. These values are all real numbers: it takes three real numbers (x, y, z) to precisely specify the location of a particle, each of which counts as the value of a different dynamical variable. The notion of a dynamical variable is inherited from classical physics.

In Newtonian mechanics we find a dynamical variable corresponding to each of the three components (p_x, p_y, p_z) of a particle's *momentum* (here understood as a vector **p** equal to the product of its mass m and velocity **v**, so $\mathbf{p} = m\mathbf{v}$); to its kinetic energy $T = 1/2\, mv^2$; total energy E; and each component (L_x, L_y, L_z) of angular momentum **L**. In classical electromagnetism we find a dynamical variable corresponding to each of the three components (E_x, E_y, E_z) of the electric field at each point, and also of the magnetic field (B_x, B_y, B_z) at that point. In classical physics light is identified with a freely propagating electromagnetic wave whose general polarization state is specified by two numbers, each of which can be taken as the value of a dynamical variable: just one of these is needed to specify a state of linear polarization—the orientation of the fixed axis along which the electric field varies in a plane orthogonal to the light's direction of propagation.

By assigning a suitable state vector to model the passage of a beam of matter through an interferometer, for each dynamical variable characterizing that matter one can predict the probability that its measurement will yield a value in any range of real numbers. More briefly, what this quantum modeling technique generates are the probabilities for each possible outcome of a measurement of any such magnitude. It would be natural to assume that a measurement outcome simply reveals the property it records—that (barring measurement errors) if the outcome records that a particle's x-position lies between x_1 and x_2 then that is because the particle *had* a determinate x-position immediately prior to its measurement and that position did indeed lie between x_1 and x_2. But reflection on what happens inside an interferometer should at least subject this assumption to reasonable doubt. For example, while two partial beams in a neutron interferometer are separated by up to several centimeters each individual neutron passing through the interferometer is sensitive to conditions along both paths. This would most naturally be explained by the hypothesis that its location is not restricted to just one path within the device, even though a *measurement* of its position would never record its presence on both paths at once. Moreover the structure of the full quantum algorithm that generalizes the modeling technique I have introduced supports strong arguments against the assumption that *every* dynamical variable of a physical system *always* has a precise real value that measurement would simply reveal. Two such arguments are presented in Appendix B.

To use a modeling technique that specifies probabilities of measurement outcomes one has to know what counts as a measurement and to be able correctly to identify its outcome. What qualifies a device as a detector, and what is the significance of the "clicks" it emits? The unwritten lore passed down among experimental physicists may have come to embody rules of thumb adequate to answer such questions when they arise in day-to-day laboratory practice. But a fundamental theory should be capable of supplying explicit answers when applied to the device and its interaction with whatever it might be supposed to measure.[5] John Bell wrote a famous paper entitled "Against 'measurement' " in which he argued [2004, p. 215] that the word 'measurement' is the worst of several words that "however legitimate and necessary in practice, have no place in a *formulation* [of any serious part of quantum mechanics] with any pretension to physical precision". He there noted the unsatisfactory character of a formulation that merely predicts probabilities *of measurement outcomes*. In Chapter 6 we shall locate among Dirac's classic formulation of the principles of quantum mechanics another place where the term 'measurement' appears—in a principle on which Bell focused open sarcasm as well as some of his strongest objections! But it is now time to consider what this modeling technique tells us about what is actually going on inside the interferometer.

2.7 What's Really Happening?

Quantum theory is often taken to be radical because of the strangeness of the world it portrays. It would certainly be strange if the theory were to portray a world in which a subatomic particle could be in two places at once. Does the quantum modeling technique I have described show that a neutron (say) is in two places at once as it passes through an interferometer? No such conclusion may be drawn simply from the success of the technique in predicting the probability that the particle will subsequently be detected by one device rather than another. If the sole function of the state vector

---

[5] Heisenberg [1971] reports a conversation in which Einstein put it this way: "It is the theory which decides what we can observe".

assigned to particles in the beam is to predict this probability, then its assignment has no implications about any particle's location either inside or outside the interferometer. But the superposition principle at least suggests that the quantum state plays an additional role by describing properties of a system to which it is assigned. Indeed, it was a key tenet of the Copenhagen interpretation of quantum theory that a system's quantum state provides a complete description of that system. If that's right, then how do (Superposition) and (Linearity) describe the position of a particle as it passes through the interferometer?

Here is what Dirac [1930] says about this:

> There will be various possible motions of the particles or bodies [composing an atomic system] consistent with the laws of force. Each such motion is called a *state* of the system.... A state of a system may be defined as an undisturbed motion that is restricted by as many conditions or data as are theoretically possible without mutual interference or contradiction. (p. 11)

> The general principle of superposition in quantum mechanics applies to [these] states... It requires us to assume that between these states there exist peculiar relationships such that whenever the system is definitely in one state we can consider it as being partly in each of two or more other states. The original state must be regarded as the result of a kind of *superposition* of the two or more other states, in a way that cannot be conceived on classical ideas. (p. 12)

He clearly views assignment of a state vector to a system as describing its physical (dynamical) state. Consistent with this view, he thinks the superposition principle represents a physical relationship among these states. If that's right, we should say that a neutron passing through an interferometer is partly in a physical state represented by $|high\rangle$ and partly in a physical state represented by $|low\rangle$. That does not mean that it follows both paths at once, but rather that it partly follows the high path and partly follows the low path ... "in a way that cannot be conceived on classical ideas"! Notice that by adjusting the values of the coefficients (h, l) one can construct a continuously infinite set of different vectors of the form (Superposition), each of which would then represent a physically different way for a neutron to encounter the detectors after partly following the high path and partly following the low path. Notice also that we have tacitly assumed here that the vector $|high\rangle$ represents a physical state of following the high path. Dirac [1930] says it is all right for us to speak this way:

> The expression that an observable[6] [such as *being in the high beam*] 'has a particular value' for a particular state is permissible in quantum mechanics in the special case when a measurement of the observable is certain to lead to the particular value... (p. 46)

---

[6] Dirac [1930] used the term 'observable' to refer to a dynamical variable corresponding to a particular kind of mathematical object in quantum theory. It is now used to refer to a dynamical variable corresponding to a self-adjoint operator—see Appendix A. A magnitude that takes value +1 for a particle in the high beam but −1 for a particle in the low beam is an observable in either usage.

Notice finally that since we can rewrite (Superposition) as

$$|high\rangle = \frac{1}{h}(|\psi\rangle - l|low\rangle)$$

it would be equally correct to say that a neutron following the high path is partly following the low path and partly following the particular superposition $|\psi\rangle$ of the high and low paths. I have no idea what any of this means. It certainly does not mean that a neutron can be in two places at once!

2.8 Summary

It is time to take stock. I described a number of experiments in which a beam of matter was caused to interfere with itself, and different parts of the beam were then detected by different detectors. I showed how one could predict the relative intensities these would detect using a model in which the state of the beam in one scenario was represented by a vector that may be expressed as a linear combination of vectors representing it in other scenarios. A more careful examination revealed the need to take these intensities as reflections of the relative probabilities for a measurement of some magnitude on each individual physical system in some sense associated with the beam to yield one rather than another outcome. But I pointed out that if this modeling technique were to be generalized into an acceptably formulated physical theory then some way would have to be found either to use that theory to provide a non-circular analysis of this notion of measurement or to reformulate the theory without using any term like 'measurement'. After noting that Dirac's formulation of quantum theory did neither, I went on to critically examine his interpretation of the crucial principle of superposition of states. While some such principle must figure in any acceptable formulation of quantum theory that generalizes the modeling technique I have illustrated, there is no need to interpret it as representing a physical relationship among dynamical states of physical systems. Nor need a quantum state be taken to represent the dynamical condition or behavior of the system to which it is assigned.

For future reference I state here five principles of the quantum modeling technique illustrated in this chapter.

1. Each *quantum system* is assigned a *quantum state*, a vector $|\psi\rangle$ in an abstract vector space.
2. Any *superposition* of state vectors $|\alpha\rangle, |\beta\rangle$ is also a state vector $|\psi\rangle = c_1|\alpha\rangle + c_2|\beta\rangle$, where c_1, c_2 are complex numbers.
3. State vectors evolve *linearly*: if states $|\alpha\rangle, |\beta\rangle$ evolve from t_1 to t_2 as $|\alpha(t_1)\rangle \implies |\alpha(t_2)\rangle, |\beta(t_1)\rangle \implies |\beta(t_2)\rangle$, then $|\psi\rangle$ evolves as $|\psi(t_1)\rangle = c_1|\alpha(t_1)\rangle + c_2|\beta(t_1)\rangle \implies |\psi(t_2)\rangle = c_1|\alpha(t_2)\rangle + c_2|\beta(t_2)\rangle$.
4. This evolution is *unitary*: that is, it preserves the norm $||\psi||$ of every vector $|\psi\rangle$ as well as the "angle" $\langle\varphi|\psi\rangle$ between every pair of vectors.

5. If a, b are two possible values of a dynamical variable A, then the relative *probability* of recording a rather than b as the outcome of a measurement of A on a system in state $|\psi\rangle$ is determined by the ratio of two numbers—$|c_1|^2 : |c_2|^2$, where c_1, c_2 are the coefficients appearing when $|\psi\rangle$ is written as a linear superposition of a *special* set of vectors determined by A.

For many types of system it is the Schrödinger equation which specifies exactly how the state vector evolves linearly and unitarily in accordance with principles 2–4. Principle 5 is a special case of what is known as the Born rule: Appendix A gives details of how a dynamical variable comes to be associated with the special set of vectors to which the principle refers. Quantum theory arises as a natural generalization of these five principles to apply to a wider class of systems.

3

Entanglement

3.1 What is Entanglement?

Schrödinger noticed something peculiar about the way we represent the states of quantum systems and called it 'entanglement' ([1935], p. 555):

> When two systems, of which we know the states by their respective representatives, enter into temporary physical interaction due to known forces between them, and when after a time of mutual influence the systems separate again, then they can no longer be described in the same way as before, viz. by endowing each of them with a representative of its own. I would not call that one but rather *the* characteristic trait of quantum mechanics, the one that enforces its entire departure from classical lines of thought. By the interaction the two representatives (or ψ-functions) have become entangled.

In this chapter I begin to explore the significance of this peculiarity and to evaluate Schrödinger's claim about its importance. Chapter 4 assesses the implications of entanglement for non-locality.

By the word 'representative' Schrödinger means a mathematical object representing a quantum state. It is now customary to follow Dirac in representing quantum states mathematically by state vectors, as I did in Chapter 2. Schrödinger refers to a ψ-function: this is often called a wave function. The wave function $\psi(x, y, z)$ of a single (spinless, massive) particle is a complex-valued function of the spatial coordinates x, y, z—magnitudes whose values represent distance from some chosen point along each of three perpendicular directions. Such wave functions may be added and multiplied by (complex) numbers. After defining an appropriate inner product, the set of wave functions forms a vector space (an infinite-dimensional Hilbert space—see Appendix A, § A.1). A wave function is then a vector in that space—a state vector.

In this quote Schrödinger took quantum systems to be localized, individual physical systems capable of interacting with each other. This is a natural way of thinking of the electrons, protons, neutrons, atoms, and molecules generally taken to compose beams of matter like those we considered in the previous chapter. It is not so clear we can think of photons this way. Photons do not interact with each other directly through forces; they have zero (rest) mass; and despite what Einstein said when introducing the concept it has turned out to be very difficult to describe them as localized objects. Schrödinger may have been thinking only of systems of massive particles here: after all, his eponymous equation (an explicit form of principle 4 at the end of Chapter 2) can't be straightforwardly applied to light. In any case it is common to model light using state vectors associated with multiple photons in close analogy with those used to model two or more massive particles, and those vectors may also be entangled in very much the same way.

Just as for Dirac's superposition principle, entanglement is here introduced as a relation among state vectors assigned to quantum systems, not among the systems themselves. Only if state vectors described the condition or behavior of physical systems to which they are assigned would this entanglement relation imply any corresponding physical relation among the systems themselves. In Chapter 2 we failed to make sense of the idea that a superposed state vector describes any physical system as itself either wholly or in part in some associated combination of physical conditions. This should prompt one to be wary of the suggestion that state entanglement describes or represents some peculiar connection between or among physical systems. But before evaluating that suggestion we need to understand this entanglement relation among states.

Consider the state vector (2.19):

$$|\Psi'(t_f)\rangle = \frac{1}{\sqrt{2}} \left(|D\rangle \otimes |45°\rangle + |S\rangle \otimes |135°\rangle \right) \tag{3.1}$$

assigned in Chapter 2 to the beam emerging from the interferometer there described whose input beam had been assigned state vector

$$|\Psi(t_i)\rangle = |\psi(t_i)\rangle \otimes |V\rangle. \tag{3.2}$$

Dirac would allow one to say that the input beam was vertically polarized (V), since it would (with probability 1) pass a polarizer whose axis was vertical: and he would also allow one to assign it some property P_{in} corresponding to the fact that its full intensity would be detected by a device set to measure a magnitude that determined a special set of vectors including $|\psi(t_i)\rangle$. But he would not permit one to say that the output beam had any determinate polarization (neither 45° nor 135°): and he would permit one to say neither that it had some property P_A corresponding to the fact its full intensity would be detected by a device set to measure a magnitude A determining a special set of vectors including $|D\rangle$, nor that it had some property P_B similarly associated with vector $|S\rangle$. All Dirac would permit one to say about the output beam assigned state vector $|\Psi'(t_f)\rangle$ is that it manifested a peculiar kind of *correlational* property—partly P_A polarized at 45° and partly P_B polarized at 135°.

Assuming the beam is composed of particles (perhaps neutrons or photons), one could assign each individual particle the vector $|\Psi(t_i)\rangle$ before and $|\Psi'(t_f)\rangle$ after it passed through the interferometer. But only $|\Psi(t_i)\rangle$ could be thought to represent each of the particle's translational and polarization states separately (by $|\psi(t_i)\rangle$, $|V\rangle$ respectively). The state vector $|\Psi'(t_f)\rangle$ is not just the product $\otimes$ of a translational and a polarization state vector. Indeed it cannot be expressed as a product $|\Psi'(t_f)\rangle = |\varphi\rangle \otimes |\pi\rangle$

of *any* pair consisting of a translational state vector $|\varphi\rangle$ and a polarization vector $|\pi\rangle$. A state vector that cannot be expressed as a product of vectors in two such separate spaces is called *non-separable* (for details see Box 3.1).

Box 3.1 Some Non-separable State Vectors

If the states of two distinct quantum systems are represented in Hilbert spaces $\mathcal{H}_1, \mathcal{H}_2$ respectively, then the state of the compound system they compose is represented in the *tensor product* Hilbert space $\mathcal{H}_{12} = \mathcal{H}_1 \otimes \mathcal{H}_2$. These systems need not be *physically* distinct. The spin and translational states of a massive particle are represented in distinct Hilbert spaces, and so are the polarization and translational states of light photons. In each case, the total state is represented in the tensor product of these two Hilbert spaces.

A set of vectors $\{|\phi_i\rangle\ i = 1, 2, \ldots, n\}$ in a Hilbert space $\mathcal{H}$ is *linearly independent* if and only if the linear sum $\Sigma_i c_i |\phi_i\rangle = 0$ if and only if $c_i = 0$ for all i. The dimension of $\mathcal{H}$ is the maximum number of linearly independent vectors in $\mathcal{H}$: any vector in $\mathcal{H}$ may be expressed as a linear sum of *basis vectors* from a linearly independent set. If $\mathcal{H}_1, \mathcal{H}_2$ have dimensions m, n respectively, then $\mathcal{H}_{12}$ has dimension $m \times n$: if $\{|\phi_i\rangle\}, \{|\psi_j\rangle\}$ are basis vectors for $\mathcal{H}_1, \mathcal{H}_2$ respectively, then $\{|\phi_i\rangle \otimes |\psi_j\rangle\ i = 1, 2, \ldots, m; j = 1, 2, \ldots, n\}$ are basis vectors for $\mathcal{H}_{12}$. This means that any vector in $\mathcal{H}_{12}$ maybe expressed as a linear sum $|\Psi\rangle = \Sigma_{i,j} c_{ij} |\phi_i\rangle \otimes |\psi_j\rangle$ for some $\{c_{ij}\}$.

However, a typical vector in $\mathcal{H}_{12}$ cannot be expressed in the form $|\Psi\rangle = |\phi\rangle \otimes |\psi\rangle$, with $|\phi\rangle \in \mathcal{H}_1, |\psi\rangle \in \mathcal{H}_2$: states that cannot be written this way are called *non-separable*. Here are some relevant examples:

1. Consider the spin state $|o\rangle = 1/\sqrt{2}(|\uparrow\rangle \otimes |\downarrow\rangle - |\downarrow\rangle \otimes |\uparrow\rangle)$ of two spin $1/2$ particles. $\{|\uparrow\rangle, |\downarrow\rangle\}$ are basis vectors for the two-dimensional spin space of each particle. Suppose that $|o\rangle = |\phi\rangle \otimes |\psi\rangle$, with $|\phi\rangle \in \mathcal{H}_1, |\psi\rangle \in \mathcal{H}_2$: expressing each as a linear sum of basis vectors

$$|\phi\rangle = c_1 |\uparrow\rangle + c_2 |\downarrow\rangle$$
$$|\psi\rangle = d_1 |\uparrow\rangle + d_2 |\downarrow\rangle.$$

Hence

$$|o\rangle = |\phi\rangle \otimes |\psi\rangle$$
$$= (c_1 |\uparrow\rangle + c_2 |\downarrow\rangle) \otimes (d_1 |\uparrow\rangle + d_2 |\downarrow\rangle)$$
$$= c_1 d_1 (|\uparrow\rangle \otimes |\uparrow\rangle) + c_1 d_2 (|\uparrow\rangle \otimes |\downarrow\rangle) + c_2 d_1 (|\downarrow\rangle \otimes |\uparrow\rangle) + c_2 d_2 (|\downarrow\rangle \otimes |\downarrow\rangle)$$

from which it follows that

$$c_1 d_1 = c_2 d_2 = 0;$$
$$c_1 d_2 = -c_2 d_1 = 1/\sqrt{2}.$$

But these equations have no solution. Hence $|o\rangle \neq |\phi\rangle \otimes |\psi\rangle$.

> 2. Neither of the polarization states $|\chi\rangle, |\eta\rangle$ of two photons can be expressed as a tensor product of polarization states in their two-dimensional polarization Hilbert spaces, for very similar reasons.
>
> 3. The state of a single particle $|\Psi'(t_f)\rangle \neq |\varphi\rangle \otimes |\pi\rangle$ for *any* pair consisting of a translational state vector $|\varphi\rangle$ and a polarization vector $|\pi\rangle$. Even though the translational Hilbert space is infinite-dimensional, the linear independence of the set $\{|S\rangle, |D\rangle\}$ is enough to prove this.

Non-separability is a necessary formal condition for entanglement of the state vectors of physical systems: but, as the example in equation 3.1 shows, it is not sufficient. We have a case of entanglement only if the state vector assigned to a compound pair (triple, quadruple, etc.) of physical systems cannot be expressed as a product $\otimes$ of state vectors individually assigned to its component subsystems. Even if the state assigned to a system is not separable into a translational state vector and a polarization vector, this does not count as a case of entanglement unless that system is composed of a pair of systems to which such vectors might have been assigned. But if quantum states are assigned only to *physical* systems then there is no reason to suppose it is a physical component rather than the particle itself that bears either its polarization or its translational properties. So the non-separable state vector (3.1) does not represent entanglement between the state vectors of two physical systems.

Bohm ([1951], sections 22.15–22.18) described an important example of entanglement between the state vectors assigned to two clearly physical systems—a pair of atoms formed by the dissociation of a diatomic molecule such as hydrogen. In this case it is the spin angular momentum states of the atoms that are entangled if the molecule is assigned a state vector that implies with probability 1 that a measurement of any component of angular momentum will yield outcome zero. A measurement on either atom of its spin component along any direction yields just one of two values which we can for convenience designate +1 (or *up*) and −1 (or *down*) with respect to that direction. If an atom were assigned spin state $|\uparrow\rangle$ then it would certainly (with probability 1) yield outcome *up* for a measurement of spin along the $\uparrow$ direction, while if it were assigned spin state $|\downarrow\rangle$ then it would certainly (with probability 1) yield outcome *down* for that same measurement. The normalized spin state vector assigned to the pair of atoms after dissociation may be written as

$$|o\rangle = \frac{1}{\sqrt{2}}(|\uparrow\rangle \otimes |\downarrow\rangle - |\downarrow\rangle \otimes |\uparrow\rangle) \quad (3.3)$$

where the vector before (after) each $\otimes$ lies in the vector space associated with the first (second) atom. It doesn't matter which order we take them in because assigning $-|o\rangle$ rather than $|o\rangle$ would yield the same probability for any measurement outcome. Nor does it matter which direction we choose to indicate as $\uparrow$: for example, after making this choice the state $|o\rangle$ may equally well be written with respect to a different direction $\rightarrow$ as

$$|o\rangle = \frac{1}{\sqrt{2}}\left(|\rightarrow\rangle \otimes |\leftarrow\rangle - |\leftarrow\rangle \otimes |\rightarrow\rangle\right). \tag{3.4}$$

The spin state $|o\rangle$ of the pair of atoms is non-separable, and the individual atoms' spin states are entangled. Dirac would allow one to say "the pair has spin-component zero along every direction" but not to say anything about the spin components of the individual atoms, despite the strong anticorrelations apparent in a joint measurement of their spin components along any one direction which would (with probability 1) yield opposite outcomes: *up* on one if and only if *down* on the other. This same state may be assigned to a very different system.

Positronium is like a miniature atom composed of an electron and a positron (its antimatter "twin"; a particle with the same mass and equal but opposite charge). The lowest energy state of a positronium "atom" has spin zero. This is unstable because as particle and antiparticle the electron and positron can mutually annihilate, releasing energy in the form of a pair of γ-rays (high-energy photons). It is hard directly to confirm the anticorrelations predicted by $|o\rangle$ for a direct joint measurement of the same spin component on both electron and positron. Instead one can measure a corresponding anticorrelation between the γ-rays emitted when these annihilate. One can even do this at home![1] Though the experiment does demonstrate entanglement, this is strictly an example of entanglement between the two γ-ray photons assigned polarization state

$$|\chi\rangle = \frac{1}{\sqrt{2}}\left(|H\rangle \otimes |V\rangle + |V\rangle \otimes |H\rangle\right). \tag{3.5}$$

It is nice to start with this case of photon entanglement because the actual discrete clicks emitted at random by the Geiger counters cry out for an interpretation in terms of detection of individual particles, and the way their rates vary as the counters are moved around strongly suggests these particles follow definite paths away from their source. But in quantum optics laboratories entangled lower-energy photons can now be produced with much greater control, then manipulated and detected with vastly greater precision. I shall describe a few experiments involving photon entanglement, beginning with an experiment that has been taken as an even more convincing demonstration that a light beam is composed of discrete photons.

Figure 2.2 depicted a beam incident on a beam splitter. Suppose this is a very weak beam of light. According to the classical wave theory, no matter how low its intensity, part of the light wave will be transmitted to detector D_1 and part will be reflected to detector D_2. It might take some time for a detector to collect enough energy from the wave to register its detection by a "click". But when the detectors did start clicking there is no reason why the detectors should not sometimes both click at once. That would

---

[1] See George Musser's description of how he did so quite cheaply in his blog entries at https://blogs.scientificamerican.com/critical-opalescence/how-to-build-your-own-quantum-entanglement-experiment-part-1-of-2/ and https://blogs.scientificamerican.com/critical-opalescence/how-to-build-your-own-quantum-entanglement-experiment-part-2-of-2/

happen sometimes, for example if when D_1 clicks is independent of when D_2 clicks—if their clicks are uncorrelated.

Very weak light may be detected by a device called a photomultiplier. Light entering a photomuliplier induces a rapid avalanche of electrons in a high-voltage tube, causing a brief electrical pulse strong enough to be easily recorded. Such recording can be arranged to occur only if a pulse is received during a carefully controlled, very short interval of time called a gate. One could study correlations between clicks of D_1 and D_2 just by opening the gate every so often and seeing which, if either, of them generated a detection pulse that passed the gate. But the recording efficiency would be greatly improved if one knew the best times to open the gate. That is where photon entanglement comes in.

Suppose a beam of light is composed of individual photons. If we knew when to expect a photon we could arrange to open the gate just after it reached the beam splitter to determine whether D_1 and/or D_2 then detected any light. If this photon was part of an entangled pair emitted at the same time but in opposite directions, we could arrange to open the gate just when its twin is detected. A more practical arrangement is to set up an electronic coincidence circuit that automatically registers coincidences between a pulse from a third detector placed to intercept light emitted in the opposite direction and one or more pulses from D_1, D_2. Such an experiment [Grangier, Roger, and Aspect, 1986] revealed strong *anti*correlations between clicks of detectors D_1 and D_2. This confirms the hypothesis that the light incident on the beam splitter was composed of photons, each of which is either transmitted or reflected as a whole. It disconfirms the classical wave theory of light. But it does more: the experimental design incorporated the auxiliary assumption that each of these photons was one of a pair emitted in an entangled polarization state

$$|\eta\rangle = \frac{1}{\sqrt{2}} \left(|H\rangle \otimes |H\rangle + |V\rangle \otimes |V\rangle \right). \tag{3.6}$$

Not only does the experiment provide a persuasive reason to believe a light beam is composed of photons: it shows the value of thinking of the state of each of these as entangled with that of a second photon.

3.2 Entanglement and Incompleteness

The polarization state $|\eta\rangle$ is a close analog of the entangled spin state $|o\rangle$. Bohm used the latter to illustrate a famous argument against the claim that a quantum state completely describes a system to which it is assigned. This is known as the EPR argument after the initial letters of its coauthors' last names [Einstein, Podolsky, and Rosen, 1935]. The argument rests on two explicit assumptions. The first states a necessary condition for a physical theory to offer a complete description of physical reality, while the second states a sufficient condition for the existence of an element of that physical reality.

46   ENTANGLEMENT

Completeness Assumption
Every element of the physical reality must have a counterpart in the physical theory.

Reality Criterion
If, without in any way disturbing a system, we can predict with certainty (i.e. with probability equal to unity) the value of a physical quantity, then there exists an element of physical reality corresponding to this physical quantity.

EPR's argument is unnecessarily tortuous (in private correspondence Einstein himself later wrote that its essential point was smothered by the formalism). But in outline it goes like this. Consider a pair A, B of atoms or other physical systems whose joint spin state at some time t is represented by $|o\rangle$. These systems may be supposed to be far apart and not interacting through any forces at t. Nevertheless, quantum theory implies that by measuring the spin component of A in direction d at t one can predict with probability 1 the outcome of a measurement of the same component of spin on B at t: its value will certainly be opposite but equal. But a measurement on A cannot instantaneously disturb B since A, B are far apart and noninteracting. By the reality criterion it follows that at t there is an element of physical reality corresponding to the spin component of B in direction d. But the measurement on A could not have instantaneously brought this into being, since A, B are far apart and noninteracting. So whether or not one chooses to measure A there is an element of reality corresponding to the spin component of B in direction d at t.

Repeating this argument for *any* direction d yields the conclusion that for *every* component of B's spin, an element of physical reality corresponds to that quantity at t. But there is no counterpart in quantum theory of these elements of physical reality. In the state $|o\rangle$ there is no counterpart of *any* component of spin of B. Even if a measurement on A were to instantaneously change their joint quantum state to make it separable—of the form $|\alpha\rangle \otimes |\beta\rangle$—there is no counterpart for *every* component of spin on B in any state $|\beta\rangle$. At most (following Dirac's prescription) a state $|\beta\rangle$ could be taken to represent B's spin component as ± 1 in a single direction (and $\mp$ in its opposite), while failing to represent its spin component in any other direction. So description of physical reality by a quantum state is incomplete: quantum theory fails to meet the necessary condition spelled out by the completeness assumption.

Besides the two assumptions they made explicit, EPR's argument also rested on two implicit assumptions noted by Einstein later after giving a version of his own:

One can escape from this conclusion only by assuming that the measurement [on A] telepathically changes the real situation of [B], or by denying independent real situations as such to things which are spatially separated from each other. Both alternatives appear to me entirely unacceptable.   (Einstein, in Schilpp, ed. [1949], p. 85).

I'll look more closely at EPR's assumptions in Chapter 4. We will see there how profound theoretical analyses by Bell prompted an explosion of further work on entangled states, including many experiments exploring the consequences of quantum entanglement such as those I'll now describe.

3.3 Entanglement Exhibited

I already explained why an experiment by Aspect and collaborators has been taken to provide a vivid demonstration of the particulate nature of light. While that experiment used an entangled polarization state, it was not intended as a detailed investigation of correlations predicted by that state. But in an earlier experiment Aspect and colleagues [1981] had used state $|\eta\rangle$ to demonstrate and study the EPR correlations themselves. Their test results displayed statistical patterns in close conformity to those expected by assigning state $|\eta\rangle$ to photon pairs when the linear polarization of each photon was measured by detectors placed some forty feet apart.

Two detectors were placed at each location, and an event recorded when a pair of photons simultaneously triggered one detector at each location. Which detector was triggered at a location indicated the outcome of a measurement of linear polarization with respect to some axis. That axis was chosen independently at each location: what these axes were could also be varied from one run of the experiment to the next. When the axis at one location coincided with that at the other location the outcomes were indeed almost perfectly correlated, as the state $|\eta\rangle$ would lead one to expect after taking account of experimental imperfections. The precise way in which the statistical correlations observed were found to depend on the difference between the axes with respect to which the polarization was measured at the two locations is just what assignment of state $|\eta\rangle$ would lead one to expect. This fact will prove crucial to Chapter 4's discussion of non-locality.

At the end of the second paper in the two-part series where he first wrote about entangled states, Schrödinger seriously entertained the idea that it might not be correct to assign such a state to a pair of systems once they became separated. Of a plausible alternative, non-entangled state assignment, he said:

I would call it a possible one, until I am told, either why it is devoid of meaning or with which experiments it disagrees.   [1936]

The results of Aspect's experiments on polarized pairs of photons certainly disagree with Schrödinger's alternative, while agreeing with entangled state $|\eta\rangle$. Subsequent experiments have confirmed predictions based on assignment of entangled polarization states to photon pairs when the measurements of the photons' polarization occurred even further apart.[2]

There are experiments on pairs of widely separated particles involving entanglement of neither polarization nor spin, but of states corresponding to two different paths each particle might have been thought to take through an interferometer. While similar in spirit to Chapter 2's discussion of Mach–Zehnder interferometers, such

---

[2] At the time of writing I believe the record separation is 143 kilometers [Herbst et al., 2014]. A Chinese satellite, Micius, was launched on August 15, 2016 to perform Quantum Experiments at Space Scale (QUESS). If successful, this is expected to demonstrate polarization entanglement between photons, one detected in Vienna, the other in Beijing.

48 ENTANGLEMENT

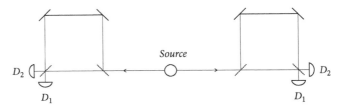

Figure 3.1 Franson interferometer, balanced

experiments also provide a vivid illustration of additional features of quantum interference involving multiple particles. The Mach–Zehnder interferometer was invented long before the development of quantum theory, to study interference of classical waves. While a classical wave model can account for gross features of single-particle quantum interference, it cannot be naturally extended to apply to multiparticle quantum interference phenomena like those displayed in the experiment of Gisin and collaborators [Tittel et al., 1998].

Gisin's group used what is called a Franson interferometer to verify quantum predictions based on assignment of an entangled two-photon state. The predictions concerned correlations between coincident detections of photons at two sites over 10 kilometers apart. Two detectors were present at each site, and the simultaneous detection of a photon by one rather than the other detector at each location was recorded as a coincidence event. Figure 3.1 is a highly schematic representation of a Franson interferometer, which I'll use to explain the logic of the experiment.

When a photon pair is emitted by the source one may be detected on the left while the other is detected on the right, in each case either by a detector labeled D_1 or by a detector labeled D_2. The detectors at each location click at random intervals, but at each location D_1 detects photons just as often on average as D_2. If the two wings of the interferometer are arranged with exact left–right symmetry, it is said to be balanced. For this balanced configuration the entangled state assigned to the photon pairs predicts a perfect correlation in each coincidence event—either both D_1 detectors will click, or both D_2 detectors will click. The experimental results bear this out, within the limits of error to which it is understood to be subject. One can represent this entangled state as a superposition of two components in a way that may seem familiar from Chapter 2:

$$|\psi\rangle = l(|long\rangle \otimes |long\rangle) + s(|short\rangle \otimes |short\rangle). \qquad (3.7)$$

But now each component represents a *two-photon* translational state. The first component is a state that Dirac would allow one to say represents both photons taking the long paths to their detectors, while the second component corresponds in this way to their both taking the short paths. Following his suggestion about how to understand the physical significance of superpositions, one could also consider $|\psi\rangle$ to represent a situation in which both photons partly follow their respective long paths and partly

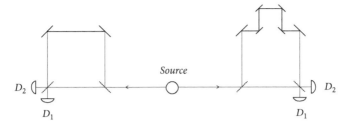

Figure 3.2 Unbalanced Franson interferometer

follow their respective short paths. But this cannot be a matter of each photon partly taking each path independent of the other, for that would be to neglect the *correlations* between their (partial!) behaviors.

Chapter 2 despaired of making sense of such talk of partial behaviors. Consider the alternative proposal that each photon wholly follows one path or the other here and that those paths are definitely correlated. I noted there that one could hold on to this view of a superposed state like this only if one were to admit that you can detect interference only when you have removed any possibility of determining which paths the detected particles took. At first sight it seems one *could* tell which paths the photons take in this case by timing how long it takes them to reach their respective detectors: this will be longer if they both take their respective long paths than if they take the short paths. But it turns out that this is impossible because the nature of the source is such that one cannot determine the precise moment at which it emits a pair of photons. Indeed, photons emitted by a source that did permit such precise timing of pair emission would yield no two-particle interference in a Franson interferometer!

As in Chapter 2, one can study interference by making small, carefully controlled changes in the length of one of these four paths so the interferometer is no longer balanced. Figure 3.2 illustrates one way of doing so.

To model the effect of such a change one must assign a modified entangled state to each photon pair. The modification consists in altering the complex argument of l and/or s without changing its modulus so that the new entangled state becomes

$$|\psi'\rangle = l'(|long\rangle \otimes |long\rangle) + s'(|short\rangle \otimes |short\rangle). \tag{3.8}$$

The argument of a complex number is represented mathematically by an angle, so this has the effect of changing only the relative phase angle between the two terms in the superposition by a known amount, depending on the alteration in relative path lengths. Just as altering the angle between the axes with respect to which linear polarization is measured in the preceding photon experiments shows how an entangled polarization state of two photons correctly predicts the corresponding correlations among measurement outcomes, so in this case altering the relative phase angle in the entangled state assigned to photon pairs in the Franson interferometer correctly predicts the corresponding variations in correlations among distant detector clicks.

Note that modifying the Franson interferometer in this way has no effect on the relative frequency with which the two detectors located in just one wing of the experiment click: there is no single-particle interference. That is why even the gross features of the experimental results cannot be understood using a classical wave theory of light. We saw that quantum state vectors may be represented by wave functions. But if one tries to model two-particle interference by representing an entangled state such as $|\psi\rangle$ or $|\psi'\rangle$ as a wave, this cannot be a wave in ordinary three-dimensional space. This is one way entanglement enforces departure from classical thought. Each of Chapters 4 and 6 will highlight a further, different, way that Schrödinger may well have had in mind when introducing this concept.

In introducing the notion, Schrödinger restricted his discussion of entanglement to the states of just two systems. But the idea naturally generalizes to entangled states of more than two systems. Entangled states of up to ten photons have been studied as I write this, and the number will very likely be higher by the time you read it.[3] There is one particular entangled state of three photons whose implications will prove especially interesting later:

$$|GHZ\rangle = \frac{1}{\sqrt{2}} \left(|H\rangle |H\rangle |H\rangle + |V\rangle |V\rangle |V\rangle \right). \tag{3.9}$$

This is a non-separable polarization state vector that may be assigned to three photons. I follow custom in labeling it $|GHZ\rangle$ in honor of the three physicists who first brought out some of its striking features [Greenberger, Horne, and Zeilinger, 1989].[4] As the notation indicates, it is a superposition of two states, in each of which quantum theory predicts perfect (probability 1) correlations between the outcomes of a joint measurement of linear polarization on each of three photons, all with respect to the same axis V orthogonal to their common plane of propagation: either all three photons will pass a polarizer with axis set along that axis, or all three will be blocked by it.

There are different ways of writing the state $|GHZ\rangle$. Interestingly, it may be written as a superposition of states involving not only linear but *circular* polarization.[5] Here is one way of doing this, in which a photon assigned state $|R\rangle$ (respectively $|L\rangle$) will certainly be measured to be right (respectively left) circularly polarized.

$$|GHZ\rangle = \frac{1}{2} \left(|R\rangle |L\rangle |45°\rangle + |L\rangle |R\rangle |45°\rangle + |L\rangle |L\rangle |135°\rangle + |R\rangle |R\rangle |135°\rangle \right). \tag{3.10}$$

---

[3] Construction of a working quantum computer is currently a major long-term technological goal. Since this is believed to require controlled entanglement of a much larger number of systems there is a strong incentive to increase this number.

[4] To simplify notation I have omitted the occurrences of the symbol '⊗' previously used to indicate the (associative) operation of forming a (tensor) product of vectors in lower-dimensional vector spaces.

[5] In a classical wave model, if light is linearly polarized the electric field always varies along a fixed axis orthogonal to the direction of propagation; if it is circularly polarized then this axis rotates as the wave moves along.

This form of the state $|GHZ\rangle$ predicts that if two photons are measured to have the same circular polarization then the third will certainly be found to be linearly polarized parallel rather than orthogonal to the 135° direction, while if they are measured to have opposite circular polarizations then the third will certainly be found to be linearly polarized parallel rather than orthogonal to the 45° direction.

It is not easy to produce triples of photons correctly assigned state $|GHZ\rangle$ and to use them to test these quantum theoretical predictions for strict (probability 1) correlations and anticorrelations between the outcomes of the relevant polarization measurements. But this has been done and the observed correlations were closely in accord with quantum theoretical predictions based on assignment of state $|GHZ\rangle$ [Pan et al., 2000]. If one asks what are the actual linear and circular polarizations of the three photons in state $|GHZ\rangle$, then that state provides no answer and Dirac would not permit one even to express one. Chapter 4 shows how perhaps the obvious answers to these questions are in serious tension with the rejection of action at a distance by both everyday thought and the rest of physics. Nor can these answers be reconciled with structural features of quantum models apparently critical to their successful application (see Appendix B.2).

Despite the suggestive evidence provided by experiments in which devices called photon detectors respond with discrete clicks, and experiments like that of Grangier, Roger, and Aspect [1986] supporting the hypothesis that a beam splitter acts independently on individual photons, one can remain skeptical.[6] Our best current quantum theory of light is a quantum field theory, a detailed examination of which shows how problematic it is to suppose that quantum theory describes light as composed of particles. I will have more to say about the ontology of quantum field theories (what things they are about!) in Chapters 12 and 13. But we have already seen reasons to remain skeptical about the particulate character of light.

Even if light does consist of photons, these don't display all the characteristic features we expect of particles. We saw the difficulties associated with the idea that a photon follows one path rather than another through a Mach–Zehnder interferometer: if all there is to light is a bunch of photons, it is hard to understand how the behavior of each individual photon depends on conditions on *all* available paths. Also, I already

---

[6] It is worth quoting (without endorsement!) the authors of a prominent recent article celebrating the centenary of Einstein's introduction of the photon hypothesis:

> When analysing quantum interference we can fall into all kinds of traps. The general conceptual problem is that we tend to reify—to take too realistically—concepts like wave and particle. Indeed if we consider the quantum state representing the wave simply as a calculational tool, problems do not arise. In this case, we should not talk about a wave propagating through the double-slit setup or through a Mach–Zehnder interferometer; the quantum state is simply a tool to calculate probabilities. Probabilities of the photon being somewhere? No, we should be even more cautious and talk only about probabilities of a photon detector firing if it is placed somewhere. One might be tempted, as was Einstein, to consider the photon as being localized at some place with us just not knowing that place. But, whenever we talk about a particle, or more specifically a photon, we should only mean that which a 'click in the detector' refers to. [Zeilinger et al., 2005]

mentioned that photons have zero mass. In consequence, a photon always travels through empty space at the same enormous speed (the speed of light, obviously!) and cannot be brought to rest. This is another sense in which photons cannot be localized. But a collection of massive particles *can* all be brought to rest together, and I conclude this chapter by describing a kind of experiment in which multiple massive particles so localized have been successfully assigned entangled states.

It is possible to trap and store atomic ions (atoms that have become positively charged through loss of outer electrons) for long periods and to manipulate them to produce entangled states.[7] Such atomic ions can be stored nearly indefinitely, each localized in space to within a few billionths of a meter. Entangled states of large groups of stored ions have been produced in this way, and their production verified by showing that suitable correlation measurements on individual ions and subsets of the group yield statistics expected from assignment of the entangled state the experiment was designed to produce.

Blatt's group [2011] verified production of states of up to fourteen calcium ions assigned generalizations of the GHZ state of the form

$$|\psi\rangle = \frac{1}{\sqrt{2}} (|\uparrow\rangle |\uparrow\rangle |\uparrow\rangle .... + |\downarrow\rangle |\downarrow\rangle |\downarrow\rangle ....) \tag{3.11}$$

in which both components of the superposition are products of n vectors ($n \leq 14$), each in a vector space pertaining to one of the ions whose states are thus entangled. The notation suggests that it is the spin states of the ions that are entangled, as in state $|o\rangle$. The actual states do typically involve spin, but the arrows need not simply correspond to two states that differ only insofar as a measurement of spin of that ion along the $\uparrow$ direction would certainly yield outcome *up* or *down*. The states $|\uparrow\rangle, |\downarrow\rangle$ differ with respect to energy, and so one often says they constitute a two-level system. It turns out that a quantum model of any such two-level system is formally identical to that used to model a physical system with spin 1/2. Here we see a usage in which the term 'quantum system' refers to an abstract mathematical model distinct from any particular physical system to which it may be applied.

---

[7] Blatt and Wineland [2008] give a good introduction to such work focused on its potential applications in quantum computing.

4
"Non-locality"

4.1 Spooky Action at a Distance?

Quantum entanglement is popularly believed to give rise to spooky action at a distance of a kind that Einstein decisively rejected:[1]

> But on one supposition we should, in my opinion, absolutely hold fast: the real factual situation of the system [B] is independent of what is done with the system [A], which is spatially separated from the former.   (Einstein, in Schilpp, ed. [1949], p. 85)

He offered this alternative formulation, which allows that a distant action's influence may be transmitted indirectly:

> The following idea characterizes the relative independence of objects far apart in space (A and B): external influence on A has no *immediate* ("unmittelbar") influence on B; this is known as the 'principle of local action', which is used consistently only in field theory.
>
> ([1948], pp. 321–2)

I will argue that this popular belief is false: quantum entanglement implies no such instantaneous action at a distance. But first we need to ask what lies behind the popular belief. Here I identify arguments of two kinds. We owe both ultimately to the work of John S. Bell.[2]

Surprisingly, the first kind of argument need never mention entanglement or even quantum theory. It starts with an idealized description of patterns of statistical correlation observed among macroscopic phenomena and asks what could possibly account for them. Einstein's principle of local action and other very plausible assumptions about the operation of causes are used to justify constraints on any probabilistic model proposed as an explanation of these patterns. The argument proceeds to show that since every model meeting these constraints predicts patterns very different from

---

[1] In a 1947 letter to Max Born [Born, 1971] Einstein rejected a view of quantum theory because it could not be reconciled with the idea that physics should represent a reality in time and space, free from "spukhafte Fernwirkung"—a phrase translated by Born's daughter as "spooky action at a distance". In Chapter 3 we saw that Einstein also used the term 'telepathic' ('telepathisch') to characterize violation of the quoted supposition. In each case his intention appears to have been to dismiss the possibility as supernatural, that is, in violation of any kind of natural order we could capture in physics. Just before each quoted passage Einstein made it clear that he was considering the situations of *A, B at a specified time*.

[2] Most of Bell's relevant papers may be found in his collected papers on quantum philosophy [2004].

those observed, the only possible explanation of the observed patterns is that they involve instantaneous action at a distance. As an afterthought one adds that these are exactly the patterns of correlation quantum theory correctly predicts for outcomes of measurements on systems assigned particular entangled quantum states.

The second kind of argument starts from quantum theory, and a dispute about how it should be understood. Recall that Einstein's principle of local action was implicit in Chapter 3's version of the EPR argument that a quantum state does not completely describe a system to which it is assigned: this concludes, for example, that every component of the spin of each particle in the entangled state $|o\rangle$ must be determinately either up or down. To Einstein this suggested a way of understanding quantum theory that might account for the peculiar patterns of correlation predicted by entangled quantum states *without* any direct action at a distance: in state $|o\rangle$, for example, each particle might acquire an equal and opposite value for an arbitrary spin component after their separation, so a joint measurement of a spin component on each of them would simply reveal these preexisting values. In this chapter we will see how Bell's deeper exploration of those patterns showed that they cannot be explained by taking Einstein's suggestion without abandoning his principle of local action.

Einstein offered his suggestion ([Schilpp, ed. 1949], p. 83) as an alternative to what he considered the orthodox way of understanding quantum theory, according to which the value of a dynamical variable like a position, spin, or polarization component arises only through the act of measurement itself. Interference experiments lend some credence to this orthodox conception of the role of measurement. The EPR argument was intended to refute the orthodox view of measurement when considered part of a package that also includes the assumption that a quantum state offers a complete description of the system to which it is assigned. But that argument, too, rested on Einstein's principle of local action. Given the principle of local action, the EPR argument apparently refutes this orthodox package, while Bell's argument refutes Einstein's suggested alternative. The success of quantum theory in accounting for the peculiar patterns of correlation predicted by entangled quantum states then seems to leave us with only one option—to abandon the principle of local action.

After presenting an argument of each kind I show how to block their common conclusion that quantum phenomena display instantaneous action at a distance. The key is to realize that the function of a quantum state is neither to completely describe the system to which it is assigned nor to describe the statistical properties of a collection of similar systems, but to offer authoritative advice on the significance and credibility of statements about the values of dynamical variables on physical systems. It will take a while to explain and defend this last understanding of a quantum state's function.[3] We shall see that it leads to a way of understanding quantum theory that

---

[3] The quantum state's role in offering advice on the credibility of such claims will prove important in this chapter. Its role in offering advice on their significance will prove important in Chapter 6's discussion of measurement in quantum theory.

differs both from what Einstein took as orthodoxy and from his suggested alternative. This offers a perspective from which one can see why the principle of local action and other assumptions about causes fail to justify the probabilistic constraints that the first kind of argument shows no model of the observed patterns of correlation could meet. And it leads to a way of explaining those patterns using quantum theory that is in complete conformity with Einstein's principle of local action.

Since at least 1935 quantum theory has been known to admit striking, reproducible correlations among types of events localized arbitrarily far apart from one another, even when individual events of these types occur essentially simultaneously. In science, as in ordinary life, when faced with such events we generally seek a causal explanation. The physicist Bernard d'Espagnat put it this way:

> Whenever a consistent correlation between such events is said to be understood, or to have nothing mysterious about it, the explanation offered always cites some link of causality. Either one event causes the other or both events have a common cause. Until such a link has been discovered the mind cannot rest satisfied. Moreover, it cannot do so even if empirical rules for predicting future correlations are already known. A correlation between the tides and the motion of the moon was observed in antiquity, and rules were formulated for predicting future tides on the basis of past experience. The tides could not be said to be understood, however, until Newton introduced his theory of universal gravitation.   ([1979], p. 160)

The philosopher Hans Reichenbach shared this intuition. To quote Arntzenius [2010]:

> It seems that a correlation between events A and B indicates either that A causes B, or that B causes A, or that A and B have a common cause. It also seems that causes always occur before their effects and, thus, that common causes always occur before the correlated events. Reichenbach was the first to formalize this idea rather precisely. He suggested that when $\Pr(A\&B) > \Pr(A) \times \Pr(B)$ for simultaneous events A and B, there exists an earlier common cause C of A and B, such that $\Pr(A/C) > \Pr(A/\neg C)$, $\Pr(B/C) > \Pr(B/\neg C)$, $\Pr(A\&B/C) = \Pr(A/C) \times \Pr(B/C)$ and $\Pr(A\&B/\neg C) = \Pr(A/\neg C) \times \Pr(B/\neg C)$. (See [Reichenbach, 1956], pp. 158–159.) C is said to 'screen off' the correlation between A and B when A and B are uncorrelated conditional upon C. Thus Reichenbach's principle can also be formulated as follows: simultaneous correlated events have a prior common cause that screens off the correlation.

Bell took a condition he called local causality to embody a precise formulation of a requirement of no action at a distance. From this he derived a probabilistic independence condition closely akin to Reichenbach's principle, and used it to argue that quantum theory is neither locally causal nor embeddable in a locally causal theory. He concluded that certain patterns of distant correlation predicted by quantum theory cannot be explained without action at a distance.

> [W]e will argue that certain particular correlations, realizable according to quantum mechanics, are locally inexplicable. They cannot be explained, that is to say, without action at a distance.
> ([Bell, 1981]; [2004], pp. 151–2)

Many careful experimenters have now exhibited such correlations with great precision, and only a conspiracy theorist or philosophical skeptic seriously contemplates future observation of their failure. I will give a schematic description of such an experiment after presenting a simplified version of Bell's argument.[4]

Bell ([2004], p. 239) begins that argument by stating an intuitive principle closely related to Einstein's principle of local action:

> The direct causes (and effects) of events are near by, and even the indirect causes (and effects) are no further away than permitted by the velocity of light.

Bell's intuitive principle extends to causes and effects other than those involving what is done to just two systems. It also allows that what is done to a distant system B may well affect the real factual situation of a nearby system A, but only after a lapse of time sufficient for light to get to A from B, since this is the minimal time any causal process would take to get from B to A (unless it were to travel faster than light!). Einstein's principle of local action and Bell's intuitive principle could both be taken to rest on the common assumption that for a distant event to cause a nearby event its influence must be continuously transmitted through the intervening space. In his formulation Bell takes relativity theory to impose a constraint on any process capable of transmitting such influence, and his argument in Bell ([1990]; [2004]) involves explicit consideration of the structure of space and time according to relativity theory. But the basic logic of the argument may be understood without such a detour.[5]

4.2 Mermin's Demonstration

Suppose a source emits systems of some kind every so often, and soon after a system is emitted two widely separated devices both respond simultaneously, each in one of two readily observable ways—suppose either its red light flashes or its green light flashes. Call the device located to the left of the source L, and the device located to the right of the source R. Each of L, R is equipped with a switch with three different settings labeled 1, 2, 3: the settings of the switches may be freely chosen independently of one another, making a total of 9 different possible combinations of settings (see Figure 4.1).

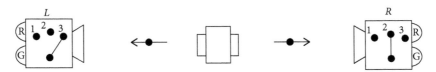

Figure 4.1 A possible experimental arrangement

---

[4] Chapter 10 will explain the simplification and show how removing it leads to a deeper understanding of probability and causation.
[5] My exposition closely follows that of Mermin ([1981]; [1985]).

"NON-LOCALITY" 57

For each combination, a record is kept of simultaneous flashes at L, R, indicating in each case which colored light on each device flashed. After a while a long list of records is available, permitting a statistical analysis for every combination of settings of how often both lights flashed the same color, and how often they flashed different colors. This shows significant regularities. No matter what the setting, the red light attached to L flashed just about as often as the green light: the same was true of R. If the settings were the same (11, 22, or 33), then the lights at L, R always flashed the same color. Considering all the results, and ignoring device settings, the same colored lights flashed on L, R (both flashed red, or both flashed green) just about half the time.

These results display rich patterns of correlation: how might they be explained? My description assumed that the devices were responding to systems emitted from the source—an assumption that may be confirmed by altering or removing the source and noting a change in the devices' responses. It did not assume that the source emits pairs of particles, with one particle traveling to each device and influencing its subsequent behavior, though one could certainly try to explain the correlations by making that assumption. According to Einstein's principle of local action, neither setting its switch nor any other external influence on the distant device can directly influence the outcome at the nearby device. Take it that the devices were set up to prevent any such influence being indirectly conveyed from one device to the other by eliminating all known mechanical or other physical connections, shielding them from all known kinds of radiation, etc. Suppose the switch at each device was set independently and randomly immediately before its light flashed, and the devices were placed so far apart that not even light could travel from one to the other in the time it took to set the switches and record the flashing of the lights. Then Bell's intuitive principle implies that this has eliminated any possibility that the distant switch setting influences the nearby outcome. So it seems the only option is to follow both d'Espagnat and Reichenbach in seeking to explain the correlations by locating their common cause, and the obvious place to look is in the systems emitted by the source.

The lights always flash the same color when the switches on the devices are turned to the same settings (11, 22, or 33): about half the time both flash red, the rest of the time both flash green. Take a case 11 in which they are both set to 1. To achieve perfect coordination, a common cause at the source must determine on this occasion whether they will both flash green or both flash red: introduction of any randomizing element would spoil their perfect correlation of colors. Call a common cause of this kind R_1 if it guarantees both will flash red, and G_1 if it guarantees they'll both flash green. For the 22 combination of settings we need common causes of types R_2, G_2; and for 33 we need R_3, G_3 type common causes.

Now on each occasion the systems are emitted by the source before the device settings are randomly chosen. Unless they are interfered with on their way to the devices, they must effectively carry instructions from the common cause at the source to the devices that will guarantee perfect correlations if the devices happen to end up at the same setting: and so each must effectively carry one of eight different instruction sets, which we can label as (X,Y,Z). Here the letters are to be replaced by labels for colors so, for example, (R,G,R) instructs both devices to flash red if they happen to be set in the combination 11, green if they happen to be set to 22, and red if they're at setting 33. (Note that the order of the three letters inside the parentheses has taken over the role of the subscripts on common cause types.) Even if they are interfered with on the way to the devices, each system must reach the devices with one of these eight instruction sets.

What happens if a system with instruction set (R,G,R) encounters devices set to 12? By Bell's intuitive principle (or Einstein's principle of local action), the outcome at L cannot depend on the setting at R, and the outcome at R cannot depend on the setting at L. So L must still flash red and R must still flash green, just as it would if both devices were set to the same settings (11, 22 respectively). So if a system reaches the devices with an instruction set like (R, G, R) this wholly determines the responses of the devices, no matter how they end up being set, and (even if it is not already fixed at the source) that instruction set is itself independent of the device settings. All that remains is to specify the distribution of instruction sets among systems that make the lights flash. But it is now easy to show that no distribution of instruction sets can be reconciled with the assumed statistics!

There are nine different combinations of settings, and our common cause model correctly predicts perfect correlations for setting combinations (11, 22, 33). Instruction sets (R, R, R), (G, G, G) predict perfect correlations for all combinations of settings. Instruction set (R, R, G) predicts perfect correlations also for combinations (12, 21), but perfect *anti*correlations for settings (13, 31, 23, 32). Each of the other five remaining instruction sets also predicts perfect correlations in five and perfect anticorrelations in four combinations of settings. The instruction sets may vary from one occasion to another: given a pair of switch settings, different sets will specify different pairs of colored flashes. But irrespective of the switch settings, and no matter how these specifications vary from one occasion to the next, the model predicts that in at least 5/9 of the cases the lights on the detectors will flash the same color. That prediction is falsified by the data we assumed, according to which the lights flash the same color only about half the time.

But what entitled us to assume that there are devices and a source that behave in the way I have described? Certainly quantum theory admits exactly these striking, reproducible patterns of correlation. But the goal of the first style of argument is to avoid any reliance on quantum theory while appealing directly to correlations that have actually been observed. After presenting a minor variant of the patterns I have described, Mermin [1981] admitted that they had *not* actually been observed. The apparatus necessary to display these correlations, he said, "could in fact be built with an effort almost certainly less than, say, the Manhattan project" (p. 398). This puts the assumed correlations rather far from the realm of actuality! But actual experiments have been performed leading to related patterns of correlation that also cannot be predicted by any probabilistic model conforming to a mathematical local causality condition that Bell took to follow from his intuitive principle.[6]

The patterns differ in two respects from those I assumed above. Schematically, each involved not three but only two switch settings at each of the devices L, R, so the pattern consisted of only four (rather than nine) sets of correlations: and for no combination of settings were the correlations required to be perfect. Even with these differences, Bell was able to derive an inequality from his local causality condition that was not satisfied by the actual data in the experiments of Aspect *et al.* [1982], Weihs *et al.* [1998], and Giustina *et al.* [2015] on photon pairs; Hensen *et al.* [2015] on electron pairs; and in many similar experiments.

4.3 A Simplified Bell Argument

At this point I'll use a simplified version of Bell's local causality condition to show how this leads to a restriction called Conditional Factorizability on any probabilistic common cause model intended to predict the results of such experiments. Bell's own [1990] statement of his condition will be the subject of detailed scrutiny in Chapter 10. Conditional Factorizability implies an inequality that Bell argues must be satisfied by any proposed locally causal model of those results: the derivation appears in Box 4.1. The results of many careful experiments have now been convincingly shown to violate this inequality. By 1990 Bell was already sufficiently convinced to conclude that nature violates local causality, and that the results of such experiments show that there is action at a distance, in violation of his own intuitive principle.

Box 4.1 Proof of a Bell Inequality, and its Experimental Violation

We wish to use Conditional Factorizability to prove an inequality I'll call (CHSH). The CHSH inequality is violated by probabilities derived by application of the Born rule to certain quantum states. Experimental statistics are in close conformity with these quantum probabilities but differ very significantly from what one would expect from any probabilities that satisfy (CHSH).

To prove the CHSH inequality, we assume a probabilistic theory that considers the experimentally accessible probabilities $Pr_{a,b}(A\&B)$, $Pr_a(A)$, $Pr_b(B)$ not as fundamental, but as arising from the underlying *conditional* probabilities $Pr_{a,b}(A\&B|\lambda)$, $Pr_a(A|\lambda)$, $Pr_b(B|\lambda)$ it specifies. λ is the complete earlier state, according to this probabilistic theory: it may be thought to supplement the quantum theoretic state assigned to the pair.

(*continued*)

---

[6] The experiment of Hensen *et al.* [2015] comes close to realizing Mermin's demonstration, differing mostly in its use of so-called entanglement-swapping to produce the experimental correlations.

Box 4.1 Continued

The experimentally accessible probability is then supposed to arise as an average over all the various possible values of λ, weighted by the relative probabilities $\rho(\lambda)$ that the theory also relies on. Because the value of λ is to be specified prior to the free choices of a and b, $\rho(\lambda)$ does not depend on these parameters.

$$Pr_{a,b}(A\&B|\lambda) = Pr_a(A|\lambda) \times Pr_b(B|\lambda). \qquad \text{(Conditional Factorizability)}$$

We will need a purely algebraic lemma:

Lemma 1 *If q, r, q', r' are any numbers, each lying between -1 and $+1$ inclusive, then*

$$-2 \leq qr + q'r + qr' - q'r' \leq +2.$$

Proof Any linear function $aq + b$ of q (where a, b are each independent of q) takes on its maximum and minimum values when q does. But $f(q) = qr + q'r + qr' - q'r'$ takes this form, with $a = (r + r')$, $b = q'r - q'r'$. So $qr + q'r + qr' - q'r'$ takes on its maximum and minimum values when $q = -1$ and $+1$. By similar reasoning, $qr + q'r + qr' - q'r'$ takes on its maximum and minimum values when each of $r, q', r' = \pm 1$. Now $qr + q'r + qr' - q'r' = (q + q')r + (q - q')r'$. This takes on its maximum and minimum values when $q, q' = \pm 1$, that is either $q + q' = \pm 2$ and $q - q' = 0$, or $q + q' = 0$ and $q - q' = \pm 2$: since it also takes on maximum and minimum values when $r, r' = \pm 1$ it follows either way that the maximum and minimum values of $(q + q')r + (q - q')r'$ are ± 2. Hence $-2 \leq qr + q'r + qr' - q'r' \leq +2$.

Each possible outcome of a measurement of A, B may be represented by one of two values that we take to be ± 1 (in a polarization measurement, $A = +1$ might indicate that a photon is measured to be polarized parallel to a particular axis, with $A = -1$ indicating perpendicular to that axis). The expectation value of a magnitude is what one would expect to be the average of the values obtained if it is measured repeatedly. Now define the conditional expectation value of the *product* $A.B$ of the two outcomes A, B with settings a, b by

$$E(a, b|\lambda) \equiv \sum_{A,B} A.B.Pr_{a,b}(A\&B|\lambda)$$
$$= \sum_{A,B} A.B.(Pr_a(A|\lambda) \times Pr_b(B|\lambda)) \text{ by Conditional Factorizability}$$
$$= \sum_A A.Pr_a(A|\lambda) \times \sum_B B.Pr_b(B|\lambda)$$

and similarly for $E(a', b|\lambda), E(a, b'|\lambda), E(a', b'|\lambda)$.[7]

Now let $q = \sum_A A.Pr_a(A|\lambda)$, $q' = \sum_A A.Pr_{a'}(A|\lambda)$, $r = \sum_B B.Pr_b(B|\lambda)$, $r' = \sum_B B.Pr_{b'}(B|\lambda)$: note that each of these defined quantities satisfies the conditions of the lemma. It follows that

---

[7] The notation $\sum_A x_A$ here means the sum of the two terms $\{x_{A=+1}, x_{A=-1}\}$, and $\sum_{A,B} x_{A,B}$ means the sum of $\{x_{A=+1,B=+1}, x_{+1,-1}, x_{-1,+1}, x_{-1,-1}\}$. In the sum $\sum_{A,B} A.B.Pr_{a,b}(A\&B|\lambda)$ I have used dots (.) rather than the usual sign $\times$ for multiplication to make the notation more readable.

$$-2 \leq E(a,b|\lambda) + E(a',b|\lambda) + E(a,b'|\lambda) - E(a',b'|\lambda) \leq +2. \quad (4.1)$$

Since $\rho(\lambda)$ is independent of the settings a, b, a', b' we may average over it to derive an inequality that relates experimentally observable expectation values

$$-2 \leq E(a,b) + E(a',b) + E(a,b') - E(a',b') \leq +2. \quad \text{(CHSH)}$$

This is the inequality first derived differently by Clauser, Horne, Shimony, and Holt [1969] with the declared aim of rendering Bell's original inequality amenable to experimental testing.

For comparison, consider the quantum probabilities predicted by assignment of polarization state $|\eta\rangle = 1/\sqrt{2}(|H\rangle|H\rangle + |V\rangle|V\rangle)$. In this state application of the Born rule gives probabilities

$$Pr^{|\eta\rangle}_{a,b}(+1,+1) = 1/2 \cos^2 \angle ab = Pr^{|\eta\rangle}_{a,b}(-1,-1)$$
$$Pr^{|\eta\rangle}_{a,b}(+1,-1) = 1/2 \sin^2 \angle ab = Pr^{|\eta\rangle}_{a,b}(-1,+1)$$

and hence $E^{|\eta\rangle}(a,b) = \cos^2 \angle ab - \sin^2 \angle ab = \cos 2\angle ab$. If we rotate the polarization axes through the following angles from some fixed initial direction

$$a = 45°, a' = 0°, b = 22.5°, b' = 67.5°$$

then

$$E^{|\eta\rangle}(a,b) = 0.707 = E^{|\eta\rangle}(a,b') = E^{|\eta\rangle}(a',b)$$
$$E^{|\eta\rangle}(a',b') = -0.707$$

in which case $E^{|\eta\rangle}(a,b) + E^{|\eta\rangle}(a',b) + E^{|\eta\rangle}(a,b') - E^{|\eta\rangle}(a',b') = 2.828$, which is substantially greater than the maximum value of 2 allowed by the CHSH inequality. Experiments support the quantum prediction and disconfirm any theory whose probability assignments satisfy the inequality.

Bell begins by envisaging a situation in which a theory specifies that what happens in a short period of time T in each of two widely separated regions 1, 2 depends probabilistically on what happened earlier. He thinks of such probabilistic dependence as causal, and local causality is intended to localize any probabilistic cause of events at 1 in T earlier than T and in reach of 1, and analogously for events at 2 in T. Local causality does this by requiring of a theory that the probability it predicts for any event occurring at 1 in T be unaltered by taking account of any event that occurs at 2 in T, assuming a sufficiently complete specification of what occurred in the vicinity of 1 just before T. If this specification were insufficiently complete, what happened at 2 in T might be correlated with *unspecified* events near 1 just before T that go on to affect what happens at 1 in T, but only because something even earlier caused both those unspecified events and what happened at 2 in T. If that were so, the probability of an event at 1 in T might also depend on an event at 2 in T despite their lack of any direct causal connection. To say exactly what local causality requires we need to spell out what it is for a specification to be sufficiently complete.

So Bell considers a time immediately before T at which experimenters are supposed to be able freely to choose a parameter a in the vicinity of 1 and (independently) a parameter b in the vicinity of 2. Then in T an outcome event A occurs at 1 and an outcome event B occurs at 2. A probabilistic theory is required to predict the probability of each of several possible outcomes at 1 and at 2, and also the joint probability for each possible pair of outcomes: it is required to do this for each possible choice of parameters (a, b). Write the joint probability of outcomes A, B with parameters set to (a, b) as $Pr_{a,b}(A\&B)$, and the corresponding individual probabilities as $Pr_{a,b}(A), Pr_{a,b}(B)$. To predict these probabilities, a probabilistic theory appeals to its specification of what occurred before T. Let λ label an arbitrary specification of this kind: this is assumed to be fixed independently of both a and b, which were freely chosen immediately before T. It is λ that is required to give a sufficiently complete specification of what occurred in the vicinity of 1 and also of what occurred in the vicinity of 2 just before T—sufficiently complete to remove any probabilistic interdependence between what happened at 2 in T and what happened at 1 in T.

So a probabilistic theory will consider the probabilities $Pr_{a,b}(A\&B)$, $Pr_{a,b}(A)$, $Pr_{a,b}(B)$ not as fundamental, but as arising from the underlying *conditional* probabilities $Pr_{a,b}(A\&B|\lambda)$, $Pr_{a,b}(A|\lambda)$, $Pr_{a,b}(B|\lambda)$ it specifies.[8] Each corresponding unconditional probability is then an average over all the various possible values of λ, weighted by their relative probabilities $\rho(\lambda)$. Because the value of λ is to be specified prior to the free choices of a and b, $\rho(\lambda)$ does not depend on these parameters. The requirement that the specification of what occurred just before T be sufficiently complete, not only in the vicinity of 1 but also in the vicinity of 2, is now to be explicated by conditions on the conditional probabilities supplied by the theory. The first such condition is that each individual probability should not depend on the setting of the distant parameter:

$$Pr_{a,b}(A|\lambda) = Pr_a(A|\lambda) \qquad \text{(Parameter Independence)}$$
$$Pr_{a,b}(B|\lambda) = Pr_b(B|\lambda)$$

Parameter Independence is intended to exclude the possibility that the probability of a nearby outcome causally depends on how one chooses to set a distant parameter: in this scenario, any such dependence would be in violation of Einstein's principle of local action as well as Bell's intuitive principle. The second condition is that each individual probability should be probabilistically independent of what distant outcome actually occurs:

---

[8] It may help to think of each value of λ as like one of section 4.2's instruction sets, except that rather than specifying what will happen it specifies only probabilities for various possible outcomes—but these probabilities are more fine-grained than the unconditional probabilities that arise from averaging over different values of λ.

$$Pr_{a,b}(A|B,\lambda) = Pr_{a,b}(A|\lambda) \qquad \text{(Outcome Independence)}$$
$$Pr_{a,b}(B|A,\lambda) = Pr_{a,b}(B|\lambda)$$

Outcome Independence is intended to exclude the possibility that the probability of a nearby outcome causally depends on a distant outcome in violation of Bell's intuitive principle.

Together, these two conditions imply that the theory's joint conditional probability for A and B can be expressed as a simple product of two separate individual probabilities:

$$Pr_{a,b}(A\&B|\lambda) = Pr_a(A|\lambda) \times Pr_b(B|\lambda). \qquad \text{(Conditional Factorizability)}$$

Bell's condition of local causality then incorporates this Conditional Factorizability condition. The derivation in Box 4.1 shows that any theory meeting the Conditional Factorizability condition and requiring that $\rho(\lambda)$ be independent of a, b must predict probabilities $Pr_{a,b}(A\&B)$ that are constrained by what is called the CHSH inequality. Actual experimental data concerning what was readily observable at widely separated locations yielded relative frequencies of outcomes A, B to which probabilities meeting this constraint would assign extremely low probability. No theory meeting Bell's local causality condition can correctly predict these data.

One can get insight into the situation by considering a simple game between two players, Alice and Bob.[9] The players are allowed to meet and to devise a strategy to try to win the game. They will then be separated so that they can no longer have any means of communicating with each other. At preset times each player will then be asked one or other of two questions: Alice may be asked either a_1 or a_2 on each occasion, and Bob may be asked either b_1 or b_2. On every occasion a player is asked a question, he or she must answer either "yes" or "no". At each distant location a fair coin will be tossed to determine which of the two questions the player there will be asked. Alice and Bob win the game on a particular occasion if and only if their answers satisfy these three conditions:

(1) If they are both asked their first questions (a_1, b_1) they give the same answer (both "yes" or both "no");
(2) If one is asked his or her first question while the other is asked his or her second question (i.e. (a_1, b_2) or (a_2, b_1)), they also give the same answer; but
(3) If they are both asked their second questions (a_2, b_2), they give *different* answers.

Before reading on, put yourself in the position of Alice or Bob and see if you can devise a strategy by which you and your partner can win this game. This would be easy if you could know what question your partner was asked before answering your own question. But on each occasion what question your partner is asked can't be predicted

---

[9] This is due to Popescu and Rohrlich [1994], who invented it to suggest the possibility of patterns of non-localized correlation even more puzzling than those predicted by quantum theory.

ahead of time because it is determined only at the last moment by the flip of a distant coin. Waiting to see how the coins land won't help since you must both answer your questions at the same time, when neither of you has any way to find out how the other's coin just landed.

After several fruitless attempts you will probably be ready to admit that no strategy can guarantee success on every occasion in this game. But there are strategies that give you a better than even chance of succeeding on a particular occasion: for example, agreeing to say "yes" all the time to whatever question you're asked should give you a long-term success rate of about 75%, while if Alice always says "no" and Bob always says "yes" the expected success rate is only 25%. In fact, while no strategy is *guaranteed* to yield a success rate greater than 0%, the best strategy is very likely to yield a 75% success rate over the long term.

What does this have to do with local action? Here's the analogy with patterns of correlation revealed in experiments like those of Aspect *et al.* [1981].[10] In such experiments, you can think of a choice of question as implemented by a choice of orientation for a polarizer at one of two distant locations, with a "yes" answer corresponding to the "click" of a device that detects polarization parallel to that axis, and a "no" answer corresponding to the "click" of a device that detects polarization perpendicular to that axis. The choice of question is then made at the last moment not by flipping a coin, but by a device that sets one of a pair of polarization axes at random, with an independent device at each location. With appropriate pairs of axes, the resulting patterns of detector "clicks" can be compared to the demands of the game (suitably translated). The actual patterns generated by employing photon pairs assigned the entangled state $|\chi\rangle$ correspond to a success rate of about 85%—significantly greater than the 75% maximum reachable with any strategy devisable without using such an entangled pair. It is interesting to note that this enhanced success rate is achieved for each of the four combinations of settings: since polarizers are never aligned along the same axis at both locations there is no combination for which one should expect perfect correlations.

The game provides a very simple instance in which entanglement provides a *resource* that can be used to do things we have no way of doing without it. Here it enhances our means of distant coordination even though it does not permit instantaneous communication. In this and many other instances, entanglement provides an informational resource. Quantum information theory continues to develop new means of exploiting this resource. Here we see more vindication for Schrödinger's view of the significance of entanglement.

Bell's ([1990], [2004]) argument shows that the predictions of every probabilistic model satisfying his local causality condition are incompatible with the predictions of a quantum theoretic model in which an entangled state is assigned to a certain type of quantum system. Those quantum predictions are strongly supported by actual patterns

---

[10] The analogy with those reported in Aspect *et al.* [1982] is even closer, as we shall see.

of statistical correlations that have been experimentally observed. The local causality condition is a natural generalization of Reichenbach's common cause principle, itself a precise statement of what D'Espagnat demanded of a kind of explanation needed before a mind faced by such correlations could rest satisfied. But both D'Espagnat and Reichenbach allowed for the possibility of something else that could explain these correlations—a direct causal connection between distant correlated events. Examination of the local causality condition reveals two ways this might fail because of such a direct causal connection. The probability of a nearby outcome might depend on the distant outcome because one is a direct cause of the other; or the probability of a nearby outcome might directly causally depend on the distant setting. In either case, this would involve direct action at a distance between events occurring at basically the same time, in violation of Bell's intuitive principle.[11]

If the probability (conditional on λ) of a nearby outcome were to directly causally depend on the distant setting, perhaps one could use this dependence to signal directly and instantaneously from one location to the other? Whether or not one could do so would depend on how this causal dependence worked.[12] In any case, the probability quantum theory assigns to a nearby outcome (conditional on state $|\eta\rangle$) is 1/2, independent of the distant setting, so if $|\eta\rangle$ by itself were the common cause the question would not arise. But there are certainly instances of distant causal dependence that don't afford us the possibility of signaling: take, for example, the causal dependence between the supernova explosion of a star and the incidence of gravitational waves on one of its planets. If the peculiar correlations exhibited in experiments like those of Aspect ([1981], [1982]) can't be used to signal directly and instantaneously from one place to another, that doesn't show they have no causal explanation in terms of some direct, instantaneous causal dependence between what happens at one location and what happens at the other. Indeed, if these correlations cannot be explained by appeal to a common cause, it's hard to see how else to account for them.

4.4 Two Views of a Quantum State

I shall argue that quantum theory itself can be used to explain the patterns of correlation among events at distant locations exhibited by experiments like those of Aspect [1982] in a way that conforms to Einstein's principle of local action. The relevant quantum probabilities do not conform to Bell's local causality condition. But this does not entail a violation of the intuitive principle on which he based that condition, since these probabilities function in a way that renders his local causality condition inadequate as an expression of that principle. To complete the argument I'll need to

---

[11] I have inserted the word 'basically' here to acknowledge the need for a brief period at each location between choice of setting and detector "click". No such qualification is needed when the scenario is as described by Bell, assuming the space and time of relativity. I'll return to this point in Chapter 10.

[12] In the Interlude we will encounter a specific proposal of this kind which permits no such signaling.

defend some assumptions I make in this chapter. These assumptions rest on more general views of probability, explanation, and causation that deserve the more detailed examination they will receive in Chapters 9 and 10. But first I need to introduce a way of understanding the significance of quantum states while explaining how it differs both from what Einstein took as the orthodox view and from his own suggested alternative.[13]

In common parlance as well as classical mechanics the term 'state' is generally used to connote the *condition* of something, either at one moment or as this varies over a period of time: a system's state says how it is—what the properties and relations are of the things that make it up. In this usage, to specify the state of a system is to describe it. In classical mechanics, a system's state is typically specified by a canonical mathematical representation of properties and relations of its elements. The state of a system of n Newtonian particles may be specified by a point x in a $3n$-dimensional configuration space, whose first three coordinates represent components of the first particle's position (its distances from some specified point in three-dimensional physical space along each of three mutually perpendicular directions), whose next three coordinates represent the second particle's position components, and so on.

An alternative specification in a $6n$-dimensional phase space is often convenient. Each point of this space represents not just where every particle is but also how it is moving. The representation specifies not just its three position coordinates (x, y, z) in physical space but also the three corresponding components of its momentum (the product of its mass and velocity) (p_x, p_y, p_z). The point in phase space representing the momentary state of a system of classical particles completely fixes its dynamical condition (including its energy, angular momentum, etc.), and the trajectory traced out by this point through time fixes its dynamical behavior. Phase space is often also used in statistical mechanics, but here the momentary state of a system of n classical particles is specified by a *probability distribution* ρ over the space: this specification is incomplete, insofar as representing the state of a system by ρ is compatible with its also being represented by one of many different points in phase space: ρ specifies the probability that the actual point lies in any "nice" (the mathematical term is 'measurable') set S of the space.

Applying the usual connotation of the word 'state', one may be led to think that ρ describes some physical feature associated with systems of n classical particles—a physical probability. The name "statistical mechanics" itself suggests that this feature has something to do with statistical regularities in the behavior of large collections of n-particle systems, each represented by state ρ. Certainly assignment of state ρ to each such system leads one to expect the relative frequency of systems represented by phase

---

[13] It is not clear how seriously Einstein took his own suggestion. He certainly thought quantum theoretic description was incomplete, and gave this suggestion of how to complete it. But the resulting description would still involve concepts "which on the whole have been taken over from classical mechanics", and would "offer no useful point of departure for future development" (Einstein, in [Schilpp, ed. 1949], p. 87).

"NON-LOCALITY" 67

space points $\omega \in S$ closely to match ρ (S): indeed, the success of statistical mechanics is measured by how well that expectation is borne out. But even if ρ does represent something physical, it is not clear that this is a feature of any individual n-particle system. A tradition stemming from the work of the nineteenth-century American physicist J. Willard Gibbs takes ρ to represent the physical properties of a collection of similar systems called an *ensemble*. But these systems are not considered real—the word 'virtual' is often used to describe either them or the ensemble itself. There is an alternative perspective from which ρ is not a complete description or representation of the physical condition of a real system or collection of them, but serves rather to represent uncertainty as to the real state of some particular system, as specified by a point in phase space.[14]

Should we think of a quantum state $|\psi\rangle$ by analogy to a phase space point ω, in which case it should be understood to offer a complete description of the real situation of the system to which it is assigned? Or is a quantum state more like ρ—a way of specifying the probability that the system to which it is assigned is in one kind of situation rather than another? These are basically the alternatives that Einstein offers in his intellectual autobiography (Schilpp ed. [1949], pp. 83–5). There he presents a version of the EPR argument to rule out the first (supposedly orthodox) alternative, while leaving open the second alternative as his own suggestion as to how we might think of quantum states. We can easily now use earlier results of this chapter to rule out Einstein's suggested alternative. Fortunately these aren't the only ways of thinking of quantum states. By adopting a third approach we will be able to evade the arguments of both EPR and Bell and see how to explain the puzzling patterns of non-localized correlations with no spooky action at a distance.

According to Einstein's suggestion, each individual system, just before a measurement, has a definite value for every one of its dynamical variables. Moreover, this is the value that would be recorded as the outcome of a (competently conducted) measurement of the variable. Now consider a photon pair assigned polarization state $|\eta\rangle$. A measurement of a component of polarization with respect to axis θ on either photon always yields one of two values which we can call ± 1, where $+1$ is recorded if the photon is detected with polarization parallel to θ after emerging from a two-channel polarizer, and -1 is recorded if the photon is detected with polarization perpendicular to θ after emerging from that polarizer. Consider a joint measurement of polarization on photons L, R in a pair with respect to axes a, b respectively, where the two polarization measurements are performed far apart at the same time. The outcome of the measurement is then in each case locally determined by that preexisting value of the component of polarization which it records as $+1$ or -1. Let each of a, b be selected at random at the last moment between one of just three axes making equal 60° angles with each other in a plane perpendicular to a line between a source of photon pairs

---

[14] One who adopts this perspective may go on to take ρ to offer an *incomplete* description of a system by taking it to have at least those properties that ρ assigns probability 1.

assigned state $|\eta\rangle$ and each polarization measurement apparatus. Selecting one of these three axes at a location may be thought to be accomplished by setting a switch there to one of three different positions, as in Figure 4.1.

However one might suppose preexisting polarization components to be distributed among a large collection of these photon pairs, the argument given in section 4.2 now shows that Einstein's suggestion leads to predictions that are incompatible with those of quantum theory itself. So Einstein's suggestion is ruled out as a way of understanding the quantum state. Note that this argument does not appeal to observation: it shows that Einstein's suggestion is unacceptable with no need to look at the world. The argument of section 4.2 assumed a rich pattern of correlations between distant flashing lights. These are exactly the correlations predicted by quantum theory when a source produces systems assigned polarization state $|\eta\rangle$ if each light is correctly hooked up to polarization detectors and the switch at each location appropriately sets the axis for a polarization measurement there to one of these three axes. The argument I gave there (following Mermin) that they could not be predicted by any common cause satisfying Einstein's principle of local action depended on the fact that the correlations were perfect when the switches on the devices were set the same way. Any difficulty in experimentally producing such perfect correlations is irrelevant to the disproof of Einstein's suggestion. It wouldn't matter even if setting up Mermin's demonstration turned out to be harder than the Manhattan project!

4.5 Quantum States Offer Prescriptions

I'll introduce a third way of thinking of the role of a quantum state $|\psi\rangle$, beginning with the homely example of an individual coin—say some quarter or euro you owned ten years ago. When tossed in normal circumstances and allowed to fall on a flat surface, you don't know whether it will fall heads or tails uppermost. We often think of such a coin toss as a random event and may use it to decide some unimportant question in a way that is acknowledged to be fair because we come to agree that each of these two outcomes has probability 1/2. We may believe the coin and its surroundings when tossed have some detailed physical state that determines which way up it will land, but we certainly don't know that state: the details of any such primary descriptive state don't help when it comes to forming reasonable expectations as to the outcome. What we do instead is to abstract from any such unknown details in assigning a secondary state that functions to prescribe a rational degree of belief 1/2 to each possible outcome: in this case, the state is just a probability assignment to each of heads and tails.

Now suppose we are also interested in a quite different question: Will this coin be inside a bank at any time over the next ten years? Again, no detailed primary descriptive state of the coin (and its much larger surroundings) will help answer this question.[15]

---

[15] Of course, you could have *settled* the answer by taking it to the bank yourself, or destroying it. But you didn't then and can't now unless you know you still have it!

But it seems reasonable to suppose that statistics of currency circulation provide some basis for assignment of a probability to this coin's being inside a bank at some time in the next ten years. If so, we can take that as another secondary state assignment when forming a reasonable expectation with regard to that possible outcome. If we like we can lump these two secondary state assignments together, but there is no unified representation of such a joint secondary state since the circumstances with respect to which we are being asked to form expectations here are quite unrelated.

The phase space probability measure ρ does provide a unified representation of a secondary state in statistical mechanics, since it helps form expectations about many different features of the system to which it is assigned, including what fraction of the n particles are in the left half of their container, what their distribution of velocities is, and so on—assuming the system is either isolated, or else in other precisely specified circumstances. The system has a primary descriptive state specified by the exact position and momentum of every particle, but again this is epistemically inaccessible and so cannot play the prescriptive role of determining rational expectations in such matters. ρ does play this role, but indirectly.

To specify the probabilities of some class of alternative outcomes at a time t one must partition phase space into regions, each corresponding to a different outcome from that class whose probability is then given by the measure $\rho(t)$ assigns to that region. To use $\rho(t_0)$ at time t_0 as a prescription for forming expectations about the outcome at a later time t_1, one needs to know $\rho(t_1)$. That knowledge is provided by a specification of the evolution of secondary state $\rho(t)$, conditional on the circumstances in which the system finds itself. Thinking of ρ as a measure of uncertainty as to the primary state ω, and its evolution as tracking the evolution of ω, one could say that ρ has an *epistemic* function, while the function of ω is *descriptive*. But it would be misleading to think of the evolution of ρ as itself representing the dynamical behavior of the system. As in the case of the probabilities in the coin example, this secondary state is used not to describe the system but to prescribe rational degrees of belief concerning it.

Finally, consider once more the quantum state $|\psi\rangle$. The orthodox view is that this provides a complete description of a system to which it is assigned. So stated, the view is ambiguous. Einstein took it to mean that $|\psi\rangle$ has a primarily descriptive function, like the phase space point ω in classical mechanics. But it might be taken instead to deny that quantum theory needs any such descriptive state—that the role of $|\psi\rangle$ is not to describe a physical system but to prescribe rational expectations about the outcome, when a physical system $|\psi\rangle$ is used to model enters into specified physical circumstances. ρ plays that prescriptive role in statistical mechanics. But there ρ could also be thought to play an epistemic role; to represent uncertainty as to the primary state ω. By denying that quantum theory requires any primary descriptive state, one insists on this important disanalogy with the state ρ of statistical mechanics. However, another analogy is preserved: just like that of ρ, the evolution of $|\psi\rangle$ does not represent the dynamical behavior of the system it is used to model.

$|\psi\rangle$ functions somewhat like the probabilities in the coin example: they both function prescriptively, in setting rational expectations about outcomes involving a system in various circumstances. But $|\psi\rangle$, like ρ in statistical mechanics, generates multiple probability measures over different classes of outcome in a unified way. Here are two related *disanalogies* between the ways $|\psi\rangle$ and ρ function in their respective theories. While ρ is itself a probability measure $|\psi\rangle$ is not, even though it is used to generate such measures, through a general algorithm called the Born probability rule, simple instances of whose operation we met in Chapter 2.[16] The failure of Einstein's suggestion shows that this is not a superficial dissimilarity but an indication of a deep difference between quantum theory and classical statistical mechanics. Only in trivial cases can Born probabilities generated by a quantum state be recovered from any underlying joint probability distribution for all dynamical variables at once.

Once we accept that the constitutive function of a quantum state is neither descriptive nor epistemic but *prescriptive*, we can ask what degrees of belief are prescribed by state $|\eta\rangle$ for the recorded outcomes of polarization measurements at 1, 2 in T. In order to answer this question we must first consider the situation of a hypothetical agent whose degrees of belief are to be prescribed. Long after T, any agent in the vicinity has ready access to the actual outcomes at 1 and 2 and so has no need to consult $|\eta\rangle$ in forming degrees of belief about them. Consider instead the situation of a hypothetical agent Alice located at 1 just before T who wonders what will be recorded as the outcome of a polarization measurement at 1 in T. Alice is in a position to know with respect to what axis a that measurement will be conducted: but because 2 is so far away and axes are chosen and measurements made at the last moment she is not in a position to know either with respect to what axis b a polarization measurement will be conducted in 2 or what will be recorded as its outcome. $|\eta\rangle$ prescribes degree of belief in outcome A equal to the Born probability $Pr_a^{|\eta\rangle}(A)$ for anyone in Alice's situation. This is equal to 1/2, independent of a. Similarly, $|\eta\rangle$ prescribes degree of belief in outcome B equal to the Born probability $Pr_b^{|\eta\rangle}(B)$ (= 1/2) for anyone in Bob's situation immediately before the polarization measurement at 2 in T. In each case this represents the rational degree of belief for anyone to hold, if he or she is in that position and accepts quantum theory, since it is based on all accessible information.

Application of the Born rule to $|\eta\rangle$ also yields a joint probability $Pr_{a,b}^{|\eta\rangle}(A\&B)$ for outcomes A and B in T at 1, 2 respectively, with axis set to a at 1 and b at 2. Since Born probability $Pr_a^{|\eta\rangle}(A)$ is independent of b and Born probability $Pr_b^{|\eta\rangle}(B)$ is independent of a, the conditional Born probabilities $Pr_{a,b}^{|\eta\rangle}(A|B), Pr_{a,b}^{|\eta\rangle}(B|A)$ can be calculated in the usual way as

---

[16] I give an explicit statement of the Born rule in Appendix A, and additional arguments refuting Einstein's suggestion in Appendix B.

$$Pr_{a,b}^{|\eta\rangle}(A|B) \equiv \frac{Pr_{a,b}^{|\eta\rangle}(A\&B)}{Pr_{a,b}^{|\eta\rangle}(B)} = \frac{Pr_{a,b}^{|\eta\rangle}(A\&B)}{Pr_{b}^{|\eta\rangle}(B)}$$

$$Pr_{a,b}^{|\eta\rangle}(B|A) \equiv \frac{Pr_{a,b}^{|\eta\rangle}(A\&B)}{Pr_{a,b}^{|\eta\rangle}(A)} = \frac{Pr_{a,b}^{|\eta\rangle}(A\&B)}{Pr_{a}^{|\eta\rangle}(A)}$$

Each of these conditional Born probabilities generally differs from the corresponding unconditional probability. If a, b are aligned along the same direction, for example, $Pr_{a,b}^{|\eta\rangle}(A\&B) = Pr_{a}^{|\eta\rangle}(A) = Pr_{b}^{|\eta\rangle}(B) = 1/2$, while $Pr_{a,b}^{|\eta\rangle}(A|B) = Pr_{a,b}^{|\eta\rangle}(B|A) = 1$ or 0, depending on whether B is taken to indicate the same or opposite polarization result to that indicated by A.

If Alice at 1 just before T knew the distant setting b and outcome B, she could use this knowledge, together with the conditional Born probability $Pr_{a,b}^{|\eta\rangle}(A|B)$, to adjust her degree of belief in A (in this example, changing it from wholly undecided (1/2) to certainty (1 or 0)). But she would be in no position to do so, since her physical situation prevents any access to timely information about either b or B in these circumstances. $Pr_{a,b}^{|\eta\rangle}(A|B)$ does not represent a better estimate of the real chance of A just before T than $Pr_{a}^{|\eta\rangle}(A)$: there is no such unique real chance. Each of $Pr_{a,b}^{|\eta\rangle}(A|B), Pr_{a}^{|\eta\rangle}(A)$ plays a role in providing wise guidance to one in Alice's situation at 1 just before T: anyone in that situation who accepts quantum theory should have a degree of belief in A equal to $Pr_{a}^{|\eta\rangle}(A)$, and a degree of belief in A conditional on b and B equal to $Pr_{a,b}^{|\eta\rangle}(A|B)$. Similarly, anyone at 2 just before T who accepts quantum theory should have a degree of belief in B equal to $Pr_{b}^{|\eta\rangle}(B)$, and a degree of belief in B conditional on a and A equal to $Pr_{a,b}^{|\eta\rangle}(B|A)$. All these probabilities are prescribed by assignment of state $|\eta\rangle$ but none of them should be singled out as the objective chance, either of A or of B.

Quantum theory is often taken to be an indeterministic theory that specifies the objective chance, or physical probability of an event. But quantum theory specifies no such unique probability if the constitutive function of quantum states is to prescribe objectively rational degrees of belief concerning events to (actual or hypothetical) physically situated agents. A quantum probability of an event is objective, in the sense that it provides an authoritative prescription for the degree of belief to hold about that event to anyone in a specific physical situation that precludes more direct access to information about its occurrence. But since the prescribed degree of belief depends on this physical situation, a quantum probability of an event localized to region R is not itself something localized in or near R, but must be relativized to arbitrarily distant places and times. Acknowledging this relativity of quantum probabilities prompts a re-evaluation of the condition of local causality Bell assumed in his argument that certain non-localized correlations (successfully predicted by quantum theory) cannot be explained without action at a distance.

Recall the Conditional Factorizability condition incorporated in Bell's local causality condition

$$Pr_{a,b}(A\&B|\lambda) = Pr_a(A|\lambda) \times Pr_b(B|\lambda). \qquad \text{(Conditional Factorizability)}$$

Quantum theory specifies no parameter that functions like Bell's λ. Instead its Born rule specifies these joint and single probabilities, given a quantum state $|\psi\rangle$: $Pr_{a,b}^{|\psi\rangle}(A\&B), Pr_a^{|\psi\rangle}(A), Pr_b^{|\psi\rangle}(B)$. Placing $|\psi\rangle$ as a label on these probabilities rather than conditionalizing on $|\psi\rangle$ marks the fact that it is to be considered a parameter like a, b and not a random variable that comes with its own probability distribution. The closest significant quantum analog to Conditional Factorizability is

$$Pr_{a,b}^{|\psi\rangle}(A\&B) = Pr_a^{|\psi\rangle}(A) \times Pr_b^{|\psi\rangle}(B). \qquad \text{(Quantum Factorizability)}$$

We have already seen that Quantum Factorizability fails for the case $|\psi\rangle = |\eta\rangle$. This is how quantum theory is able successfully to predict patterns of correlation that violate the inequality to which any theory satisfying Conditional Factorizability is subject. But hasn't Bell's argument now shown that quantum theory cannot *explain* these correlations without action at a distance? To answer this question we must look more closely at Bell's local causality condition in the light of the prescriptive function of the quantum state and the consequent relational character of quantum probabilities.

4.6 Exorcising the Spook

Recall that Bell proposed that his local causality condition be applied to any theory intended to explain non-localized correlations while conforming to his intuitive principle. He supposed that any such theory would do so by specifying a unique probability for each relevant event e at 1 in T and a unique probability for each relevant event f at 2 in T. He thought the theory should explain the correlations in terms of their probabilistic causes, where local causality was intended to localize the cause of e within reach of 1 and f within reach of 2 earlier than T. A probabilistic cause of an event was to act by affecting its (unique) probability of occurrence, so if an event f at 2 in T could alter the probability of an event e at 1 in T even after allowing for the influence on this probability of all probabilistic causes prior to T, then that would make f a non-local probabilistic cause of e in violation of his intuitive principle.

The relational character of quantum probabilities presents a problem when trying to apply Bell's condition of local causality to quantum theory. Local causality presupposes what quantum theory denies—that a unique probability attaches to an event at 1 in T (or 2 in T). There is no single probability that is altered by specifying setting b and conditionalizing on outcome B at 2 to replace $Pr_a^{|\eta\rangle}(A)$ by $Pr_{a,b}^{|\eta\rangle}(A|B)$: these are just different objective probabilities, each of which tells an agent in a specified physical situation how strongly to expect each possible outcome A. The proffered advice may be based on all information accessible in that situation, or it may be conditional on receipt of additional information not so accessible. Putting yourself in this situation (even in thought) and taking this advice, you will set your degrees of belief and/or conditional degrees of belief accordingly. These unconditional and conditional degrees of belief should often differ. So while quantum probabilities do violate Quantum Factorizability this does not imply that quantum theory violates local causality, and so Bell's argument does not show that quantum theory implies action at a distance.

But Bell left open another option—that the non-localized correlations successfully predicted by quantum theory were simply inexplicable. That presents a challenge: can quantum theory itself explain the non-localized correlations it so successfully predicts? One can't hope to show that it can without saying quite a lot about what more is required of an explanation than mere prediction of what is to be explained. Such philosophical work will have to be postponed to Part II of this book (see Chapter 9). For now I'll just offer a narrative in terms of quantum states and probabilities that connects such correlations to physical circumstances in which they are manifested.

Prior to T Alice near 1 and Bob near 2 each take the polarization state of the system to be $|\eta\rangle$. What justifies this quantum state assignment is their knowledge of the conditions under which the system was produced—perhaps by parametric down-conversion of laser light by passage through a non-linear crystal. Such knowledge depends on information about the physical systems involved in producing the pair that may be provided without using the terminology of quantum theory. This state assignment is backed by claims about the values of physical magnitudes on such systems—if anything counts as a descriptive statement acknowledged by quantum theory, these do. Then Alice measures polarization of subsystem L along axis a, and Bob measures polarization of subsystem R along axis b. What happens at the detectors licenses both of them to treat the Born rule measure corresponding to a quantum polarization state assignment as a probability distribution over statements about the values of dynamical variables that may be taken to describe possible outcomes of Alice's (and/or Bob's) polarization measurements.[17]

Alice and Bob should then both assign state $|\eta\rangle$ and apply the Born probability rule to find the joint probability $Pr^{|\eta\rangle}_{a,b}(A\&B)$ for polarization recorded along both a, b axes, and use this to calculate the (well-defined) conditional probability $Pr^{|\eta\rangle}_{a,b}(A|B)$. Each will then expect the observed non-localized correlations between the outcomes of polarization measurements at 1, 2 in T when the detectors are set along the a, b axes. They will expect analogous correlations as these axes are varied, and so they will expect violations of the key inequality derived from Quantum Factorizability in such a scenario.

Immediately after recording one or other polarization B for R, Bob should use that knowledge to update his unconditional degree of belief for Alice to record polarization A of L with respect to the a-axis to match the conditional probability specified in the previous paragraph: But Alice, lacking this knowledge, should continue to base her

---

[17] I'll make this more precise in section 2 of Chapter 5.

degree of belief in outcome A on the unconditional probability $Pr_a^{|\eta\rangle}(A)$ that she will record polarization A of L along the a-axis. In this way each forms expectations as to the outcome of Alice's measurement on the best information available to him or her at the time. Alice's statistics of her outcomes in many repetitions of the experiment are just what her degree of belief (based just on $|\eta\rangle$, a) led her to expect, thereby helping to explain her results. Bob's statistics for Alice's outcomes (in many repetitions in which his outcome is B) are just what his updated degree of belief (based on $|\eta\rangle$, a, b, B) led him to expect, thereby helping him to explain Alice's results. Though objectively prescribed by quantum theory, neither Alice's nor Bob's degree of belief corresponds to any unique physical propensity or chance of Alice's outcome.

Anyone can apply quantum theory from the perspective provided by the local situation of Alice or of Bob to show that the correlations of outcomes recorded at 1, 2, as well as the statistics separately recorded at each of 1, 2, were just as expected for an agent in that situation. These applications also make it clear what both these non-localized correlations and the localized outcomes that constitute them physically depend on. By showing that they were to be expected and what they depend on, quantum theory helps us to explain the non-localized correlations it successfully predicts. The correlations depend on whatever physical conditions make $|\eta\rangle$ the correct quantum polarization state to assign to the systems involved; the settings a, b; and the physical conditions at 1, 2 in T necessary for events localized in each region to count as recordings of the outcome of the relevant linear polarization measurement there.

The outcome at 1 is also probabilistically correlated with the outcome at 2 (and vice versa). But this kind of dependence on a distant physical event is not in violation of Einstein's principle of local action: since quantum theory allows for no external influence on the *outcome* of a polarization measurement at 1(2), there is no way this could have any influence (direct or indirect) on the outcome at 2(1). Note especially that while it is common to refer to the systems involved here as photon pairs (as I did myself earlier), nowhere in this quantum narrative was it necessary to refer to a photon or its actual linear polarization.

5
Assigning Values and States

Chapter 4 proposed that the function of a quantum state is neither descriptive nor epistemic but prescriptive. Many people first learn about prescriptions when they go to see a doctor. Three questions arise in that context: what is your prescription, exactly what is to be expected if you follow it, and what are the grounds for that expectation? Parallel questions arise for quantum states: what quantum state should you assign, exactly what expectations follow from this assignment, and what are the grounds for those expectations? I'll address these questions one by one in the three sections of this chapter. In answering them I'll need to introduce a third important quantum idea besides superposition and entanglement, namely decoherence.

5.1 Making Quantum State Assignments

Recall how Schrödinger introduced the notion of entanglement:

> When two systems, of which we know the states by their respective representatives, enter into temporary physical interaction due to known forces between them, and when after a time of mutual influence the systems separate again, then they can no longer be described in the same way as before, viz. by endowing each of them with a representative of its own.... By the interaction the two representatives (or ψ-functions) have become entangled. ([1935], p. 555)

We saw in Chapter 3 that this notion extends to more than two systems: indeed, the quantum state vectors of arbitrarily many systems may be or become entangled. When entanglement is a result of interactions it would still persist even if these interactions were to cease altogether. Moreover, entangled states are now commonly assigned to systems that have never interacted.[1] This may make one wonder how useful it can be to assign a vector state to a system, and whether this is even possible given the ubiquity of entanglement. These concerns may be effectively addressed if one acknowledges the prescriptive function of quantum state assignment introduced in Chapter 4.

As we have seen, to account for correlations between outcomes of measurements it is often essential to assign an entangled quantum state vector Ψ to a system composed of multiple subsystems. But if you only want to know what outcome to expect from a measurement involving a single subsystem, then you don't need to consult that state

---

[1] This occurs in so-called entanglement-swapping experiments like those of Hensen et al. [2015].

vector. It is possible to form one's expectations by assigning a mathematical object of a new kind as quantum state $\hat{\rho}$ to just that subsystem, and then applying a slight generalization of the Born probability rule to $\hat{\rho}$.[2] Applying this generalization to $\hat{\rho}$ can easily be shown to yield the same probability for any outcome on that subsystem as applying the original Born rule to Ψ. $\hat{\rho}$ wears a hat to mark it as an *operator* on vectors—a function from vectors to vectors: it is known as a *density operator*. Whether or not a subsystem can be assigned a "representative" (i.e. state vector) of its own, no formal condition of ordinary quantum mechanics prohibits assignment of a density operator as its quantum state.

As a function from vectors to vectors, a density operator has a domain over which it is defined and a range of values that result from its application to vectors in its domain. While every density operator $\hat{\rho}$ has the entire vector space as its domain, the range of a density operator is restricted to a subset of the space. Since all complex linear sums of vectors in the range of a density operator are also vectors in its range, together they also form a vector space wholly contained within the original space. This vector *subspace* is often of lower dimension than the original space. If the range of a density operator is confined to a one-dimensional subspace it is said to represent a *pure state*; otherwise the state is said to be *mixed*. Quantum states represented by vectors are pure. Indeed, an equivalent vector representation of a pure state is provided by *any* of an infinite set of vectors forming the one-dimensional range of a pure state density operator—if $|\psi\rangle$ represents a pure state, then $c|\psi\rangle$ represents that same pure state, for arbitrary (non-zero) complex number c.

The introduction of this more general density operator representation of quantum states opens up the possibility of assigning a quantum state to a system irrespective of whether there is some larger system of which it is a subsystem. Because of the pervasive nature of entanglement, a state so assigned will typically be mixed rather than pure. So this has not yet resolved the question of when, if ever, it is possible to assign a *pure* state vector to a system. But a change in an agent's situation may give him or her access to information that will make that possible, as we can see by looking again at an example from the previous chapter.

Recall the scenario in which Alice and Bob assign entangled polarization state $|\eta\rangle$ to a system on which they will each independently record the outcome of a polarization measurement in T, Alice at 1, Bob at 2. Consider Bob's situation. Just before T he would be correct to assign pure state $|\eta\rangle$ to the compound entangled system, and mixed state $\hat{\rho} = 1/2\,\hat{1}$ to the subsystem L on which Alice will make a polarization measurement.[3] This mixed state is quite uninformative: application of the generalized Born rule yields probability $1/2$ for each of her two possible outcomes no matter what polarization

---

[2] Appendix A states the generalization (see § A.5) and gives some mathematical details of density operators and mixed states.

[3] When applied to any vector representing a pure polarization state of L the operator $\hat{1}$ leaves that vector unchanged.

measurement Alice chooses to make at 1. But right after making his polarization measurement at 2 Bob is in a position to update his degrees of belief concerning the possible outcomes of Alice's measurement to take account of the outcome of his own measurement.

One way for Bob to do this is by using the conditional probability $Pr_{a,b}^{|\eta\rangle}(A|B)$ calculated by applying the Born rule to entangled state $|\eta\rangle$. Suppose his outcome was recorded as polarization B with respect to the b axis, and he wonders what outcome to expect if Alice has measured polarization with respect to that same axis. Since the conditional probability $Pr_{b,b}^{|\eta\rangle}(B|B) = 1$, he should be certain (degree of belief 1) that Alice will get the same outcome he did. But Bob also has an entirely equivalent way to arrive at the same expectation for Alice's outcome if she chooses to measure along the same axis as Bob. If Bob assigns pure state $|B\rangle$ to subsystem L then $Pr_b^{|B\rangle}(B) = 1$ also, where the subscripted b now indicates the setting of Alice's polarization measurement device along the same axis as Bob's, and B now indicates the outcome of Alice's measurement.[4]

More generally, since $Pr_a^{|B\rangle}(A) = Pr_{a,b}^{|\eta\rangle}(A|B)$, Bob can reliably set his degrees of belief regarding the possible outcomes of any polarization measurement by Alice simply by updating the quantum state he assigns to L from mixed state $1/2\,\hat{1}$ to pure state $|B\rangle$ as soon as he has a record of the outcome B of his own polarization measurement at 2. Similarly, Alice can reliably set her degrees of belief regarding the possible outcomes of any polarization measurement by Bob by updating the quantum state she assigns to the other subsystem R from mixed state $1/2\,\hat{1}$ to pure state $|A\rangle$ as soon as she has a record of the outcome A of her own polarization measurement at 1.

Many cases in which experimenters claim to prepare a pure quantum state may be understood in a similar way. Recall the situation illustrated in Figure 2.8 when the second beam splitter is polarizing. Such a polarizing beam splitter is usually said to split a single monochromatic input beam of light into two output beams with orthogonal polarizations. But in order to account for interference effects if the two output beams are suitably recombined (say, in a Mach–Zehnder interferometer) one must assign an entangled state to light output by the device of the form

$$\Xi = c_1 |\psi_1\rangle |\updownarrow\rangle + c_2 |\psi_2\rangle |\leftrightarrow\rangle \tag{5.1}$$

where $|\updownarrow\rangle, |\leftrightarrow\rangle$ are pure polarization states of light certain to pass a polarizing filter with axis oriented along $\updownarrow, \leftrightarrow$ respectively, and $|\psi_1\rangle, |\psi_2\rangle$ are pure states corresponding to certain detection of light by detectors *Different*, *Same* respectively.

Suppose Alice instead wishes to perform an experiment on a beam of light to which she can correctly assign pure polarization state $|\updownarrow\rangle$. All she has to do is to replace detector *Different* with her experimental apparatus. This apparatus will record a result only for light passing the beam splitter that would have been detected by *Different*,

---

[4] The notation $|B\rangle$ is intended to represent this certainty about Alice's measured value B of polarization with respect to the b axis.

corresponding to state $|\psi_1\rangle$: light corresponding to state $|\psi_2\rangle$ would certainly not have been detected by *Different* or her experimental apparatus. But any light passing the beam splitter then encountering her apparatus would certainly pass a polarization detector with axis $\updownarrow$. That is why Alice would be correct to assign polarization state $|\updownarrow\rangle$ to the light on which she performs her experiment.

Alice's justification for assigning pure state $|\updownarrow\rangle$ to a subsystem of an entangled system here parallels her justification for assigning pure state $|A\rangle$ to R in the previous example. In this new example, Alice is interested in forming expectations regarding the possible outcomes recorded in her planned experiment. For this purpose, she is entitled to restrict attention to events whose outcomes she will record—those resulting from light corresponding to state $|\psi_1\rangle$, not to state $|\psi_2\rangle$. So she should conditionalize the Born probability derived from entangled state Ξ on the occurrence of such an event. The degrees of belief in each possible outcome of her experiment she will arrive at by assigning polarization state $|\updownarrow\rangle$ are just the same as those she would arrive at by assigning Ξ and conditionalizing on recording an outcome of her experiment. That is why she is justified in assigning pure polarization state $|\updownarrow\rangle$ to the light she will use in her experiment.

This prescription for state assignment may be generalized. Suppose you want to be able to assign a (possibly unknown) pure state $|\psi\rangle$ to a system, but you are confident that this state is somehow entangled with the state of another system (in something like the way that equation (5.1) indicates a polarization state entangled with a translational state). Even with very little information relevant to assigning the exact entangled state, you may still be justified in assigning state $|\psi\rangle$ to the system of interest for the purpose of guiding your expectations as to the recorded outcome of a measurement on just that system. For example, one would need no details of either the coefficients c_1, c_2 or the states $|\psi_1\rangle, |\psi_2\rangle, |\Leftrightarrow\rangle$ in equation (5.1) in order to justifiably assign polarization state $|\updownarrow\rangle$ to a system subjected to measurement by a device that would yield an outcome only for a pure state component that included $|\updownarrow\rangle$ as one factor.

It is important to stress that assignment of pure state $|\psi\rangle$ to a subsystem s of a compound system S does not exclude but is perfectly consistent with assignment of entangled state $|\Psi\rangle$ to S. The two state assignments are based on different assumptions that give them different functions. State $|\Psi\rangle$ is assigned to guide expectations as to the outcome of any measurement on the compound system, including arbitrary joint or single measurements on any of its component subsystems. With no further assumption about the outcome of a measurement on a subsystem, application of the Born rule to $|\Psi\rangle$ will yield the same probability for any outcome of a measurement on s as application of the generalized Born rule to the appropriate *mixed* state of s.

But if one is in a position to make a suitable assumption about the outcome of a measurement on a subsystem of S, then one can assign *pure* state $|\psi\rangle$ to s solely to guide expectations as to the outcome of any measurement on s alone. Because of its limited function, an agent may be in a position to assign state $|\psi\rangle$ to s but not to assign state $|\Psi\rangle$ to S. If one were also in a position to assign $|\Psi\rangle$ to S one could instead use this assignment to guide expectations as to the outcome of any measurement on s alone. In this case one would conditionalize the probability for that outcome generated by applying the Born rule to $|\Psi\rangle$ on the information supplied by whatever additional assumption justified assignment of $|\psi\rangle$ to s: this would lead one to adopt just the same degrees of belief in the possible outcomes as one would arrive at by simply assigning $|\psi\rangle$ to s.

We have seen how it is possible consistently to assign a quantum state (pure or mixed) to a subsystem of an entangled system. But what makes such a state assignment correct? Experience in the laboratory has taught physicists what they need to do to prepare a particular quantum state, so one could appeal to a rule implicit in such experience as constitutive of the correctness of the corresponding state attribution. But more needs to be said, since a state attribution is subject to independent testing: are expectations based on application of the Born rule to the candidate state borne out by the statistics of outcomes in repeated measurements on systems assigned that state? There is an analogy with the assignment of a probability for a tossed coin to fall heads uppermost. One may appeal to rules of thumb for counting a coin fair, but if it comes up heads roughly nine times out of ten in a long sequence of repeated trials, this would undermine reliance on those rules to assign heads probability 1/2.

When they are available, statistics of measurement outcomes may trump other considerations in justifying assignment of a particular quantum state to a system. But if the question is what makes a quantum state assignment correct, then we cannot rest content with an answer in terms of measurement outcomes. There are two reasons why not. As we learnt from Bell in Chapter 2, we cannot arrive at a satisfactory understanding of quantum theory by appealing to the unexplained notion of a *measurement outcome*, since the notion of measurement has no place in a satisfactory formulation of a fundamental physical theory. Moreover, a state may be correctly assigned to a system that is never measured. So what makes it correct must be something else that accounts for what *might* be expected to happen if it *were* measured. The next section explains how to use the notion of decoherence to eliminate talk of measurement from the statement and application of the Born probability rule: in the third section I show how this notion may be used to license claims stating the grounds not only of correct quantum state assignments but also of a host of other familiar statements about the physical world.

5.2 Measurement-free Quantum Probabilities

To remove talk of measurement from a precise formulation of quantum theory we need to state the Born rule without using a term like 'measurement' and to say how that rule may be applied without reintroducing any such term. Only if we can do this will it become clear how quantum theory can be applied to account for what happens in the big world outside the laboratory, and not just used successfully to predict the statistics of physicists' experiments.

Actual applications of the Born rule to generate probabilities require the detailed mathematical formulation I give in Appendix A. But since our present focus is rather on understanding what it is that these probabilities concern, we may abstract from these details. All we need here is the schematic form of the underlying algorithm. This is usually expressed as follows:

(Usual Born rule schema)
For a system s assigned quantum state $|\psi\rangle$, the probability that the outcome of a measurement of dynamical variable M will yield a value in set Δ of real numbers is the value of a function $\Pr(M, \Delta, |\psi\rangle, s)$.

As we saw, Einstein suggested an alternative schematic formulation of the Born rule:

(Alternative Born rule schema)
For a system s assigned quantum state $|\psi\rangle$, the probability that the value of dynamical variable M lies in set Δ of real numbers is the value of a function $\Pr(M, \Delta, |\psi\rangle, s)$.

For future reference I'll call a statement of the form *The value of dynamical variable M on physical system s lies in* Δ a *canonical magnitude claim* and abbreviate it as $M_s \in \Delta$. It is because in this alternative formulation the Born rule assigns probabilities to magnitude claims of this form that I call them canonical.[5] So formulated, the Born rule explicitly concerns probabilities of canonical magnitude claims—not of the outcomes of measurements.

But we saw that the specific probabilities prescribed by the Born rule for certain quantum systems s assigned some state $|\psi\rangle$ cannot be recovered from any joint probability distribution over real-numbered values of all dynamical variables at once.[6] I will briefly dismiss two ways one might try to reconcile adoption of the alternative Born rule schema with this result before proposing and elaborating a third alternative.

An obvious response to this problem is to drop the assumption of faithful measurement. It is perfectly consistent to maintain that every dynamical variable has a precise real value, but to deny that this is always the value that would be revealed by its properly conducted measurement. However, mere consistency is not enough. By severing the connection between the actual value and the measured value of a dynamical variable this approach would render the alternative Born rule schema useless in predicting or explaining any observed phenomenon. That would be to cure quantum theory by killing it!

Another idea is to hang on to the faithful measurement assumption and to maintain that in each quantum state the precise real values of all dynamical variables have some

---

[5] Other magnitude claims are not of this form, such as the claim that the outcome of a measurement of M on s lies in Δ, and the claim that if the value of s's x-position lies in Δ then the value of s's x-momentum lies in Γ.

[6] The interested reader is referred to Appendix B for simple proofs of this claim based on analysis of the entangled polarization states of three photons. The statistics of outcomes for compatible polarization measurements confirm their Born probabilities even when these take the extreme values 0,1.

well-defined joint probability distribution, but to deny that probabilities derived from this distribution equal the corresponding Born probabilities. The Born rule prescribes joint probabilities only for certain specific subsets of dynamical variables, which are said to be compatible.[7] If there were a joint probability distribution over the values of all dynamical variables this would automatically yield joint distributions over all compatible subsets. One could deny the equality of the latter joint distributions to corresponding Born rule probabilities by supposing that measurement, though faithful, was inevitably conducted on systematically biased samples of the total population. But then this idea would also render the alternative Born rule schema useless in practice.

The usual Born rule schema escapes these criticisms by restricting its application to compatible dynamical variables—those for which joint Born probability distributions are well defined. This restriction lines up nicely with its explicit talk of measurement provided one makes the traditional assumption that arbitrarily precise joint measurements of the values of a set of dynamical variables are possible only if those variables are compatible. And that assumption squares with an understanding of the uncertainty principle as prohibiting arbitrarily precise simultaneous measurements of (a single component of) the position and momentum of a particle, even though arbitrarily precise simultaneous measurements of distinct components of position (and/or of momentum) are possible.

But we can have our cake and eat it too! Quantum theory itself provides a way to restrict the application of the alternative Born rule schema. This involves no explicit mention of measurement. It does, however, make it clear why the alternative Born rule schema is applicable in a wide variety of circumstances, including those where we take ourselves to be performing measurements: and it also explains why arbitrarily precise joint measurements are possible only for compatible sets of dynamical variables.

The Born rule should be used to specify the probability for a dynamical variable P on s to have a value in Δ only when s is in a situation where application of quantum theory would justify one in concluding that s should be assigned an appropriately decohered quantum state. It will take a while to explain what this means. But before I explain, let me emphasize that the explanation will at no point depend on referring to something as a measurement or observation, an apparatus, an environment, a classical or macroscopic system, or irreversible amplification. Once the explanation has been given, you will see why it is natural to use such terminology when commenting on it: but this commentary should not be confused with what is being commented on.

The easiest way to introduce the idea of decoherence is to begin with the superposition principle. Consider a superposed quantum state like

$$|V\rangle = \frac{1}{\sqrt{2}}\left(|45°\rangle + |135°\rangle\right). \tag{5.2}$$

---

[7] Compatible dynamical variables are represented by pairwise commuting self-adjoint operators on the system's vector space: see Appendix A.

When one first encounters such a state it is tempting to suppose that a system to which it is assigned is either in state $|45°\rangle$, linearly polarized at 45° to the vertical, or (with equal probability) in state $|135°\rangle$, linearly polarized at 135° to the vertical. That supposition is certainly consistent with the fact that on encountering a polarizer with axis set to 45° the system has a 50–50 chance of passing. But that is true also of an infinite number of other states, including

$$|H\rangle = \frac{1}{\sqrt{2}}\left(|45°\rangle - |135°\rangle\right) \tag{5.3}$$

and the mixed state $\hat{\rho} = 1/2\,\hat{1}$.

What distinguishes superpositions like $|V\rangle$ and $|H\rangle$ from the mixed state $1/2\,\hat{1}$ is the interference between the two components of the superpositions, as manifested by the fact that a system assigned $|V\rangle$ is certain to pass a polarizer with axis vertical while a system assigned $|H\rangle$ is certain to be absorbed. Only the state $1/2\,\hat{1}$ seems capable of representing a system either linearly polarized at 45° to the vertical, or (with equal probability) linearly polarized at 135° to the vertical: and even that is brought into question once one realizes that with equal justification $1/2\,\hat{1}$ may be taken to represent a system either vertically polarized, or (with equal probability) horizontally polarized.

This example may be generalized. Consider a measurement of dynamical variable Q on a system assigned superposed state

$$|\psi\rangle = c_1\,|P_1\rangle + c_2\,|P_2\rangle. \tag{5.4}$$

If $|Q_1\rangle$, $(|Q_2\rangle)$ is a state certain to yield outcome Q_1, (Q_2) in a measurement of Q, the probability that $|\psi\rangle$ will give outcome Q_1 is

$$\begin{aligned}
Pr_Q^{|\psi\rangle}(Q_1) &= |\langle Q_1|\psi\rangle|^2 \\
&= |c_1\langle Q_1|P_1\rangle + c_2\langle Q_1|P_2\rangle|^2 \\
&= |c_1|^2|\langle Q_1|P_1\rangle|^2 + |c_2|^2|\langle Q_1|P_2\rangle|^2 + 2\,\mathrm{Re}(c_1^*c_2\langle P_1|Q_1\rangle\langle Q_1|P_2\rangle) \\
&= |c_1|^2 Pr_Q^{|P_1\rangle}(Q_1) + |c_2|^2 Pr_Q^{|P_2\rangle}(Q_1) + 2\,\mathrm{Re}(c_1^*c_2\langle P_1|Q_1\rangle\langle Q_1|P_2\rangle)
\end{aligned} \tag{5.5}$$

Notice that the first two terms on the right give the weighted sum of the probability of getting result Q_1 in state $|P_1\rangle$ and getting result Q_1 in state $|P_2\rangle$. If it weren't for the third "interference" term,[8] we could have computed $Pr_Q^{|\psi\rangle}(Q_1)$ by assuming that a system in state $|\psi\rangle$ is either in state $|P_1\rangle$ (with probability $|c_1|^2$), or in state $|P_2\rangle$ (with probability $|c_2|^2$). In certain circumstances, these interference terms all vanish, or at least become extremely small compared to the non-interference terms.[9] When that happens, terms like $|P_1\rangle$, $|P_2\rangle$ in a superposition like $|\psi\rangle$ are said to *decohere*. In that case, the state may be regarded as a mixture rather than a superposition of these

---

[8] Notation: Re z means the real part x of the complex number $z = x + iy$: $z^* \equiv x - iy$.
[9] Equation 2.19 is an example in which the interference terms vanish for any measurement insensitive to polarization because of the orthogonality of the polarization states $|45°\rangle, |135°\rangle$.

components—at least for the purpose of computing probabilities of getting one result rather than another when measuring dynamical variable Q on systems in that state.

In Appendix C I give details of a simple model introduced by Zurek [1982] in which the joint evolution of the state of quantum systems α and ϵ induces decoherence in the state assigned to α. This can be used to model a simple physical system interacting with a compound physical system made up of a huge number of simple components. In the model the initial state assigned to α may be expressed as a superposition

$$|\psi\rangle_\alpha = (a|\Uparrow\rangle + b|\Downarrow\rangle) \tag{5.6}$$

of two states, for each of which the usual statement of the Born rule would imply that a measurement of P is certain to yield the corresponding value ($\Uparrow$ or $\Downarrow$). Interaction between physical systems is modeled as entangling the state of α with that of ϵ so the state $\hat{\rho}_\alpha(t)$ of α becomes mixed. But $\hat{\rho}_\alpha(t)$ extremely quickly comes to very closely approximate a mixture of the pure states $|\Uparrow\rangle, |\Downarrow\rangle$ with probabilities $|a|^2, |b|^2$ respectively. The interference terms between these states continue to fluctuate but remain extremely small for an enormously long time.

It is this *stability* that entitles one to infer that P has acquired a value as a result of the interaction, and to regard rival claims about that value as empirically significant. Suppose we apply this model to a physical spin 1/2 system s, identifying P with the spin component of s in the z direction (S_z) and Q with the spin component of s in the x direction (S_x). Here are some statements about s that we represent in the simple model by α: each of them is readily recast as a canonical magnitude claim (e.g. (5.7) abbreviates *The spin component of s in the z direction (S_z) lies in unit set $\{+\hbar/2\}$*, where $\hbar$ is Planck's constant [$\approx 1.055 \times 10^{-34}$ Joule seconds]); and (5.11) abbreviates *The spin component of s in the x direction (S_x) lies in unit set $\{-\hbar/2\}$*).

$$S_z = +\hbar/2 \tag{5.7}$$
$$S_z = -\hbar/2 \tag{5.8}$$
$$S_z \in \{+\hbar/2, -\hbar/2\} \tag{5.9}$$
$$S_x = +\hbar/2 \tag{5.10}$$
$$S_x = -\hbar/2 \tag{5.11}$$
$$S_x \in \{+\hbar/2, -\hbar/2\} \tag{5.12}$$

Anyone applying the simple model to s should regard statements (5.7) and (5.8) as empirically significant; and (5.9) as both empirically significant and true even with no information bearing on the question as to which of (5.7) or (5.8) is true. But s/he should regard none of the statements (5.10), (5.11), (5.12) as empirically significant.

Application of the alternative Born rule schema is restricted to empirically significant magnitude claims. We have just seen an illustration of how to apply a model of decoherence to assess empirical significance. By applying this model one can conclude that the Born rule is applicable to (5.7), (5.8), and (5.9), and so assign these probabilities 1/2, 1/2, 1 respectively; and one can also conclude that the Born rule is not applicable to

(5.10), (5.11), or (5.12). Notice that it is by applying quantum theory itself in a model of decoherence that we encounter such limits on the applicability of the Born rule: it is the form of decoherence, not the initial state of α, that privileges the "preferred basis" $|\Uparrow\rangle, |\Downarrow\rangle$ by effectively decohering just these states. There is no need to call upon an obscure notion of measurement that has no place in a precise formulation of a fundamental physical theory. Understanding the Born rule in accordance with the alternative schema removes any talk of measurement from the formulation of that rule, and the restrictions on its application may be explained without reintroducing such talk.

Decoherence is not perfect, even in this simple model. As it evolves, the interference terms between $|\Uparrow\rangle, |\Downarrow\rangle$ in state $\hat{\rho}_\alpha(t)$ of α will almost always be extremely close, but not equal, to zero. Consider a statement (5.13) about the component of spin S_θ on s, where θ is some angle to the z-axis:

$$S_\theta \in \{+\hbar/2, -\hbar/2\}. \tag{5.13}$$

If θ is close enough to zero then, very rapidly and quite stably, the evolution of $\hat{\rho}_\alpha(t)$ will almost as effectively remove interference terms between states associated with definite values of spin component in the θ direction (as can be shown by re-expressing $\hat{\rho}_\alpha(t)$ in a basis of these states). This will endow statement (5.13) with almost as much empirical significance as (5.9).

More generally, the empirical content of a statement of the form (5.13) here is a function of θ, varying continuously from its maximum value for $\theta = 0, \pi$ to zero for $\theta = \pi/2$. Empirical significance has no natural cut-off here, or in any application of quantum theory. So this restriction on application of the Born rule does not yield a precise selection criterion. This is a classic case of vagueness. The Born rule is clearly applicable to statements (5.7), (5.8), (5.9) in the simple model, and clearly inapplicable to statements (5.10), (5.11), (5.12). But the limits of applicability of the Born rule to a statement of the form (5.13) may be set anywhere within a narrow (but equally indeterminate) range of values of θ in the neighborhoods of $0, \pi$. Does this vagueness matter?

Some take quantum theory to be fundamental because it provides our most accurate descriptions of nature—call this fundamental$_a$. But Bell criticized contemporary formulations of quantum theory on the grounds that these are fundamentally approximate and intrinsically inexact. "Surely", he asked in 1989, "after 62 years we should have an exact formulation of some serious part of quantum mechanics?" ([2004], p. 213). Bell denied that quantum theory is fundamental$_a$, because contemporary formulations are not in terms of what he called 'beables':

It is not easy to identify precisely which physical processes are to be given the status of "observations" and which are to be relegated to the limbo between one observation and another. So it could be hoped that some increase in precision might be possible by concentration on the beables, which can be described "in classical terms", because they are there. ([2004], p. 52)

Bell would surely have rejected the view of quantum theory presented in this book. He would have taken the vagueness inherent in the conditions of applicability of the Born rule to introduce an unacceptable imprecision into the theory, and regarded quantum theory viewed this way as not a serious theory, that is, not a serious candidate for the job of truly describing nature. But in this view quantum theory achieves its unprecedented success *without* itself describing nature, either vaguely or precisely. An agent does not use a quantum-theoretic model to describe physical systems: a quantum model is not itself in the business of representing physical reality. Since quantum theory does not itself yield descriptions of nature, it is clearly not a fundamental$_a$ theory. But it is fundamental in another sense: it gives us our best and only way of predicting and explaining a host of otherwise puzzling phenomena. We do this using quantum models—not to describe reality but to advise us on how to describe it and what to believe about it. Any such use depends on application of the Born rule. So the predictive and explanatory successes we achieve using quantum theory depend on judicious application of that rule.

Now one can see why any vagueness associated with application of the Born rule does not matter. Application of a theory or rule always requires judgment, and this is no exception. In applying any physical theory one must first decide how to model the part or aspect of the physical world on which the application is targeted. This model of decoherence was called simple because it has few if any real world targets—a wise agent would rarely if ever decide to apply it. When applying a quantum-theoretic model, an agent must make a further decision about which statements are apt for application of the Born rule. Here, too, good judgment is called for.

Models of quantum theory are not inherently imprecise. Their specification need contain none of Bell's ([2004], p. 215) "proscribed words": 'measurement', 'apparatus', 'environment', 'microscopic', 'macroscopic', 'reversible', 'irreversible', 'observable', 'information', or 'measurement', though one may use any of these words harmlessly in commenting on the model with a view to its intended applications.[10] Any element of imprecision or inexactness can enter only when a quantum model is *applied* to a specific physical situation.

The Born rule itself is in no way imprecise or inexact: specifically, a statement of the rule should not contain 'measurement' or any other similarly problematic term. The Born rule simply assigns a *mathematical* probability measure to every statement about any spin component of s in the simple model. In a model in a higher-dimensional vector space for α, the Born rule also assigns a *mathematical* joint probability measure to sets of statements concerning compatible dynamical variables. In applying the model, an agent needs to judge which of these mathematical measures should be taken

---

[10] It does seem necessary to use a word like 'system' to say what is ascribed a quantum state in a quantum model. But a quantum system in a model (such as α) need not be identified with any *physical* system prior to or independent of a decision to apply the model. And, with no mention of any apparatus, the model cannot enshrine the "shifty split" between system and apparatus of which Bell complained.

to govern credence and which lack empirical significance in this application. Previous experience, as filtered through categories such as those criticized by Bell, may improve this judgment. But, just as in classical physics, quantum theory can help structure the agent's deliberation by making available extended models that take account of the interaction of the target system with its environment, whether this is thought of as an experimental arrangement or just the natural physical situation in which the target system finds itself. We may *apply* an extended model to the target system, its environment, and their physical interaction: but none of these is an element of, or described by, the model itself. In application, quantum theory is no more inexact than classical physics.

What is the relation between the simple model of decoherence, the possibility of measuring the value of a magnitude M represented in the simple model, and what it takes to have a piece of apparatus which is capable of determining the value of a magnitude M? The model itself makes no mention of any apparatus or measurement. But in *applying* the model to a physical system s, an agent is effectively committed to counting the interaction it models as itself a potential measurement of S_z on s that excludes measurements of other magnitudes S_θ with θ far from $0, \pi$ while simultaneously serving as a somewhat less reliable measurement of magnitudes S_θ with θ very close to $0, \pi$.

An agent makes this commitment by taking it for granted that the process being modeled had or will have a determinate outcome the agent could come to recognize as indicating the value of S_z by examining either s or some part of e—a physical system modeled by ϵ. The agent is thereby committed to regarding the whole system being modeled as effectively including an apparatus capable of determining the value of S_z, but excluding any apparatus capable of determining the value of any magnitude S_θ with θ far from $0, \pi$. This commitment is acknowledged when referring to a preferred basis like $|\Uparrow\rangle, |\Downarrow\rangle$ as a *pointer basis*, and a preferred dynamical variable (or "observable") such as S_z as a *pointer position*.

Notice that one will never be committed in this way to regarding a system modeled by decoherence as effectively including an apparatus capable of determining the values of incompatible dynamical variables with arbitrary precision. This is a consequence of the fact that no element of the preferred basis of states with respect to which a system's quantum state decoheres is associated with exact values of incompatible dynamical variables.

5.3 The Grounds of Quantum State Assignments

Section 5.1 explained how it is possible to assign a pure quantum state to a system despite the pervasiveness of quantum entanglement. I also gave a couple of examples of situations in which one would be justified in assigning a pure state to a quantum system for certain purposes, while acknowledging that for other purposes it would be necessary to assign an entangled state to a system of which it is one component.

But what is it about such situations that provides the backing for assignment of one quantum state rather than another? Without an answer to this question, assigning a quantum state and basing one's expectations on probabilities generated by the Born rule would be groundless: following the quantum state's prescription would just be like basing one's beliefs on the pronouncements of an oracle with a reliable track record.

One may be justified in assigning a particular quantum state while ignorant of its grounds, much as you may be justified in following a doctor's prescription even though you don't know what makes the prescription work. But there is an important disanalogy. It is following the prescription that brings about recovery, whereas using the assigned quantum state to set one's expectations does not bring about events that fulfill them. One is justified in basing one's expectations on assignment of a particular quantum state because whatever physical conditions provide the backing for that assignment are reliably correlated with the statistics it leads one to expect.

Experimental physicists talk of "preparing" quantum states. Such talk may be understood as a way of specifying the physical grounds of the resulting quantum state assignment. For example, to prepare a beam of C_{60} molecules in a pure (center of mass) quantum state $|p_x\rangle$ certain to yield value p_x in a measurement of x-component of momentum, one can pass the beam through spinning disks that pass only molecules with the corresponding velocity: to prepare a state of a collection of rubidium atoms corresponding to a Bose–Einstein condensate, one inserts a few into a high vacuum, cools them to within one millionth of a degree of absolute zero using carefully designed laser beams and evaporative cooling, and traps them with magnetic fields.[11]

But there are two reasons why we can't remain content with such a way of specifying the physical grounds of quantum state assignment. First, if quantum theory is to be applied outside the laboratory it must be possible to assign quantum states to systems that no one has prepared. More fundamentally, this kind of specification makes free use of the language of classical physics, while we have just seen that statements in such language don't always have well-defined empirical content. So why are we entitled to use such language in specifying the grounds of a particular quantum state assignment?

Fortunately, we can appeal once more to quantum decoherence to quiet these worries. In section 5.2 we used decoherence to show how to state and apply the Born rule with no mention of measurement. But models of decoherence may be much more widely applied to provide reassurance that a host of magnitude claims have well-defined empirical content. Some of these truly state the grounds for assignment of quantum states. What determine the correct quantum state assignment are the empirically significant and true magnitude claims that state its grounds.[12] The same thing applies to an enormous number of statements about the physical world and not only those stated in the language of classical physics, such as the statement that a

---

[11] See https://web.archive.org/web/20051220114441/http://www.colorado.edu/physics/2000/bec/how_its_made.html for a more complete elementary description.

[12] Philosophers like to say the quantum state assignment supervenes on these statements.

Bose–Einstein condensate was first prepared in 1995. I'll have more to say in Chapter 12 about how acceptance of quantum theory impacts our ordinary talk about the world.

We noted in Chapter 4 that if the Born probability of an outcome is to appropriately guide the credence of an agent, then it must be relativized to the agent's situation. Since this is the Born probability generated by the correct quantum state, the correct quantum state must also be relativized to the agent's situation: a system is not *in* a unique quantum state, and does not *have* a unique quantum state. But it is important to stress that this does not make quantum state assignments or Born probabilities subjective—they are merely relational. My debtor cannot escape the objectivity of his debt by pointing out that others owe me different amounts!

A quantum state assignment is objective because it is relative to the *physical*, not merely epistemic, situation of an actual or hypothetical agent. However this physical situation is described, it is true, empirically significant, magnitude claims that make the description correct—importantly including claims about the location of such an agent. In Chapter 4's scenario we saw how Alice and Bob each correctly assigned different quantum states at the same time because they were in different places. An agent's physical situation may influence what information is physically accessible, so the ideal epistemic state of an agent is also relative to its physical situation. But this introduces no element of subjectivity into quantum states or Born probabilities.

The grounds of a quantum state assignment relative to the physical situation of an actual or hypothetical agent always concern events to which such an agent has access. Such events typically lie in the past of the agent's present situation:[13] in Chapter 10 we shall see what this comes to in the space-time of relativity. The quantum state assignment these back is useful insofar as it governs the agent's beliefs about other unknown events to which s/he presently lacks access. While some of the grounds of the quantum state may be thought to concern causes of such unknown events, others may not—perhaps they occur far away at the same time or even later. It is always a mistake to assume that the grounds of a quantum state assignment *cause* a system to be in that quantum state, and many such grounds should not be thought of as causes of events unknown to the agent who wishes to use that state to form expectations about them. In Chapter 4's scenario, for example, Bob's outcome is not a cause of Alice's (or vice versa) and there is no instantaneous action at a distance.

I'll conclude this chapter by summarizing its main points. Though objective, both quantum state assignments and Born probabilities must be understood as relativized to the physical situation of an actual or hypothetical agent for whom they offer advice on the empirical significance and credibility of magnitude claims about physical systems. The use of a quantum model for this purpose is subject to two kinds of restrictions: a canonical magnitude claim must be empirically significant before such an agent is licensed to apply the Born rule to match degree of belief to Born probability;

---

[13] Typically, but perhaps not always. An agent may be said to have knowledge of events he can, and presently commits himself to, bring about.

and the quantum state the agent assigns must be backed by other true magnitude claims. Neither quantum states nor Born probabilities bring about the events described by true magnitude claims. Instead, a quantum state helps a physically situated agent track the correlations between accessible events and inaccessible events the agent is warranted in believing to a degree given by their Born probabilities. As we shall see in more detail in Chapter 10, not all such correlations should be considered causal.

An agent may be justified by prior experience in assigning a particular quantum state and applying the Born rule even while ignorant of the physical conditions that back that state assignment and license the application of the rule. But it is the correlations among these physical conditions that undergird the successful application of quantum theory. Moreover, quantum models of decoherence can be applied to certify the empirical significance of magnitude claims stating such grounds and expressing the statements to which the Born rule may be applied.

6

Measurement

6.1 The Measurement Problem

There is no doubt that quantum theory has been enormously successful. In the form of quantum electrodynamics it has led to some of the most accurate predictions in all of science. It has helped us to understand the stability of atomic matter, the form of the periodic table of chemical elements, and why the sun shines. With applications to lasers, transistors, superconductors, and MRI scanners, one third of our economy has been said already to involve products based on quantum theory: new potential applications including quantum computers hold out the prospect of a second quantum technological revolution.

But, focusing instead on the concepts underlying the quantum revolution, some physicists and many philosophers have diagnosed what they take to be a serious weakness in the foundations of the theory. This has come to be known as the quantum measurement problem. There are various ways of stating the problem. I'll begin with this brief statement by a contemporary physicist:

> [M]ost interpretations of quantum mechanics at the microscopic level do not allow definite outcomes to be realized, whereas at the level of our human consciousness it seems a matter of direct experience that such outcomes occur. (Leggett [2005], p. 871)

If quantum theory does not allow a measurement to have a definite outcome then either the theory is falsified every time an experimenter becomes aware of its outcome or every report of what the experimenter has directly experienced is false—but in that case we have no experimental evidence for quantum theory. Either way, we should not believe quantum theory: its claimed success rests on an illusion.

Another statement of the problem by a contemporary philosopher (Maudlin [1995], p. 7) expands on reasons why one might think quantum theory does not allow a definite outcome of a measurement:

> The following three claims are mutually inconsistent.
>   A. The wave-function of a system is *complete*, i.e. the wave-function specifies (directly or indirectly) all of the physical properties of a system.
>   B. The wave-function always evolves in accord with a linear dynamical equation (e.g. the Schrödinger equation).
>   C. Measurements of, e.g., the spin of an electron always (or at least usually) have determinate outcomes, i.e., at the end of the measurement the measuring device is either in a state which indicates spin up (and not down) or spin down (and not up).

Maudlin here follows Schrödinger in calling a representative of a pure quantum state a wave function, which he then represents as a vector. To demonstrate the inconsistency, he proceeds to apply a simple model of quantum theory to an interaction between the z-component of the electron's spin and a dynamical variable (the "pointer position") of a measuring device. The thought behind this application is the following. A measurement involves a physical interaction between the measured system (here microscopic) and a second device system that correlates their states so that direct experience of the state of this second system yields reliable information about the state of the first. As a fundamental theory, quantum theory should be applicable to all interactions, including this one. If the electron's z-spin were up, then the device's pointer should indicate "UP"; if its z-spin were down, then the device's pointer should indicate "DOWN". We can assume the device begins with its pointer in some neutral "READY" position.

It follows from assumption A that the initial state of the device is a pure state $|\text{"READY"}\rangle_d$, that if the electron's z-spin were initially up its state would be $|z - up\rangle_e$, and that if the electron's z-spin were initially down its state would be the orthogonal state $|z - down\rangle_e$. To get the required correlations between initial state of electron and final state of device, the state of the combined system must evolve from either of these initial states as follows:

$$|z - up\rangle_e \otimes |\text{"READY"}\rangle_d \longrightarrow |z - up\rangle_e \otimes |\text{"UP"}\rangle_d \quad (6.1)$$
$$|z - down\rangle_e \otimes |\text{"READY"}\rangle_d \longrightarrow |z - down\rangle_e \otimes |\text{"DOWN"}\rangle_d$$

where the states $|\text{"UP"}\rangle_d, |\text{"DOWN"}\rangle_d$ are also orthogonal. This evolution is unitary, as required.

It then follows from assumption B that if the initial spin state of the electron were instead $1/\sqrt{2}\left(|z - up\rangle_e + |z - down\rangle_e\right)$ then the interaction would yield an entangled final state

$$1/\sqrt{2}\left(|z - up\rangle_e + |z - down\rangle_e\right) \otimes |\text{"READY"}\rangle_d \quad (6.2)$$
$$\longrightarrow 1/\sqrt{2}\left(|z - up\rangle_e \otimes |\text{"UP"}\rangle_d\right) + \left(|z - down\rangle_e \otimes |\text{"DOWN"}\rangle_d\right)$$

in which state assumption A implies that the device's pointer fails to indicate "UP" and fails to indicate "DOWN" (but also no longer indicates "READY"—it points nowhere, indicating nothing). Assumptions A and B imply that a quantum measurement has no determinate outcome, in contradiction to assumption C. Of course, this is a very crude quantum model of a measurement interaction. But decades of attempts to remove the inconsistency by more sophisticated and realistic models have failed to do so.

Bell ([2004], p. 201) summed up the problem in this way:

Either the wavefunction, as given by the Schrödinger equation, is not everything, or it is not right.

Taking C as a fact of immediate experience, Bell here presents us with the options of rejecting either A or B. Maudlin [1995] says these correspond respectively to additional variables theories and non-linear theories. Unlike Bell he does acknowledge the option of denying C but argues that this leads nowhere. I will rely on this useful classification of attempts to solve the quantum measurement problem in the Interlude (Chapter 7), where I briefly explain my reasons for rejecting currently influential theories or interpretations that make them. Bohmian mechanics rejects A; objective collapse theories reject B; and Everettian or many worlds views reject C.

One way to try to solve the problem is to reject B by proposing that while a system's wave function evolves deterministically and linearly except when a measurement occurs, measurement induces an indeterministic, non-linear "collapse" onto a state specifying physical properties that indicate a determinate outcome (in accordance with A). In that case quantum theory would replace the simple model of a measurement interaction (6.2) by the transition

$$1/\sqrt{2}\left(|z-up\rangle_e + |z-down\rangle_e\right) \otimes |\text{"READY"}\rangle_d \quad (6.3)$$
$$\longrightarrow \text{either } |z-up\rangle_e \otimes |\text{"UP"}\rangle_d \text{ or } |z-down\rangle_e \otimes |\text{"DOWN"}\rangle_d,$$
each with probability $1/2$.

This was how Dirac put it in his classic work, originally published in 1930:

When we measure a real dynamical variable ξ, the disturbance involved in the act of measurement causes a jump in the state of the dynamical system. ([1930], p. 36)

Also in the 1930s, the mathematician von Neumann [1932] echoed this idea in his seminal, mathematically rigorous presentation of quantum mechanics. He there proposed two different evolution laws for quantum states, which he called process 1 and process 2. Process 2 was to be described by the linear evolution of the Schrödinger equation, while process 1 applied instead in a measurement.

It would be surprising, though not necessarily inconsistent, if quantum theory were to postulate two such different dynamical laws. Dirac, von Neumann, and others gave their reasons why both were needed. But without some precise specification as to when the wave function followed one law and when it followed the other we have not arrived at a satisfactory solution to the measurement problem. This is where Bell lodged his strongest complaints against the appearance of the term 'measurement' in a formulation of quantum theory:

It would seem that the theory is exclusively concerned about 'results of measurement', and has nothing to say about anything else. What exactly qualifies some physical systems to play the role of 'measurer'? Was the wave-function of the world waiting to jump for thousands of millions of years until a single-celled living creature appeared? Or did it have to wait a little longer, for some better qualified system.... with a Ph.D.? If the theory is to apply to anything but highly idealized laboratory operations, are we not obliged to admit that more or less 'measurement-like' processes are going on more or less all the time, more or less everywhere? Do we then not have jumping all the time?   ([2004], p. 216)

While operating under the constraints provided by assumptions A and C, the challenge for an objective collapse theory is to provide a precise specification in physical terms as to how, and exactly when, assumption B fails. No such specification can be given in vague terms like 'measurement', 'observation', 'apparatus', 'environment', 'classical', 'macroscopic', or 'irreversible amplification', not to mention 'living' or 'conscious'. In the Interlude I'll critically assess attempts by some recent objective collapse theories to meet this challenge.

When faced with an intractable conceptual problem it is often a good idea to question the terms in which that problem has been posed. By undermining the presuppositions of these terms it may be possible to *dissolve* the problem rather than solving it—to show that with a clearer understanding of the matter the problem never arises in the first place. That is how I propose to dispose of the quantum measurement problem, based on the insight that the function of a quantum state assignment is not descriptive but prescriptive, and the consequent appreciation that this makes all quantum state assignments relational.

Examining assumption A with this thought in mind one is struck by two incongruities. Assumption A presupposes that it is (at least) *a* function of a wave function to specify the physical properties of a system to which it is assigned, and also that for a given system there is a unique wave function with that role. But if a quantum state assignment serves a prescriptive function for an actual or hypothetical physically situated agent, then differently situated agents may each consistently assign different wave functions to the same system, neither of which need specify any physical properties of that system. Assumption B also presupposes the descriptive function of a quantum state by assuming that the evolution of a wave function represents a dynamical process by (directly or indirectly) specifying the changing physical properties of the system to which it is assigned. There are no similarly obvious grounds for questioning the presuppositions of assumption C, although we shall see that once the measurement problem has been dissolved these presuppositions require defense.

The inconsistency between assumptions A, B, and C loses its bite once one rejects presuppositions of A and B, since this removes the grounds for believing they are true. But nor are A and B straightforwardly false: each may be said to contain an element of truth. Einstein argued that A is straightforwardly false and suggested that a quantum state plays an epistemic role with respect to a complete description of the physical properties of a system in terms of additional variables. But if assignment of a quantum state has a prescriptive function then this provides a sense in which the wave function is indeed complete: it fulfills this prescriptive function in a way that admits no supplementation by additional variables. In particular, no specification of the value of one or more dynamical variables on a system immediately before a measurement will provide a better guide to an agent's expectations than application of the Born rule to the wave function correctly assigned in that agent's physical situation.

Similarly, B is not straightforwardly false: the Schrödinger equation *is* right, when correctly understood. After assigning a wave function, an agent is correct to evolve it linearly until the agent's changing physical situation provides access to true magnitude claims that require assignment of a new wave function. Since such updating on access to new information corresponds to no physical process involving the system there is no physical "collapse" here. Von Neumann's process 1 is not a physical process, and nor is his process 2, since the Schrödinger equation is not a dynamical law.

Pace Bell, no mystery attaches to the question of when the wave function of the world collapses. The world does not have a wave function, though an agent may assign a wave function to the large-scale structure of the world in order to guide expectations about features of that structure beyond those that back the assignment.[1] The grounds for assigning or reassigning a quantum state are physical: they may be precisely specified by true magnitude claims about physical systems and the physical situation of an agent, relative to which the assignment would be correct. Since no agent need actually occupy that situation, we can talk freely about wave function assignments and reassignments relative to agent situations billions of years ago and billions of light years away.

Because each of A and B contains an element of truth it is important to revisit the argument as to why together they are inconsistent with C and to see in more detail how quantum theory allows measurements to have definite outcomes. After introducing equation (6.2) as a crude quantum model of a measurement interaction I noted that more sophisticated and realistic models have failed to remove the inconsistency among A, B, C to which its application leads. But the prescriptive function of quantum state assignment and Chapter 5's analysis of the applicability of the Born rule enable one to see how this model is too crude to serve its intended purpose.

Models like (6.2) derive from a standard example of measurement that even predates the formulation of quantum theory in anything like its present form. Stern and Gerlach [1922] prepared a beam of silver atoms, all with the same momentum, passed it through the poles of a magnet shaped to produce an inhomogeneous magnetic field, and detected the silver atoms after passing through the magnet, as illustrated in Figure 6.1.

In a modern version of the experiment individual silver atoms may be detected, each in one or other of two vertically separated regions of a screen. If the poles of the magnet are vertically oriented along the z-axis this is said to constitute a measurement of the z-spin of each of the silver atoms: for those detected in one region the measured value of z-spin is said to be $+\hbar/2$ ("UP") while for those detected in the other region the measured value of z-spin is said to be $-\hbar/2$ ("DOWN"). Since the spin of a single electron in its outer shell is responsible for the spin of a silver atom, some view this as a measurement of the spin of an electron. The z-spin of a free electron cannot be measured in this way because, unlike a silver atom, an electron is charged. The field of the magnet interacts with the charge of a free electron in a way that is difficult if not impossible to separate out from the effect on its magnetic moment associated with its spin.

---

[1] As did Hawking and Hartle [1983] in their paper "The wave function of the universe".

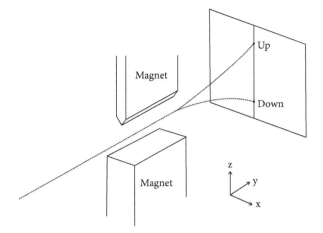

Figure 6.1 Stern–Gerlach apparatus

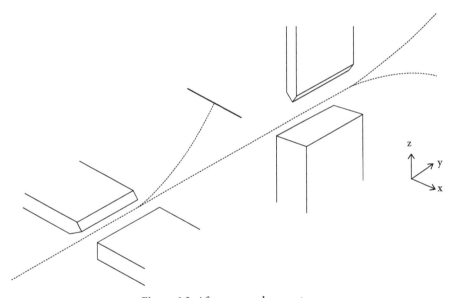

Figure 6.2 After a second apparatus

Now suppose one introduces a second Stern–Gerlach magnet, oriented along the x-axis instead of the z-axis, blocks the path of any silver atoms that would have been measured as having x-spin "DOWN", and passes the remaining atoms in the beam through the original Stern–Gerlach magnet oriented along the z-axis. This arrangement is shown in Figure 6.2.

If we model this arrangement in quantum theory we arrive at something quite close to equation (6.2), namely

96   MEASUREMENT

$$1/\sqrt{2}\left(|z-up\rangle_a + |z-down\rangle_a\right) \otimes |\psi_{\text{``READY''}}\rangle_a \quad (6.4)$$
$$\longrightarrow 1/\sqrt{2}\left(|z-up\rangle_a \otimes |\psi_{\text{``UP''}}\rangle_a\right) + \left(|z-down\rangle_a \otimes |\psi_{\text{``DOWN''}}\rangle_a\right).$$

The main difference is that instead of two physical systems, an electron e and a device d, this specifies the Schrödinger evolution of the quantum state of a single physical system (a silver atom a) that models passage between the z-oriented magnets of Figure 6.2 as entangling its spin and translational states.

We first met a similar entangled state in Chapter 2, namely the state

$$|\Psi'(t_f)\rangle = h\,|high(t_f)\rangle \otimes |V\rangle + l\,|low(t_f)\rangle \otimes |H\rangle \quad (6.5)$$
$$= \frac{1}{\sqrt{2}}\left\{\left(|D\rangle \otimes |45°\rangle\right) + \left(|S\rangle \otimes |135°\rangle\right)\right\}$$

assigned to a beam of light emerging from an interferometer. In Chapter 3 we noted that Dirac would not permit one to say that the output beam assigned state (6.5) had any determinate polarization. Further, he would permit one to say neither that it had a property corresponding to the fact it would be detected by the device *Different* nor that it had a property corresponding to the fact it would be detected by the device *Same*.

Chapter 2 already put forward a consideration in favor of Dirac's prohibitions. While the two components of (6.5) do not display interference, one can restore interference between the states $|high\rangle$ and $|low\rangle$ by using a polarizing filter with axis set to 45°. Similarly, by carefully overlapping the translational states $|\psi_{\text{``UP''}}\rangle_a$, $|\psi_{\text{``DOWN''}}\rangle_a$ of the two components of (6.4) one could (in principle, though not without enormous difficulty in practice) demonstrate interference by passing the resulting beam through an x-oriented Stern–Gerlach magnet and detecting all of the output beam in the region corresponding to x-spin "UP".

Armed with the discussion of decoherence in Chapter 5, we can now see why the entangled state in the model (6.4) *cannot* be understood as one in which passage through a Stern–Gerlach measurement has produced an outcome of a quantum measurement. The reason is simple: the two components of this state have not stably decohered. So an agent assigning this entangled quantum state to a silver atom should regard all magnitude claims of the form "The z-spin of a is $+\hbar/2$", "The z-spin of a is $-\hbar/2$", or "The z-spin of a is $+\hbar/2$ or $-\hbar/2$" as lacking empirical significance, and apply the Born rule to no such statements.

But doesn't this establish inconsistency with C? No: as Margenau [1937] long ago argued, there is not yet any definite outcome since nothing has been recorded. Only when a detector registers a silver atom in one region rather than another is there a record of an outcome. No doubt Bell and others would object to the vagueness of such talk of records, registration, and outcomes. But following the prescriptions of a correct quantum state assignment enables one to replace such talk with empirically significant magnitude claims. This requires an extension of (6.4) to include an idealized interaction with a detecting device d. In the original Stern–Gerlach experiment the detector was simply a screen, and the silver atoms adhered to different regions of it.

Interaction with the screen is then represented in a model of decoherence that licenses application of the Born rule to the positions of silver atoms on the screen. I'll discuss a similar model in Chapter 9 for large molecules adhering to a silicon surface in a recent interference experiment. Each magnitude claim of the form "The z-position of the silver atom on the screen is confined to interval Δ" is empirically significant for Δ very much less than the distance separating the regions where silver atoms can be seen adhering to the screen. Application of the Born rule to the decohered quantum state leads one to expect a particular silver atom to adhere to the screen in one of these two regions, assign equal degree of belief to the two regions, and confidently expect the symmetric distribution that can be seen to form on the screen after a large number of atoms have adhered to it.

Application of the model leading to equation (6.4) does not show how quantum theory allows the determinate outcomes demanded by C, since that equation models nothing that qualifies as a measuring device. This shows that the model is inadequate to its task. It must be extended to a model of the right kind of stable decoherence induced by interaction with a complex physical system in the atom's physical environment. This will be a model in which approximate determinate-position wave functions of the atom (not spin quantum states) decohere at the screen. Then quantum theory certainly allows an empirically significant magnitude claim about the atom's position on the screen to acquire a determinate truth value. Such a claim, if true, can be taken to express a determinate outcome, recorded in one state and not another of an apparatus, of a measurement of the z-spin component of an atom, if not of one of its constituent electrons.

In this way application of a quantum model that also treats the interaction with the screen reconciles (6.4) (the result of applying the original simple model) with the truth of C. But application of no model of quantum theory could *imply* that a measurement has a definite outcome, here or anywhere else. Even if different conditions backed assignment of state $|z-up\rangle_a \otimes |\psi_{\text{``UP''}}\rangle_a$ in a model, a legitimate application of the Born rule would merely assign probability 1 to a claim about a silver atom's position on the screen that one could take to express that outcome. Quantum theory never implies magnitude claims.

Versions of Dirac's jumps and von Neumann's process 1 persist even in contemporary discussions of quantum theory, and the measurement problem is often taken to be that of reconciling such discontinuous change in a quantum state with continuous evolution according to the Schrödinger equation or some linear variant. Albert ([1992], p. 79), for example, puts it this way:

The dynamics and the postulate of collapse are flatly in contradiction with one another... the postulate of collapse seems to be right about what happens when we make measurements, and the dynamics seems to be bizarrely wrong about what happens when we make measurements, and yet the dynamics seems to be right about what happens whenever we aren't making measurements.

Why did Dirac and von Neumann, among others, think both were needed? What has impressed such thinkers are cases in which immediate repetition of a quantum measurement is alleged certainly to yield the same result. Dirac ([1930], p. 36) argued that this follows from physical continuity, and von Neumann ([1932], pp. 212–15) took it to be illustrated by, if not a consequence of, an experimental result (observed by Compton and Simon). It is also common to appeal to examples involving Stern–Gerlach experiments on spin 1/2 systems or linearly polarized light. If one selects only whatever spin 1/2 particles emerge in the "UP" beam from a z-oriented Stern–Gerlach magnet and passes them through a second z-oriented Stern–Gerlach magnet, then no particles are found to emerge in the "DOWN" beam—all detected particles emerge in the "UP" beam from this second magnet. If the ordinary ray of light emerging from a suitably oriented crystal of Iceland spar is incident on a second similarly oriented crystal, then no light emerges in the extraordinary ray—only an ordinary ray emerges from the second crystal.

As we have seen, the outcome of a measurement of a component of spin or linear polarization is typically unpredictable, while its probability is given by the Born rule: it is certain only when that probability is 1. If the outcome of a repeated Stern–Gerlach experiment or repeated passage through a crystal of Iceland spar is certain, then the quantum state to which the Born rule is applied must make it so. It follows that after emerging in the "UP" beam of a z-oriented Stern–Gerlach magnet a spin 1/2 particle must be assigned state $|z-up\rangle$, and a photon emerging from a crystal of Iceland spar in the ordinary beam must be assigned polarization state $|V\rangle$. In each case it does seem that the first measurement has caused a jump or collapse into a new state—one in which a subsequent measurement is certain to yield the same outcome.

But appearances are deceptive. Passage of a system through a Stern–Gerlach magnet or crystal of Iceland spar does not constitute a measurement unless and until the system suitably interacts with another system. In an appropriate model, this interaction must be such as to induce stable decoherence of the quantum state assigned to a system, thereby certifying the empirical significance of a set of incompatible statements about the value of some "pointer position" magnitude. The measurement leads to a definite outcome: it is a matter of direct human experience that exactly one of these statements is true. So sequential passage through multiple Stern–Gerlach magnets or crystals of Iceland spar does not constitute a set of repeated measurements of spin component or linear polarization (respectively), but only a single measurement, whose outcome is recorded in the "pointer position" of a detector at the end of the sequence.

Someone may object that there is nothing in the model of decoherence to represent one rather than another definite outcome, and so this quantum model fails to account for our direct experience of one such outcome in each case, though possibly different outcomes in different cases. So decoherence of a quantum state through interaction with a complex environment does not solve the quantum measurement problem.

While true, this is no objection to the present approach, which aims not to solve the problem but to *dissolve* it by rejecting presuppositions of assumptions that generate the inconsistency that defines the problem. Prominent among these is the assumption that it is the function of a quantum model to represent definite properties of a physical system. But that is a mistake: quantum models always have a prescriptive, not a descriptive, function. The function of this model of decoherence is to advise an actual or hypothetical agent on the significance and credibility of magnitude claims, including some that may be reconstrued as rival claims about a determinate outcome of a measurement.

Use of the model for this purpose presupposes both that empirically significant magnitude claims have determinate truth values, and that exactly one such claim about a determinate outcome of the measurement modeled is true. This is indeed a matter of direct human experience: denying it amounts to extreme skepticism and removes the empirical foundation of quantum theory. To see that we are justified in accepting quantum theory as well as the evidence of our senses, all one has to do is to recognize that the function of quantum state assignment is not descriptive but prescriptive. Quantum theory cannot be expected to account for determinate measurement outcomes in the sense of itself implying that they occur. Our direct experience assures us that they do. If we were to deny this we could not consistently apply quantum models that presuppose that they do, and we would have no reason to rely on these models even if we could.

Even if passage through a z-oriented Stern–Gerlach magnet does not by itself constitute a measurement of z-spin, it is widely acknowledged as a way of preparing quantum states $|z-up\rangle_a$, $|z-down\rangle_a$ of atoms: if atom a emerges in the "UP" beam its quantum state is often said to have been prepared as $|z-up\rangle_a$, while emerging in the "DOWN" beam prepares state $|z-down\rangle_a$. This can be seen as a motive to adopt a collapse model of state preparation analogous to (6.3) rather than (6.2).

But as we saw in Chapter 5, state assignment involves no such physical collapse. An agent may justifiably assign spin state $|z-up\rangle_a$ to an atom to guide expectations about the possible outcomes of an experiment conducted only on atoms that would be detected in the upper beam, while acknowledging that it would be necessary to assign the entangled state $1/\sqrt{2}\left(|z-up\rangle_a \otimes |\psi_{\text{"UP"}}\rangle_a\right) + \left(|z-down\rangle_a \otimes |\psi_{\text{"DOWN"}}\rangle_a\right)$ for a different purpose, such as guiding expectations as to the possible outcomes of an experiment in which the "UP" and "DOWN" beams were carefully recombined and then passed through an x-oriented Stern–Gerlach magnet.

As an example of state preparation without collapse, consider how one might go about justifying assignment of the initial quantum state in (6.4). Here is a simple way to do so. Begin by evaporating silver in an oven and extracting silver atoms from a small hole. Pass any emerging silver atoms through a velocity selector, and then through an x-oriented Stern–Gerlach magnet. Conduct further experiments only on silver atoms that would be detected emerging from this Stern–Gerlach magnet in the "UP" beam. Then one is justified in assigning the initial quantum state in (6.4) to the silver

6.2 The EPR Argument Revisited

In the course of their argument, EPR (Einstein, Podolsky, and Rosen) made the assumption that measurement collapses a quantum state onto a state such that immediate repetition would certainly yield the same outcome. It is interesting to revisit their argument to see whether it would go through without this assumption. They also assumed this criterion for the existence of an element of physical reality:

> If, without in any way disturbing a system, we can predict with certainty (i.e. with probability equal to unity) the value of a physical quantity, then there exists an element of physical reality corresponding to this physical quantity. (Einstein, Podolsky, and Rosen [1935])

Applied to a pair of widely separated systems modeled by the entangled spin state introduced by Bohm

$$|o\rangle = \frac{1}{\sqrt{2}}(|\uparrow\rangle \otimes |\downarrow\rangle - |\downarrow\rangle \otimes |\uparrow\rangle)$$

the EPR argument is that measurement of the z-spin of a nearby system would collapse this state onto either $|\uparrow\rangle \otimes |\downarrow\rangle$ or (with equal probability) $|\downarrow\rangle \otimes |\uparrow\rangle$, and that this would enable you to predict with certainty the value of the distant system's z-spin, but without in any way disturbing that system. They used the reality criterion to conclude that, whether or not one performs a measurement on the nearby system, there exists an element of reality corresponding to the distant system's z-spin. Since this is nowhere represented in the state $|o\rangle$, quantum description is incomplete.

Moreover, since $|o\rangle$ may be re-expressed as

$$|o\rangle = \frac{1}{\sqrt{2}}(|\rightarrow\rangle \otimes |\leftarrow\rangle - |\leftarrow\rangle \otimes |\rightarrow\rangle)$$

a parallel argument reveals another element of reality corresponding to the x-spin of the distant system—indeed, by extension, there is an element of reality corresponding to *every* component of that system's spin. But *no* quantum state can be taken to represent more than one component of the spin of a spin 1/2 system. So not only does $|o\rangle$ fail to provide a complete description of physical reality, but so would any other quantum state one might suggest as an alternative to $|o\rangle$ in this situation.

Recall Einstein's (Schilpp, ed. [1949]) later remark about his own version of the EPR argument

> One can escape from this conclusion only by assuming that the measurement [on A] telepathically changes the real situation of [B], or by denying independent real situations as such to things which are spatially separated from each other. Both alternatives appear to me entirely unacceptable.

If one thinks of measurement as collapsing the quantum state $|o\rangle$ in accordance with the view that a quantum state's function is descriptive, then one seems forced to adopt Einstein's first option here, committing one to the "spooky" action at a distance that Einstein rejected. Einstein's own suggestion was to reject this descriptive function of a quantum state in favor of an epistemic conception. On that conception, every component of the distant particle's spin has the value that would certainly be revealed if it were measured, and these values are distributed among a large number of similar pairs of systems with relative frequencies very close to the Born probabilities derived from $|o\rangle$. But Bell proved that cannot be true, as we saw in Chapter 4.

We also saw in Chapter 4 how, by recognizing the prescriptive function of a quantum state assignment, one can see there is no "spooky" action at a distance in a case like this. The key point is that reassignment of the quantum state of a pair does not represent some physical collapse (involving an instantaneous change in the distant system's properties): it is necessary because access to the outcome of a measurement on a nearby system requires an agent to update the state s/he assigns not just to that system but also to the distant system, if s/he is to form justified expectations concerning the outcomes of possible measurements on either or both systems. We saw also that by adopting this prescriptive conception of quantum state assignment one can evade Bell's argument that refuted Einstein's suggested alternative, epistemic conception.

Where does this leave EPR's reality criterion? If assignment of state $|\downarrow\rangle$ or $|\uparrow\rangle$ to a distant system on learning the outcome of a z-spin measurement on a nearby system has no implications for the value of a physical quantity on the distant system, then the reality criterion cannot be applied to draw any conclusions about the distant system. In that case the EPR argument fails to show that quantum description is incomplete even if the reality criterion is correct. But a slightly modified statement of the criterion is then incorrect, namely

If, without in any way disturbing a system, we can predict with certainty (i.e. with probability equal to unity) the outcome of a measurement of a physical quantity, then there exists an element of physical reality corresponding to this physical quantity.

The point is that assignment, either of state $|\downarrow\rangle$ or of $|\uparrow\rangle$, to a distant system on learning the outcome of a z-spin measurement on a nearby system *does* enable one to predict with probability equal to unity the result of a measurement of the distant system's z-spin, and it does so without in any way disturbing that distant system. But prior to a measurement on the distant system, there exists no element of physical reality corresponding to the distant system's z-spin.

7

Interlude: Some Alternative Interpretations

This Interlude is directed primarily to readers predisposed to reject the understanding of quantum theory developed in Part I in favor of a more orthodox Interpretation of quantum theory such as those considered here. The unprejudiced reader should feel free to skip it, at least on a first reading, especially since its condensed presentation scarcely serves as a self-contained introduction to these Interpretations. In writing it I had in mind not specialists in the philosophy of physics but those, including philosophers and scientists, who take it for granted that quantum theory offers some description of the world at a fundamental level and are content to leave it to the experts to thrash out among themselves exactly how the world should be described.

Such readers are in for a shock. Rather than converging on a consensus, disagreements between proponents of radically different Interpretations of quantum theory seem to have intensified as a result of continuing research. The result has come to resemble more closely a sectarian schism than progress toward the truth about the world. But there has been progress, of a kind more characteristic of philosophy; the depth and sophistication of arguments offered has improved—both for and against each Interpretation. Interminable disputes among religious sects can promote atheism. Perhaps this interlude may motivate an analogous attitude toward Interpretations of quantum theory.

Part I offered an opinionated understanding of quantum theory: Part II will assess the philosophical significance of this revolutionary theory. There is an obvious reason to proceed in this way: no assessment of the philosophical significance of X is possible without an adequate understanding of X. But in this case the two projects—of understanding quantum theory and assessing its philosophical significance—are more intimately intertwined. Part I's attempt to convey an understanding of quantum theory was guided by a view of what such understanding requires—roughly, the view that understanding a scientific theory is manifested by the ability successfully and unproblematically to use it to further the goals of prediction, explanation, and control of natural phenomena. That view is by no means universally held by physicists and philosophers concerned with the conceptual foundations of quantum theory, many of whom have demanded something more—a precisely formulated and self-contained description of the world according to quantum theory.

An overarching theme of Part II will be that quantum theory has achieved its phenomenal success as a fundamental scientific theory without offering such a description but by other means. But that possibility is likely to be dismissed out of hand by anyone who maintains that a fundamental theory *must* offer a precisely formulated and self-contained description of the world. So before going on to develop that theme I'll pause to address proponents of three prominent rival Interpretations of quantum theory, each of which seeks to understand the theory as offering a precisely formulated and self-contained description of the world.

Each of these proposals has been extensively articulated and ably defended against many objections. This is not the place for a detailed response. My aim is more limited. It is to sketch their ideas only in broad brush strokes sufficient to make it clear why I find none of them sufficiently attractive to be worth pursuing as an alternative understanding of quantum theory. There are good scientific reasons to develop such ideas as *alternatives* to quantum theory as I understand it. We surely cannot expect the quantum revolution in physics to be the last. But we can arrive at a perfectly adequate understanding of the quantum theory we already have without casting around for alternatives like these.

7.1 Bohmian Mechanics

Especially after their promulgation by Bell [2004], ideas of de Broglie [1927], rediscovered and further developed by Bohm [1952a], [1952b], have come to figure prominently in recent discussions of the conceptual foundations of quantum theory. I'll mostly rely on an important recent book (Dürr, Goldstein, and Zanghi [2013]) advocating what its authors call Bohmian mechanics, a theory its authors defend as a well-defined theory from which one can recover all empirical predictions of (at least non-relativistic) quantum theory, which, they believe, is a less well-defined theory.[1] One important reason for their belief is that conventional formulations of quantum theory are in terms of measurement and observation: they seek to remove such vague and misleading talk, and to free quantum physics from what they call "quantum philosophy":

Quantum philosophy provides a philosophical foundation for the claim that a fundamental physics focused on an objective reality is impossible.

[T]he core of quantum philosophy is that physics is about measurement and observation, and not about an objective reality, about what seems and not about what is.

(Dürr, Goldstein, and Zanghi [2013])

This tendentious use of the term 'quantum philosophy' should be rejected because of its false insinuation that without a foundation in something like Bohmian mechanics quantum physics rests on an obscure idealism or subjectivism. Doubtless one can find

---

[1] Goldstein [2013] is a readily available introduction with additional references.

bad arguments why fundamental physics can't focus on objective reality. But Part I exhibited an understanding of quantum theory that made it clear that the theory is not about measurement and observation and that its physical applications are to objective reality. As we shall see, quantum theory helps us form true beliefs about objective reality while prompting us to re-evaluate their significance as well as the grounds of their objectivity. Bohmian mechanics is not quantum physics without quantum philosophy, but an alternative to quantum theory motivated in part by implicit philosophical prejudices that quantum theory should prompt one to question.[2]

Bohmian mechanics proposes to solve the measurement problem by supplementing the quantum wave function of n non-relativistic particles by additional variables—the positions of the particles. These variables are not hidden but manifest. Each particle i always occupies some precise location x_i in three-dimensional physical space, and a well-conducted observation or measurement will reveal just where it is. Each particle's position changes continuously (or stays the same), so a particle has a well-defined velocity and traces out a continuous trajectory. The particles' wave function $\Psi(x_1,\ldots,x_i,\ldots x_n)$ always evolves continuously according to the n-particle Schrödinger equation, with no "collapses". It also influences particle trajectories in accordance with a simple "guidance" equation: at any time, the velocity v_i of particle i at x_i is uniquely determined by its mass m_i and the value of $\Psi(X_1,\ldots,x_i,\ldots X_n)$ at and arbitrarily close to x_i, where the actual position of the jth particle is X_j.

It is useful to specify the momentary positions of all the particles by a single point $x \equiv (x_1,\ldots,x_i,\ldots x_n)$ in a configuration space of $3n$ dimensions. The initial values of $x, \Psi(x)$ determine their values at any later (or earlier) time, in accordance with the Schrödinger equation and the guidance equation. Probability enters epistemically, since these values are unknown: Ψ has a second, epistemic function as a probability density $\rho(x) = |\Psi(x)|^2$. If the (epistemic) probability density at an initial time t_0 for all the particles to lie in a small volume d^3x of configuration space around x is $|\Psi(x,t_0)|^2$, then the probability density at any later time t will be $|\Psi(x,t)|^2$. In this way Ψ specifies the probabilities for one or more particles to occupy particular regions of space at any time. These will then equal the corresponding position probabilities specified by the Born rule. This instance of the Born rule therefore follows from the assumed initial state Ψ and its role in specifying epistemic probabilities.

In Bohmian mechanics, all measurements of dynamical variables are based on measurements of particle positions. What is called a measurement of a component of spin or momentum is an experiment that must be analyzed within Bohmian mechanics in terms of the positions of all particles involved, including those of the physical apparatus interacting with the measured system. This interaction is designed so that the final particle positions in the apparatus are appropriately correlated with the initial

---

[2] I don't mean to imply that advocates of Bohmian mechanics such as Dürr et al. dismiss all work in the philosophy of quantum theory as worthless. Indeed, they have made many valuable contibutions to that enterprise themselves.

state of the system. That initial state is here specified by an *effective* wave function assigned to that system alone. The effective wave function is not the total wave function of all the particles involved in the experiment. It is based on assumptions about that total wave function and approximations about the system's relation to the rest of the particles, but not on detailed knowledge of their positions.

Suppose that the effective wave function of a system is expressed as a superposition of vector states, in each of which Dirac would permit one to say dynamical variable Q has a particular (different) value. In an experiment designed to measure the value of Q, the particles of the apparatus would assume positions following interaction with this system correlated with one or other of these different values. Moreover, the (epistemic) probability associated with each such "pointer position" would equal the corresponding Born probability specified by that effective wave function for a measurement of Q. This is how the full Born rule is recovered in Bohmian mechanics. At no point in the analysis of the experiment was it necessary to assume that Q had a value—only that the interaction correlated a final "pointer position" with each component vector state in which Dirac would permit one to say dynamical variable Q has a particular (different) value. A proponent of Bohmian mechanics would see no need to say that unless Q is a function of the positions of the particles that make up the system.

Even though the total wave function always obeys the Schrödinger equation, there is a sense in which the effective wave function of a system does effectively collapse from time to time. A system has an effective wave function ψ only when its m constituent particles move in a way that may be specified by the guidance equation as applied to a function ψ of just those $m < n$ positions. This will be the case immediately following the kind of interaction suitable for a quantum measurement, which may therefore be thought to "collapse" the system's state onto this effective state (even while Ψ continues to evolve linearly). As long as a system with effective state ψ remains suitably isolated, it will continue to have an effective state ψ that evolves linearly. But interactions may cause it to cease to have any effective state at all for a period, after which it may again acquire an effective state that is *not* the result of linear evolution of the previous effective state. So there is effective collapse in Bohmian mechanics in the absence of anything like von Neumann's process 1.

The ideas of de Broglie and Bohm have never been very popular among physicists, and many objections have been raised against them. Most of these can be seen to be based either on misunderstandings of the view, extraneous unreasonable assumptions, or aesthetic prejudices. But serious difficulties remain after such insubstantial criticisms have been dealt with. I am most troubled by these objections.

Bohmian mechanics violates Einstein's principle of local action. When the total wave function entangles the state of a distant particle with that of a particle nearby, the motion of the distant particle may be immediately influenced by influencing the nearby particle. This is a direct consequence of the guidance equation because the distant particle's velocity depends on the position of the nearby particle: so any interaction that changes the nearby particle's trajectory instantaneously changes that of the distant particle in a way that is not mediated by any process occurring in the intervening space. Proponents of Bohmian mechanics accept this, but respond that *every* "serious" theory violates locality—violation of Bell's inequality shows that the world is non-local. Chapter 4 showed the inadequacy of this response by explaining how quantum theory may be understood as conforming to Einstein's principle even though it correctly predicts violation of Bell's inequality. A proponent of Bohmian mechanics who rejects this understanding of quantum theory because it fails to portray it as a "serious" theory is relying on just the kind of unexamined philosophical prejudice that quantum theory should prompt one to question. I shall examine and reject this prejudice in Part II.

Bohmian mechanics is essentially a non-relativistic theory. Its proponents freely admit that the version I have sketched recovers the empirical content only of non-relativistic quantum mechanics. But they have made considerable progress in extending its central ideas to relativistic quantum field theory and hope for further extensions to a theory of quantum gravity.[3] Bohm already sketched one extension to field theory in which particle configurations are replaced by classical electromagnetic field configurations and the Schrödinger and guidance equations are modified accordingly. Bell sketched an alternative approach in which the positions of electrons and other particles remain basic, and this has been further developed by others, resulting in a recipe for constructing a Bell-type quantum field theory claimed to yield all the same predictions as (almost any) regularized quantum field theory. The basic ontology of such a theory consists of particles (including photons), but these may from time to time come into existence or cease to exist as a result of a stochastic process superimposed on their otherwise deterministically evolving trajectories.

This raises an objection to the de Broglie–Bohm approach. The quantum theory we have includes the Standard Model of interacting relativistic quantum gauge fields. Physicists have used this model to explain a host of phenomena in high-energy physics and to make many subsequently verified predictions of the properties and behavior of so-called elementary particles (including the existence of the recently discovered Higgs particle). The de Broglie–Bohm approach arguably offers a serious alternative to the non-relativistic quantum mechanics of a fixed number of particles. Methods have recently been developed for extending it to relativistic quantum field theories. But the approach trails behind the cutting edge of contemporary quantum theory. Rather than being an interpretation of the quantum theory we have, it proposes replacing it by an alternative theory capable of recapturing its empirical success. However, the actual construction of this theory remains a work in progress.

Even empirically successful extensions of Bohmian mechanics to the relativistic domain would be essentially non-relativistic. This is because the equations governing particle motion in such an extension still single out a preferred notion of simultaneity:

---

[3] See, for example, Bohm [1952b]; Bell ([2004], pp. 173–80); Dürr, Goldstein, and Zanghi [2013].

the velocity of a particle (or field value at one place) at an instant is a function of the positions of distant particles (or field values at distant places) at that instant, as is also any stochastic process added to accommodate the creation and annihilation of particles. The structure of space and time in relativity includes no preferred notion of the simultaneity required to define such an instant. The objection is usually expressed by saying that a Bohmian theory fails to be fundamentally Lorentz invariant. But the point is actually more general. Quantum theory may be applied in the context of the curved space-times required by general relativity, for example in the vicinity of black holes. Applications of quantum theory to such space-times are not Lorentz invariant;[4] but nor do they depend on a privileged notion of simultaneity.

It is true that the empirical content of a special-relativistic Bohmian theory (such as a Bell-type quantum field theory) can be shown to be Lorentz invariant. This involves a Bohmian analysis of experiments whose results are normally considered outcomes of measurements of a dynamical variable such as momentum- or spin-component. The statistics predicted for the final positions of particles in an "apparatus's pointer" accord with those the Born rule would lead one to expect for the corresponding outcomes of the measurement of the relevant variable. So no such experiment can be used to discover the preferred notion of simultaneity required to formulate the Bohmian theory that predicts these statistics.

This leads to perhaps the deepest objection to Bohmian mechanics and its extensions. These theories all postulate physical structures that they themselves imply to be empirically inaccessible. Relativity theory represented a great epistemological advance by showing absolute rest and absolute simultaneity to be not merely empirically inaccessible but theoretically superfluous structures. Quantum theory is epistemologically superior to Bohmian alternatives for the same reason that Einsteinian relativity is superior to a Lorentzian "conspiracy theory" in which the peculiar behavior of matter blocks all attempts to find out the true state of absolute rest.

Absolute simultaneity (in the form of a preferred space-time foliation) is not the only element of empirically inaccessible structure required by Bohmian mechanics and its extensions to encompass relativistic phenomena. It is a consequence of Bohmian mechanics that at any time a system's effective wave function ψ provides the best available guide in forming expectations concerning the outcomes of a measurement on it at that time: the predicted statistics would be the same as those predicted by the total wave function Ψ conditional on the exact positions of every particle except those composing the system. But since the motion of every particle is deterministic in Bohmian mechanics, knowledge of Ψ as well as the positions of all the particles would put one in the position of Laplace's demon, able to predict every detail of the motion of the system as well as all the other particles in the universe. Determinism, though true, remains empirically inaccessible.

---

[4] That is because these space-times are Lorentzian manifolds, sharing the tangent space structure of special relativity but not its Minkowski metric.

Even though the position x of a particle is measurable, its velocity v and its wave function ψ are not. Once again, this is a consequence of an analysis of the kind of experiment needed to correlate the system's initial value of **v** or of ψ with the final positions of particles making up some "pointer" that could record the outcome of its measurement.[5] *A fortiori*, a particle's trajectory is not measurable when specified by its position and velocity throughout some interval of time. One often sees depictions of the paths taken by Bohmian particles through the two slits of an interference experiment. Any precise measurement of the position of a particle on its way through the experimental apparatus would so alter its wave function as to radically change its subsequent motion (and so its contribution to the pattern recorded on the screen), thereby frustrating any attempt to verify this depiction.[6]

Not only are particle motions in Bohmian mechanics empirically inaccessible in these ways, empirical adequacy does not suffice to single out the form of the Bohmian trajectories. In fact, there are infinitely many distinct but empirically equivalent theories, all of which differ in the trajectories to which they give rise. Considerations of simplicity and symmetry narrow down the candidates, and the Bohmian trajectories arguably represent the most natural choice. But this situation reminds one of von Neumann's remark that the problem with Lorentzian alternatives to Einsteinian relativity is not that they exist but that there are too many of them. A similar issue arises in a Bell-type quantum field theory, for which one must choose among a variety of different options for the stochastic evolution equation on non-empirical grounds.

Finally, the total wave function Ψ which plays such a key theoretical role on a Bohmian approach is empirically inaccessible. The standard procedure of quantum tomography involves preparing many systems in the same state, performing measurements of several different dynamical variables, one on each system, tabulating the statistics of their outcomes, and inferring the quantum state from them. We can't find out about Ψ like this: there is only one universe and its state obviously can't be measured through interaction with an external apparatus.[7]

Notice that its commitment to all these elements of empirically inaccessible theoretical structure does not show that a Bohmian theory is false. What it shows is that verification of the theory's empirical consequences should leave one skeptical enough about the existence of these structures to actively seek out an alternative, empirically equivalent, theory without them. We don't have far to look: quantum theory is such

---

[5] It has been claimed that a so-called weak measurement of velocity is possible. While such an experiment is well defined and yields a value *v*, it remains controversial whether this procedure counts as a genuine measurement of the particle's velocity. I believe it does not.

[6] While not significantly disturbing the interference pattern, neither does a weak measurement directly yield information about the trajectory of an individual particle. I believe this refutes any claim to have observed Bohmian trajectories using weak measurement.

[7] Some (e.g. Valentini and Westman [2005]) have suggested a modification in Bohmian mechanics as discussed here that countenances non-equilibrium distributions of particle positions. This might render the wave function empirically accessible. But by yielding different predictions such a modified theory would become a rival to rather than an Interpretation of quantum theory.

a theory! This is something that Bohmians must deny, on the grounds that quantum theory is simply not a well-defined theory, or not a "serious" theory. Part I of this book formulated quantum theory without appeal to undefined and obscure notions like measurement and wave-collapse. Part II will respond to the challenge that, so formulated, this is not a "serious" theory.

7.2 Non-linear Theories

Largely in response to the measurement problem, a class of theories has been formulated in which the quantum state of a system does not always evolve linearly, even in the absence of interaction with other systems. They include Pearle's [1976]; a theory proposed by Ghirardi, Rimini, and Weber [1986] more as a toy model than a candidate for a fundamental physical theory; Ghirardi, Pearle, and Rimini [1990]; and proposals by Diosi [1987] and Penrose [2014] as steps toward the incorporation of gravity into a unified theory of quantum phenomena. To recover the verified predictions of quantum theory, based on linear evolution in accordance with the Schrödinger equation or some relativistic generalization, these theories typically postulate deviations whose effects are negligible for electrons and other very small, simple systems. But for systems composed of very many such components, the deviations can become significant enough to mimic the effects of von Neumann's process 1. That is why they have come to be known as collapse theories.[8]

Insofar as its predictions differ from those of quantum theory, a collapse theory is not an interpretation of quantum theory but a rival theory. But it has not yet been possible to test predictions of a well-defined alternative against those of quantum theory. Aside from experimental difficulties, the situation is not clear-cut for two reasons. On one hand, the GRW (Ghirardi, Rimini, and Weber) theory, for example, contained two parameters whose values could be adjusted within limits to ensure compatibility with the outcomes of known and foreseeable experiments predicted by quantum theory. On the other hand, Bell and others have maintained that the imprecision and ambiguity introduced into quantum theory by its talk of measurement renders quantum predictions themselves sufficiently ill-defined to vitiate any crucial experiments.

Part I showed how to understand quantum theory without relying on any problematic notion of measurement. But one can use the terminology of measurement in describing a hypothetical experiment whose outcome, as predicted by quantum theory, differs from the prediction of a well-defined collapse theory.[9] More importantly, a collapse theory views its non-linear evolution equation as describing the dynamical behavior of a physical system, while collapse is not a physical process in quantum theory. Even if a collapse theory were empirically equivalent to quantum

---

[8] Ghirardi [2016] provides a readily available introduction with additional references.
[9] One such experiment would involve interference of objects much larger then C_{60} molecules.

theory (as Bohmian mechanics is empirically equivalent to non-relativistic quantum mechanics) it would still be a different theory, not an interpretation of quantum theory.

While there are collapse theories that have not yet been experimentally refuted, all of them are of more limited scope than the quantum theory we have. Extension of non-relativistic collapse theories to the relativistic domain has proven difficult. Even the recent progress has been limited.[10] Though hailed by some as a significant advance, Tumulka's [2006] relativistic "flash" theory extends GRW only to encompass massive, non-interacting Dirac electrons. Bedingham's [2011] collapse quantum field theory promises greater generality, but because of its introduction of a novel, fundamental field for mathematical rather than physical reasons it remains highly speculative.

Even a successful relativistic collapse theory violates Einstein's principle of local action (see, for example, Bedingham *et al.* ([2014], p. 628) since a locally initiated collapse instantaneously alters the probability of a distant (i.e. space-like separated) event. A collapse theory is committed to such non-locality because it takes collapse to be a physical process and regards application of the Born rule to the collapsed state as specifying the unique chance of the event it concerns. We have already seen in Chapter 4 how quantum theory avoids such non-locality by relativizing both quantum state assignment and consequent Born probabilities to the physical situation of an actual or hypothetical agent: I'll have more to say in Chapter 10 about how this also removes even the appearance of tension between quantum theory and relativity.

Penrose's [2014] search for a non-linear theory is motivated partly as a solution to the measurement problem, but partly also by the widely perceived need to unify quantum theory with the theory of general relativity. It has not yet led to a new theory capable of comparison with quantum theory, though it has suggested (extremely difficult) experiments whose results might refute quantum theory. Diosi [1987] has proposed a universal master equation for the gravitational violation of quantum mechanics, but this has neither received experimental support nor been extended into a research program capable of displacing the quantum theory we have.

In sum, non-linear theories are not interpretations of but alternatives to quantum theory. As such they are valuable in removing complacency and suggesting severe tests of quantum theory. But none has yet been developed into a serious rival to quantum theory, and the basic motivation for such theories seems misguided: the quantum measurement problem requires no solution because it simply dissolves under critical scrutiny.

7.3 Many-outcomes Theories

Some currently influential attempts to solve the measurement problem deny that a well-conducted quantum measurement has a unique outcome, despite the fact that at the level of our human consciousness this seems a matter of direct experience.

---

[10] See, for example, Bedingham [2011]; Tumulka [2006], [2009]; Bedingham *et al.* [2014].

INTERLUDE: SOME ALTERNATIVE INTERPRETATIONS 111

Inspired by the formally elegant but conceptually elusive work of Everett [1957], such attempts are known as Everettian, or many-worlds, interpretations (though some add outcomes by multiplying minds rather than worlds). A substantial minority of physicists advocate such a view: cosmologists are especially inclined to do so, in large part because of their desire to apply quantum theory to a self-contained universe subject to no external measurements. Two recent works provide in-depth coverage of the diversity of opinion on the merits of Everettian interpretations (Saunders, Barrett, Kent, and Wallace [2010]; Wallace [2012]).[11]

Despite Everettians' ingenious defense of the approach, I have yet to be convinced by their responses to two common objections. The first objection is that we have no sufficient reason to believe that a quantum measurement has any outcome other than the one experimentalists commonly take it to have, or to accept the existence of other worlds in which all possible contrary outcomes actually occur. The second objection is that there would be no way to deploy a concept like probability (notably in the Born rule) in an Everettian deterministically evolving universe within which "our" world is just one of a vast and rapidly diverging set of branch-worlds constituting the multiverse.

In his vigorous and resourceful advocacy, Wallace [2012] claims that since quantum theory taken at face value already is an Everettian theory no further interpretation is required: in saying what the universe is like if quantum theory is true of it, (other) interpretations *add* structure (Bohmian trajectories, wave-collapse, a preferred basis, etc.) to the pure unadorned theory. That theory simply postulates a linear, deterministically evolving wave function governing the dynamics of the universe in space-time (or perhaps some more fundamental structure from which space-time itself emerges).

The initial quantum state of the universe evolves unitarily in a way that guarantees the progressive formation of independently evolving components, when expressed as a superposition in some (thereby defined) decoherence basis. Each independently evolving component defines a branch. Though rapid and robust, this decoherence remains approximate, so the number and exact identities of these branches are themselves not precisely defined. A world of objects and properties, from atoms and subatomic particles to tables, trees, tigers, and humans emerges in each branch as functionally realized by the dynamical evolution of the quantum state of that branch-component. Such objects and properties are real even though they are world-bound and do not figure in the ontology of a fundamental (quantum) theory.

Splitting or divergence of branches is a consequence of the initial quantum state of the universe and the form of its linear evolution. Branching events include occasions when experimentalists in a world perform a quantum measurement: The resulting branches then realize different outcomes, each with branch weight given by the Born

---

[11] Two good introductions by Vaidman [2014], and Barrett [2014] are available in the *Stanford Electronic Encyclopedia of Philosophy*.

rule applied to the pre-branching quantum state of that world. All outcomes with non-zero Born weight are actually realized in some resulting branch of the universal quantum state, and every one may be uniquely experienced and recorded by experimenters realized in that branch-world, each a distinct world-bound continuation or "version" of an original experimenter. Because each branch state then evolves essentially independently, "inhabitants" of one branch-world can never become aware of the existence of anything or anyone that emerges dynamically in any other branch. By conspiring to block our epistemic access to most of the multiverse quantum theory itself explains why we remain under the illusion that a quantum measurement has a unique outcome and ignorant of the presence of everything but what we are able to experience in this world.

Why should anyone believe this? The argument is straightforward: we should believe what our best scientific theories tell us, and this is what quantum theory, a phenomenally successful theory fundamental to current science, tells us. If you understand quantum theory in the way I presented it in Part I you will find this argument unconvincing because you will deny that this is what quantum theory tells us. But that is not the only reason why the argument should not convince.

Suppose one understands quantum theory as Everettian quantum mechanics (EQM), as Wallace argues one should. Does the claimed success of EQM justify belief that EQM is true? Do the results of the experiments scientists take to convincingly support quantum theory provide good, or indeed *any*, evidence for EQM? Everettians call this the evidential problem and have tried to solve it by providing an analysis of how statistical data provide evidence for a probabilistic theory that applies equally to one-world and many-worlds theories. The claim is that this analysis solves any evidential problem faced by EQM by showing just how statistical data scientists take as overwhelming evidence for probabilistic predictions of quantum theory do indeed provide strong evidence for EQM.[12]

Neither scientists nor philosophers agree on how to analyze the evidential bearing of statistical data on a probabilistic theory. Though influential among philosophers of science, a Bayesian approach remains controversial and even its supporters engage in internecine disputes. Physicists prefer to use classical statistical methods due to Fisher and to Neyman and Pearson and don't generally analyze and report their experimental results in a Bayesian framework. But Jaynes and other theoreticians have sought to answer the stock objection to Bayesian approaches—that the demands of scientific objectivity rule out any assignment of a prior probability to a hypothesis. Influenced by quantum information theory, so-called QBists have gone further by advocating the personalist approach of de Finetti, in which objective probability plays no role, even

---

[12] See Greaves [2007], Greaves and Myrvold [2010], whose work Wallace [2012] incorporates into his own extensive discussion in Part Two (especially chapter 6).

in science.[13] Fortunately, any plausible analysis of the bearing of statistical data on a probabilistic hypothesis will accommodate the following objection to Everettians' attempts to solve the evidential problem.

If EQM is true then a scientist experiences statistical data only in the branch-world in which she, her data, her apparatus, and the systems on which she is experimenting exist. If anyone in the world at the start of the experiment were to believe EQM, she would believe that after the experiment there will be a version of herself in each of many different branch-worlds, in each of which scientists have collected different data in what started out as the same experiment before these worlds diverged. Indeed, she would believe that every possible set of data will be collected in some such world, including data whose statistics have very low Born weight given the branch state prepared at the start of the experiment, as well as data with high Born weight. She believes that a distinct version of herself in each post-experiment branch-world will experience just the data collected in that world: so she believes that some versions of herself will experience statistics closely matching the Born probability predicted for that data by the initial branch state while others will experience statistics nowhere near that Born probability.

Every version of herself after the experiment will experience one such set of data, just as she then remembers her former self believing she would. This experience is consistent with EQM, but it provides no support for EQM. A version should take the data she experiences in conformity to the Born rule of EQM as evidence for that rule only if she takes her data as evidentially salient while discounting the evidential bearing of "contrary" data she believes to have been experienced only by her other versions. But if she were to believe EQM she would not be justified in so prioritizing the particular data she happens to have experienced. A believer in EQM who comes to be aware of data in conformity to the Born rule learns only that *she* is in the same world as that data—something that EQM could neither predict nor assign a Born probability because EQM has no way of independently referring to *her*, either directly or as the unique realizer of a certain dynamic pattern in some branch state.

Greaves and Myrvold [2010] developed an axiomatic approach to the evidential problem within a personalist Bayesian analysis. One of their axioms (P7) required an agent's preferences to be non-dogmatic. This was intended to secure the possibility of learning from experience—specifically, adjusting her confidence in a theory (including EQM) on acquiring new data. Anyone entertaining the hypothesis both that EQM is true and that Born branch weights correctly predict the data should accept that she acquires no new evidence on experiencing experimental data. It is only a version of herself in a world who experiences the data in that world, so strictly speaking *she* can

---

[13] See Mermin [2014] and the works he cites. Formerly referred to as quantum Bayesians, they now call themselves QBists.

learn nothing.[14] More importantly, assuming EQM, this version of her acquires merely self-locating knowledge that we just saw to be evidentially irrelevant to EQM. There is no way that newly experienced data can be evidentially relevant to Born weights in EQM, so experience can give one no reason to believe EQM.

We think statistical data support a probabilistic theory when the theory renders that data more probable than other potential data. EQM assigns higher Born weight to some data than it does to other data. So why doesn't finding oneself in a world of high-Born-weight data support EQM? It would do so only if Born weights play the role of probabilities in EQM: my second objection is that they cannot.

One reason they cannot is that there is no place for genuine uncertainty in EQM—anyone who believes EQM and knows how his branch quantum state will evolve has the resources to say exactly what will happen in each resulting branch-world.[15] This includes knowledge of what will happen to him—not which branch-world he will be in after some future divergence (he will not be in a single one of them) but which branch-worlds will contain or realize versions of himself. There is no place for subjective uncertainty prior to a branching event, but immediately afterwards a replica may be initially uncertain as to which branch-world he is in—an uncertainty at least partially resolvable by subsequent experience. But this is not the right kind of uncertainty to provide Born weights with a role as probabilities, since EQM assigns Born weights to branch-worlds and so states of affairs in a world, not self-locating propositions.

Everettians have argued that Born weights can play both the inferential and the practical roles of probabilities in EQM even in the absence of genuine uncertainty: they can be legitimately inferred from statistical data and used as a basis for rational decision-making. I have already rejected the first claim: the reason EQM derives no support from statistical data is just that one cannot legitimately infer Born weights from statistical data in EQM. I will now argue that Born weights cannot serve as a uniquely rational basis for decision-making in EQM. Since Deutsch [1999] and Wallace [2012] claim to have used decision theory to prove the Born rule in EQM—in the sense of establishing that rational action in the multiverse will maximize expected utility as calculated using Born weights—this amounts to rejecting that claim.

What is there to decide about in a deterministically evolving Everettian universe? Agents such as us are realized on branches of the universal quantum state, and I will grant that such agents can act freely despite this underlying determinism. An agent cannot decide what will be the outcome of a branching event—all branching events have all possible outcomes, though contrary outcomes occur only in different worlds. But she can decide what branching events to initiate, or at least form preferences over alternative branch events. Contemporary Everettians have adapted Savage-style

---

[14] Assuming, as seems overwhelmingly plausible, that there are no grounds for believing that she in this branch-world is the unique continuation of her former self, with imposters in all the others falsely believing themselves to be her!

[15] Just as Laplace's infinite intelligence could figure out the entire future (as well as past) state of a Newtonian world from its present classical state.

treatments of decision-making under uncertainty to the scenarios envisaged in EQM to prove theorems that purport to establish that the Born rule follows from innocuous rationality requirements in EQM in the sense that a rational agent will always act on the basis of credences (degrees of belief) equal to the Born weights of branch-worlds that realize (all) possible outcomes of each contemplated action.

This has struck many as an implausibly strong claim. In a one-world scenario decision theorists typically claim only to impose consistency constraints on rational credences and preferences while leaving open their actual values. But the claim is supported by rigorous arguments from axioms that appear very plausible, at least at first sight. I think the best way to undermine their plausibility is to begin by stepping back to gain a broader perspective on the nature of the decision problem facing an agent in EQM.[16]

EQM presents a decision theorist with a novel scenario in which an agent already knows the immediate consequence of each contemplated action: every associated branch quantum state will emerge and dynamically realize the objects, properties, and events in the corresponding branch-world. Rather than making a decision in a position of uncertainty concerning the consequences of her actions in the one world, she has to decide which set of branch-worlds she should prefer to be realized. If one focuses on the interests of people or other agents realized in a branch-world this resembles a situation of distributive justice in one world, where the goods and bads are to be (hypothetically) divided up among them in some way.

Of course there are also disanalogies. "Inhabitants" of one branch-world will have no physical or epistemic access to inhabitants of another, so there is no point in asking how they would all "get along" after a branch event: and a would-be "world-maker" may care as much or even more about the "environment" of a world than about the agents that "inhabit" that environment. A further consideration is whether the deciding agent biases her preferences toward what will happen to her own future selves in different branch-worlds. In simple illustrations of one-world decision-making under uncertainty one presents an agent with bets, assuming the agent will persist to receive the subsequent pay-off. Arguably no agent will uniquely continue through a branch event in EQM, though typically an agent will continue in multiple versions in all or many resulting branch-worlds.

Decision-making in EQM is obviously a very complicated business, involving a trade-off between concerns that threaten to be incommensurable. Without claiming to be exhaustive, Kent [2010] articulates seven different decision strategies one might try to follow in face of this complexity, including what he calls the mean-utilitarian strategy pursued by an agent who aligns her credences to the corresponding Born branch-weights. The Everettian may respond by dismissing many such strategies as either ill-defined or impracticable, but the reasons for dismissal may be queried.

---

[16] Here I rely heavily on the contributions of Price, Kent, and others to Saunders, Barrett, Kent, and Wallace [2010].

116 INTERLUDE: SOME ALTERNATIVE INTERPRETATIONS

In some ways a stronger response is to reiterate that the derivation of the Born rule proceeded from plausible axioms and to point out how alternative strategies implausibly violate one or more of these axioms. Here the debate is in danger of degenerating into a clash of intuitions, contrary to the expressed intention of an Everettian like Wallace [2012].

My own intuition is that the scenario faced by a would-be rational decision-maker in EQM is so complex and multifaceted as to defeat any attempt to impose clear and compelling rationality demands in the form of either axioms or the theorems they imply. Even if there were good reasons to believe EQM this would give one a Pascalian practical reason not to do so. I cannot believe EQM because if I came to believe it I would literally not know what to do next.

I began this chapter by casting each of three rival "interpretations" as really proposing an alternative to quantum theory. While it is always a valuable scientific project to develop alternatives to existing theories (even to an extraordinarily successful theory), this project is not well motivated as a way to rid contemporary quantum theory of alleged conceptual problems.

Wallace would dispute this characterization of EQM on the grounds that EQM simply *is* quantum theory, rather than an interpretation of or alternative to it.

[R]eally the "Everett interpretation" is just quantum mechanics itself, read literally, straightforwardly—naively, if you will—as a direct description of the physical world, just like any other microphysical theory.

[N]or do I wish to be read as offering yet one more "interpretation of quantum mechanics".

([2012], p. 2)

The problem, of course, is that it is clear neither how one can read quantum mechanics (by which Wallace means quantum theory) as a literal description of the physical world nor whether that is the right way to understand quantum mechanics.

A main thesis of this book is that one cannot understand quantum theory as offering a literal description of the physical world, though quantum theory does improve our ability to make significant and true claims about the physical world, not just about our observation and measurement of it. It is appropriate to speak of the Everett *Interpretation* precisely because it is an attempt to say what the physical world is like if quantum theory is true of it. I believe that attempt fails. But the failure does not reflect badly on quantum theory—only on a mistaken way of trying to understand it.

I argued that Bohmian mechanics is an alternative theory to quantum theory even though (non-relativistically) these theories are empirically equivalent. A related argument applies to the Everett interpretation, assuming (for now) that EQM is both coherent and empirically equivalent to quantum theory as presented in Part I. Wallace summarizes the Everett interpretation as follows:

It consists of two very different parts: a contingent physical postulate, that the state of the Universe is faithfully represented by a unitarily evolving quantum state; and an a priori claim about that quantum state, that if it is interpreted realistically it must be understood as describing a multiplicity of approximately classical, approximately non-interacting regions which look very much like the "classical world". ([2012], p. 38)

Since in the first quote Wallace denied that he was offering an interpretation of quantum mechanics he must now be read as using the 'Everett interpretation' not to refer to an interpretation of quantum mechanics but as the conjunction of this physical postulate (EQM) and the a priori claim. So understood, it postulates a quantum state of the universe in EQM and uses the a priori claim to argue for a multiplicity of approximately non-interacting physical regions (the worlds), each equipped with a physical feature—its branch-weight. As understood in Part I, while every application of quantum theory presupposes there is but one physical world, no application assigns this universe a quantum state. By its addition of what count as excess physical structures from the point of view of quantum theory as presented in Part I, the Everett interpretation counts as an alternative to quantum theory (so viewed), just as its postulation of the excess structures of particle trajectories, a total wave function, and a privileged foliation makes Bohmian mechanics an alternative to quantum theory.

This Interlude has presented three proposals for describing quantum phenomena, not as interpretations of quantum theory but as alternative theories. I gave reasons why I presently consider each of them inferior to the quantum theory we have, while stressing the scientific importance of further developing them. What they all have in common is the view that "the only way to understand a scientific theory is to understand it as offering a description of the world" (Wallace [2012], p. 26). Part I presented an understanding of quantum theory as a scientific theory that does not itself offer a description of the world. The enormous success of such a theory has much to teach philosophers; not just the "serious philosophers of science" who share the common view Wallace here elevates to a consensus, but also contemporary philosophers (including analytic metaphysicians and philosophers of language) and anyone else wishing to understand the philosophical significance of the quantum revolution.

PART II
Philosophical Revelations

8
Theories, Models, and Representation

8.1 Representation and Theories

The most significant break marked by acceptance of quantum theory is a novel, indirect use of models to further the aims of fundamental science. This, above all else, is what makes quantum theory radical: not the introduction of chance or the observer into fundamental physics, not entanglement, not non-locality, and not the revelation that our world is but one of many. Quantum theory is revolutionary not because it represents new and unfamiliar physical things and processes in the universe, but because of the way it improves our use and understanding of representations of the universe we could offer without it. To see why, we'll need to look more closely at what a scientific theory is and what it does. But first it's important to say a little about representation.

The word 'representation' may be used to refer either to an act or to the consequence of such an act. In both cases the core notion is that an entity represents another entity by standing in place of it: so 'representation' may be used to refer either to the function of standing in place of, or to the entity that performs this function (be it an object, event, person, or symbol). Through the development of theories, science has provided us with many novel ways of representing physical phenomena. We can represent a glass of water using the language of theory by describing it as containing partially ionized molecules of H_2O. But it does not follow that the only way to understand a scientific theory is to understand it as offering a description of the world. Linguistic description is one form of representation, but there are others. Many images depict and maps portray spatial aspects of the world, and graphs are often used to represent processes of change.

A distinctive feature of modern science is its use of mathematics. The logistic equation may be used to represent the growth of a population, and so may graphs of its solutions. In Chapter 10 I will use space-time diagrams to represent relations between events that could also be represented mathematically. The variety of ways that mathematics is used in formulating and applying a theory in modern physics make it unwise to assume one could express the theory's content by saying how it describes the world.

Philosophers of science have come to see that mathematics provides such a flexible expressive framework that in understanding a theory (at least in physics) it is best to begin by directing attention to the mathematical models involved in its applications. As one prominent philosopher put it:

> An empirical scientist is concerned with a certain class of phenomena, and when she presents a theory, she presents a class of models for the representation of those phenomena. So a scientific model is a representation. (van Fraassen [2014], p. 277)

He argues here and elsewhere that a scientific model is a representation only insofar as it is *used* to represent phenomena.[1] But it does not follow that the function of every element of a model of a physical theory is to represent something physical, nor that representation is the sole (or even the main) use of models of every physical theory.

I introduced quantum theory in Chapter 2 by describing a set of techniques for modeling interference phenomena. The chapter ended by stating five principles and claiming that quantum theory arises as their generalization to a wider class of systems. But what exactly *is* quantum theory? I will take it to include quantum field theory as well as the quantum mechanics of particles. According to one recent physics text (Peskin and Schroeder [1995], p. xi) "Quantum field theory is a set of ideas and tools...". This helpfully directs our attention to the theory's function but leaves us in the dark as to its structure. When faced with the analogous question "What is Maxwell's theory?" Hertz famously replied that he knew of no shorter or more definite answer than "Maxwell's theory is Maxwell's equations".[2]

Taking it as their professional duty, philosophers of science over the past century have had a lot to say about the structure and function of scientific theories. The most influential accounts were created with theories of physics in mind: only in recent years have theories of biology and other branches of natural science commanded similar attention. Perhaps conscious or unconscious adoption of one or another account has made it hard to see the quantum revolution for what it is—a new kind of science.

The logical positivists' accounts of the structure of a scientific theory were influenced by the development of modern logic and relativity theory. Rudolf Carnap attended lectures by Gottlob Frege and initially presented a dissertation on the structure of relativity theory to the physics department at his university before being advised to try the philosophy department instead! Carnap's views on the structure of a scientific theory evolved throughout his career. His [1966] book, based on lectures at the University of Chicago, gives a simple presentation of his later view.[3]

---

[1] See especially van Fraassen [2008].

[2] A sentiment endorsed by their inscription on the T-shirt of many an electrical engineering student. An unusually reflective physicist, Hertz developed quite a sophisticated view of scientific theories and this epigram does not adequately express his full view of Maxwell's theory.

[3] The book was first published under the title *Philosophical Foundations of Physics*: only when later republished by Dover was the title changed to *An Introduction to the Philosophy of Science*.

For Carnap, the task of the philosopher is to seek clarification of what a scientific theory says by means of a logical reconstruction of the theory within a precisely defined language. Ideally, this would involve a formalization in first-order symbolic logic whose theorems capture the content of the theory when supplemented by informal rules of interpretation ("correspondence rules") connecting terms of the theory to observational procedures. By endowing the theory's formal syntax with semantic content these rules would specify just what the theory had to say about the world.

This way of thinking of a scientific theory has deep historical roots, stretching all the way back to Euclid's axiomatic presentation of geometry. When Newton wrote his *Principia* [1687] he presented his theory of mechanics and gravitation in Euclidean fashion, beginning with his three laws of motion and law of universal gravitation as axioms and going on to prove propositions from them. This served as a paradigm for scientific theories and encouraged the idea that a properly constructed scientific theory should be based on laws of nature that serve as axioms from which the rest of the theory follows, given only the principles of logic and mathematics. Even today many physicists take themselves to be engaged in a quest for fundamental laws, whose successful outcome would be a theory or theories in which these laws could be formulated and used in deducing predictions and grounding explanations.

In fact, scientific theories are hardly ever formulated axiomatically,[4] and there is now a consensus among philosophers of science that the labor involved in re-expressing a significant theory in a formal language and then giving its semantics by means of correspondence rules would make this neither a practicable nor a useful technique for revealing its structure and function. Patrick Suppes proposed to take a theory as defining a set-theoretic predicate (e.g. *is a system of classical particle mechanics*, to be predicated of an ordered set of mathematical objects, namely sets and functions) and Bas van Fraassen suggested that many physical theories are best understood by reference to an associated class of state-space models—mathematical structures of a certain kind. In each case what mattered was not the syntactic structure of any canonical formulation in some language, but a class of mathematical structures associated with a theory. Because it de-emphasized language and syntax, this new orthodoxy came to be called the semantic approach to the structure of a scientific theory.

The basic idea is to present a scientific theory by means of a class of models that can be used to represent features of real or merely hypothetical physical systems: for a theory of physics these models are mathematical. In Ronald Giere's [1988] influential version, this involves formulating a theoretical hypothesis claiming that some real physical system is (more or less) faithfully represented by such a model. One might, for example, formulate the theoretical hypothesis that the motion of the pendulum that Foucault hung from the roof of the Panthéon is well represented by a particular

---

[4] Even Euclid's axiomatization did not meet today's mathematical standards, though David Hilbert and others have since provided rigorous axiomatizations of Euclidean geometry.

model of Newtonian mechanics involving a point mass freely suspended above a large, uniformly rotating, spherical mass—the source of a constant gravitational field.

Viewed this way, a scientific theory does not itself say anything about the physical world—its users make claims about the world by asserting theoretical hypotheses to the effect that certain features of some physical system or systems are more or less faithfully represented by some model associated with the theory. Box 8.1 casts Newton's distillation of Kepler's astronomical achievements as a simple theory to show how the semantic approach differs from the syntactic approach.

Box 8.1   Kepler's Laws Approached Syntactically and Semantically

Kepler's "laws" are usually stated this way:

1. The path of every planet is an ellipse with the sun at one focus.
2. Every planet traverses its path so that a line from the sun to the planet passes over equal areas in equal time intervals.
3. For every planet, the square of the time it takes to go once around its path is proportional to the cube of its average distance from the sun.

Suppose we regard them as providing the basis for a simple physical theory. On the syntactic approach, the way to reveal the structure of this theory is to begin by expressing each of these laws as a sentence in a formal language. For example, one might initially take the letter 'P' to symbolize the predicate *is a planet*, and the letter 'u' to denote the sun. Considering the letters 'X','Y','Z' as variables ranging over positions in space, one could partially symbolize Kepler's first law as follows

$$(\exists! y, z) \{Luy \& (p)(x)(Pp \& Lpx \rightarrow (\exists n)[d(x,y) + d(x,z) = n])\}.$$

In the intended interpretation, this says that the sun is located at a unique point y and there is a unique point z such that if a planet is located at any point x then its distance from the sun plus its distance from z has the fixed value n. (In a plane, an ellipse may be defined as the set of points the sum of whose distances from two fixed points are equal.[5]) Symbolizing the other two "laws" is considerably harder.

After forming the conjunction of the three symbolized laws, the next step would be to regard the resulting symbolic expression as initially devoid of any meaning: we no longer understand P to apply to *planets*, or u to the sun, but leave the interpretation of all such symbols to be specified independently at a later stage.[6] The theory is then specified purely syntactically as the formal consequences of this

---

[5] The symbolization is incomplete, since it does not symbolize the requirement that x,y,z always lie in the same plane.

[6] This is how Hilbert proceeded in axiomatizing Euclidean geometry. Unlike Euclid (who offered unhelpful definitions of 'point', 'line', etc.) Hilbert stressed that the primitive symbols of his formalization were to be regarded as devoid of any intuitive meaning. Regarded as a formal theory, Euclidean geometry could then be given a quite different interpretation as a set of truths of real analysis.

formal conjunction in accordance with precisely specified rules of transformation, designed to preserve truth no matter how the constituent symbols are interpreted.

The final step in the syntactic approach involves laying down correspondence rules intended to specify exactly how the symbols were to be applied to the world. One would specify observational procedures to identify the referent of u, to determine whether an object counts as a planet, to settle the question of whether a planet is located at a point of space, and so on. (This came to be widely regarded as the Achilles heel of the approach, as realist critics objected that such observational procedures cannot be adequately specified without relying on theories, including the theory whose meaning they were intended to give.)

Now let's apply the semantic approach to the theory of Kepler's "laws". The first step is to specify a class of mathematical models to which one can appeal in applying the theory to the motion of the planets. It is convenient to represent the position of a moving object by its distance $r(t)$ from a specially chosen geometric point x, and the angle $\theta(t)$ through which it has turned away from a straight line emanating from this point after an interval of time t. The state space is then the set of points (r, θ) and the moving point's trajectory is a curve in this space. A model of the theory takes the form $\Theta = \langle a, e, T, r, \theta, t \rangle$ where e is a real number in the interval $[0, 1]$ and a, T are positive real numbers. It is a model of this Keplerian theory K if and only if two conditions hold:

$$r = \frac{a(1-e^2)}{1-e\cos\theta}$$

$$\frac{1}{2}r^2 d\theta/dt = \frac{\pi(1-e^2)^{1/2}a^2}{T}$$

One can use these models to state theoretical hypotheses such as

H_P: The orbit of planet p is faithfully represented by a model of K where a represents the semi-major axis of an ellipse of eccentricity e with the sun at one focus and T represents the time it takes to go once around this ellipse.

H_{All}: The orbit of every planet p is faithfully represented by a model of K where a_p represents the semi-major axis of an ellipse of eccentricity e_p with the sun at one focus and T_p represents the time it takes to go once around this ellipse.

H_3: If the orbits of two planets p, q are faithfully represented by models Θ_p, Θ_q then $\frac{T_p^2}{a_p^3} = \frac{T_q^2}{a_q^3}$.

Here H_{All} expresses the content of Kepler's first and second "laws", while H_3 expresses Kepler's third "law". Notice that on this approach K treats Kepler's "laws" as referring to elements of mathematical models, and that they have no theory-independent interpretation. If laws govern anything it is not the world but the behavior of scientists in constructing models whose elements satisfy certain equations.

8.2 Applying Quantum Models

Soon after presenting his state-space version of the semantic approach to scientific theories, van Fraassen ([1972], [1973]) applied it to quantum theory. In ordinary quantum mechanics, the state space of a model includes one or more Hilbert spaces $\mathcal{H}$ (see Appendix A). Van Fraassen took a vector $|\psi\rangle$ in one such space to specify the instantaneous dynamic state of a (temporarily) isolated physical system s whose behavior is being modeled. This vector may then be supposed to represents s's physical evolution as it varies continuously with time according to the Schrödinger equation: but the equation's specific form varies from model to model. One element of a model is a linear operator $\hat{H}$ called the Hamiltonian which corresponds to the dynamical variable energy E. $\hat{H}$ is itself a function of operators corresponding to other dynamical variables (such as components of position, momentum, and spin), but not the same function in every model. So although the Schrödinger equation

$$\hat{H}|\psi\rangle = i\hbar\frac{d|\psi\rangle}{dt} \qquad \text{(Schrödinger equation)}$$

holds in all models, the linear evolution it prescribes for $|\psi\rangle$ differs from model to model, much as the motion of a body in Newtonian mechanics depends on what forces act on it even though it always conforms to the second law of motion.

Each dynamical variable A on s is uniquely represented by an operator $\hat{A}$ acting on the Hilbert space $\mathcal{H}$ containing $|\psi\rangle$: this sets up a many–one correspondence between (what I called) the canonical magnitude claim $\mathbf{A}_s \in \Delta$ (restricting the value of A on s to set Δ of real numbers) and a subspace $\mathcal{M}_\Delta^A$ of $\mathcal{H}$. There is a unique operator $\hat{P}^A(\Delta)$ that projects vectors of $\mathcal{H}$ onto $\mathcal{M}_\Delta^A$. The basic idea of van Fraassen's modal interpretation was that $A_s \in \Delta$ if s's dynamic state $|\psi\rangle \in \mathcal{M}_\Delta^A$ (i.e. $\hat{P}^A(\Delta)|\psi\rangle = |\psi\rangle$) but $A_s \notin \Delta$ if $\hat{P}^A(\Delta)|\psi\rangle = \emptyset$ (the null vector): otherwise quantum theory does not say whether or not $A_s \in \Delta$. But it does assign a probability to $\mathbf{A}_s \in \Delta$ in a context in which A is measured in an interaction with s.[7] This is given by the Born rule as:

$$\Pr{}^{|\psi\rangle}(A_s \in \Delta) = \langle\psi|\hat{P}^A(\Delta)|\psi\rangle.$$

In this view, ordinary quantum mechanics is associated with a family of state-space models. A model includes at least one Hilbert space with operators defined on it representing dynamical variables. These models may be used to make claims of the form $\mathbf{A} \in \Delta$ about physical systems as well as claims of the form $\Pr(\mathbf{A} \in \Delta) = \mathbf{p}$ about the probability of measurement outcomes on them. A model is applied by formulating

---

[7] After extending his modal interpretation to model interacting systems, van Fraassen [1991] was able to offer idealized quantum models of measurement contexts. A quantum model assigns a probability to the claim $\mathbf{A} \in \Delta$ about the measured system rather than to a claim about the "pointer position" of an "apparatus system" with which it interacts only in the case of a special kind of (von Neumann/Lüders) measurement.

a theoretical hypothesis to the effect that the behavior of some physical system or systems is faithfully represented by a particular state-space model.[8]

The behavior may be attributed to the system(s) as token or type, as in "the hydrogen atom", "the simple harmonic oscillator", "a two-level system". The class of models is diverse. Not only the Hamiltonian operator figuring in the Schrödinger equation, but also the number of Hilbert spaces and their dimensionality, vary from model to model, as do the initial and boundary conditions on $|\psi\rangle$ and even the topology of the physical space used to model the behavior of the particles or other physical systems (two- or three-dimensional? compact or non-compact?).

Applied to quantum theory, the state-space version of the semantic conception is a clear improvement on a syntactic conception that seeks to present the theory as a deductively closed formal system plus an interpretation in terms of correspondence rules. It acknowledges the central role of mathematical models—not in defining quantum theory but in forming the theoretical hypotheses involved in applying it. And it reveals a respect in which the relation between the theory and representational claims based on it is indirect.

The state-space approach nicely captures the use of models in theories of classical physics. But it requires significant modification if one is to appreciate what makes quantum theory a radical departure from theories of classical physics. As we shall see, the relation between quantum theory and representational claims based on it is even more indirect than this approach would allow.

I'll call the system(s) to which quantum theory is applied in any instance the *target* of that application. The target is physical, and any actual (rather than merely hypothetical) application takes it to be real—an element of physical reality. Recall how Bell introduced beables as what a theory takes to be physically real—as what may be "described in 'classical terms', because they are there". Any application of quantum theory assumes that its target has beable status. But it does not follow that the quantum model applied *represents* the target of its application or that the target is a beable *of quantum theory*. Bell ([2004], p. 55) lists the settings of switches and knobs and currents needed to prepare an unstable nucleus as beables "recognized in ordinary quantum mechanics", and presumably he would be prepared to add the nucleus and its α-particle decay product to that list. Things whose physical existence is presupposed by an application of quantum theory deserve a name of their own: I'll call them *assumables*.

The target of any actual application of quantum theory is clearly assumable in this sense, whether or not the model applied is taken to represent it. Whatever entities and magnitudes back assignment of a quantum state in this model are also assumables, such as experimental equipment including Bell's settings of switches,

---

[8] At least this is how a realist like Giere would regard model application. As a constructive empiricist, van Fraassen takes science here to have the more limited goal of faithfully representing all observed or measured behavior.

knobs, and currents. Chapter 6's dissolution of the measurement problem stressed that application of a quantum model *assumes* rather than implies that a measurement has a definite outcome. The readings of instruments recording such outcomes must also be counted among quantum theory's assumables. Clearly the application of a model of quantum theory assumes a lot about how the physical world can be represented. But examination will show that no element of a quantum model has the function of representing any beable that is novel to quantum theory. In that sense, quantum theory has no beables.

Just as classical mechanics may be used to model a system of any number of particles, quantum theory has models that may be applied to any number of physical systems (particles or fields). I shall use a simple example to illustrate the function of quantum models: a model that can be applied to the behavior of a single silver atom. A model of quantum theory is a mathematical structure. To preserve a clear distinction between models and their application to physical systems it is best to take a model to include only mathematical objects. I emphasize their mathematical character by adopting the convention of symbolizing mathematical objects using Greek letters like Ψ and μ or calligraphic letters like $\mathcal{H}$ or $\mathcal{A}$.

A (non-relativistic) quantum model Θ of a single physical entity like a silver atom includes a Hilbert space $\mathcal{H}$, a set T of real numbers t, a state $|\Psi\rangle$ (a vector in $\mathcal{H}$), a set $\mathcal{A}$ of operators on $\mathcal{H}$, and a measure μ on subspaces of $\mathcal{H}$: we can write $\Theta = \langle \mathcal{H}, T, \Psi, \mathcal{A}, \mu_\Psi \rangle$.[9] One operator in $\mathcal{A}$ is the Hamiltonian $\hat{H}$ that figures in the Schrödinger equation. Each model of this form will feature a specific Hamiltonian chosen with a view to its application. For example, in a quantum model of the effect of passage of a silver atom through an inhomogeneous magnetic field, this Hamiltonian will be a sum of terms including a term intended to model the free motion of the atom and another term intended to model the interaction between the field and the atom's magnetic moment associated with its intrinsic (spin) angular momentum.

How is such a model applied to the behavior of a silver atom passing through an inhomogeneous magnetic field? T is taken to represent the interval between an initial time t_i when the atom is still far from the magnet and a later time t_f at which it has been detected after passage. The prior conditions back assignment of an initial state $|\Psi\rangle_{t_i}$ to the atom. Adopting a Cartesian coordinate system in which the axis of the magnet is aligned in the z-direction, an operator $\hat{S}_z$ in the model

---

[9] As in classical physics, in an application the interval of real numbers T will represent an interval of time while t represents a moment in that interval. (Since quantum mechanics doesn't change their representational functions, I follow standard notation in writing t, T even though these are mathematical objects.) Each operator $\hat{A} \in \mathcal{A}$ is self-adjoint: in an application $\hat{A}$ will correspond uniquely to a dynamical variable A. A measure μ assigns, at each $t \in T$, a number in the interval [0,1] to each subspace $\mathcal{K}$ of $\mathcal{H}$ such that $\mu(\mathcal{H}) = 1$, and if every vector in $\mathcal{K}_i$ is orthogonal to every vector in $\mathcal{K}_j$ for $i \neq j$ then $\mu(\mathcal{L}) = \sum_i \mu(\mathcal{K}_i)$, where $\mathcal{L}$ is the smallest subspace such that, for each $\mathcal{K}_i$, $\mathcal{L}$ contains every vector in $\mathcal{K}_i$. On this occasion I have placed the subscript Ψ on element μ_Ψ of Θ to make explicit its dependence on Ψ: in future I will omit this subscript to simplify notation. In an application *some* values of μ_Ψ may yield Born probabilities for canonical magnitude claims.

corresponds to the z-component of spin S_z: but neither $\hat{S}_z$ nor anything else in the model represents the value of this z-component in the interval T, even though passage through the inhomogeneous field of a Stern–Gerlach magnet is often said to measure the value of S_z.

Suppose that prior conditions were such that the correct initial state to assign to the atom has the form

$$|\Psi\rangle_{t_i} = \frac{1}{\sqrt{2}}(|\Psi_\uparrow\rangle_{t_i} + |\Psi_\downarrow\rangle_{t_i}), \tag{8.1}$$

a superposition of two component vectors in the Hilbert space $\mathcal{H} = \mathcal{H}_1 \otimes \mathcal{H}_2$, where $\mathcal{H}_2$ is used to model the motion of the atom and $\mathcal{H}_1$ is used to model its intrinsic angular momentum. In more detail,

$$|\Psi_\uparrow\rangle_{t_i} = |\uparrow\rangle \otimes |\psi\rangle_{t_i} \tag{8.2}$$
$$|\Psi_\downarrow\rangle_{t_i} = |\downarrow\rangle \otimes |\psi\rangle_{t_i}.$$

Since this is a model of quantum theory with the Hamiltonian appropriate for the assumed situation, the correct quantum state to assign at a time t after passage through the magnet but before detection will be (compare (6.4))

$$|\Psi\rangle_t = \frac{1}{\sqrt{2}}\left[(|\uparrow\rangle \otimes |\psi_+\rangle_t) + (|\downarrow\rangle \otimes |\psi_-\rangle_t)\right]. \tag{8.3}$$

This represents neither the motion nor the spin of the atom at t.

Unlike van Fraassen's modal interpretation, in this view it is *never* the function of a quantum state to represent the physical condition of a system to which it is assigned: even if state $|\Psi_\uparrow\rangle_{t_i}$ had been correctly assigned at t_i this would not have represented the atom as having z-component of spin "up" at t_i. When a model of quantum theory is applied, the vector $|\Psi\rangle$ (or whatever alternative mathematical object specifies the quantum state in the model[10]) does not have a direct representational function of standing in place of some physical entity or magnitude. The universe does not contain new physical things represented by state vectors or wave functions, and nor is it the role of vectors or wave functions to assign truth values to statements about the values of dynamical variables, whether classical or quantum.

The only element in quantum model Θ with a direct representational function is the interval T of real numbers whose elements t serve to represent moments of time in an application of this model. Models of relativistic quantum mechanics and quantum field theory instead contain elements that serve to represent points or regions of space-time. But neither Hilbert spaces nor operators[11] serve to represent anything in the physical world when a quantum model is applied. That leaves the measure μ and its analogs

---

[10] Chapter 3 noted that the vector for a (pure) quantum state may sometimes take the form of a wave function, and we saw in Chapter 5 that a mixed state may be specified by a density operator. In algebraic quantum theory a state is specified as a linear map on an abstract algebra.

[11] Nor the elements of a C^* algebra that generalizes $\mathcal{A}$ in algebraic quantum theory.

in models of other forms of quantum theory.[12] Since these are intimately connected to the empirical content of quantum theory through the Born rule, one might expect μ and its analogs to have some representational function in applications of quantum models. But we shall see that they have at most an indirect representational function, since Born probabilities are not physical magnitudes.

In an application of a quantum model the measure μ functions by yielding certain statements of the form $\Pr(A_s \in \Delta) = \mathbf{p}$ assigning a probability p to the statement $A_s \in \Delta$ that restricts the value of dynamical variable A on s to set Δ of real numbers. In Chapter 5 I called a statement of the form $A_s \in \Delta$ a *canonical magnitude claim* about a physical system s. A quantum model neither contains nor implies any magnitude claim since it is just a mathematical structure. Even when this structure is applied to physical systems the theoretical hypothesis making that application does not imply any magnitude claim. But the fundamental point of applying a quantum model is to help one to form better beliefs about magnitude claims: though not part of quantum theory, they are its *raison d'être*.

To illustrate the application of quantum models I used the example of a simple model in non-relativistic quantum mechanics of a silver atom a passing through a (classical) inhomogeneous magnetic field. This example featured canonical magnitude claims about the values of two different kinds of dynamical variables on a—position variables, and components of (intrinsic) angular momentum (i.e. spin). As I noted in Chapter 6 (§ 6.1), claims of the form "The z-spin of a is $+\hbar/2$" and "The z-spin of a is $-\hbar/2$" lack empirical significance *at least* prior to a's interaction with the detection screen, while a claim of the form "The z-position of the silver atom on the screen is confined to interval Δ" then becomes empirically significant for Δ very much less than the distance separating the regions where silver atoms can be seen adhering to the screen. So the Born rule may be legitimately applied to yield a probability assignment to the latter but not to the former claims.

It is appropriate to call the dynamical variables featured in this example classical, since they also featured in applications of models of classical physics (though angular momentum was treated very differently in such applications). But not all canonical magnitude claims assigned Born probabilities concern dynamical variables acknowledged by classical physics, and nor do these all concern entities recognized by classical physics. In applications of a relativistic quantum field theory canonical magnitude claims may concern entities (such as the Higgs field, or a K^0 meson with its quark constituents) and dynamical variables (such as the Higgs field magnitude, and the charm of one of the K^0 meson's quarks) unknown to classical physics. So canonical magnitude claims are not all stated in the language of classical physics, and they certainly do not all concern observable objects and magnitudes. But though in this

---

[12] These include those theories of what Ruetsche [2011] calls "extraordinary quantum mechanics" whose models are typically presented without specifying any particular Hilbert space.

sense they are non-classical it is nevertheless appropriate to deny that quarks and the Higgs field magnitude are quantum beables since they are not represented by elements of the models of quantum field theory that are applied in making magnitude claims about them. It is because quantum models contain no elements whose function is to represent beables that magnitude claims are not part of quantum theory.

In applying quantum theory, the Born rule helps a scientist by advising him or her how strongly to believe each canonical magnitude claim to which it may be legitimately applied. Such degrees of belief may vary between firm belief and firm disbelief, with firm belief represented by 1, firm disbelief by 0, and intermediate numbers between 0 and 1 representing partial beliefs of different strengths. This numerical representation is common among Bayesian statisticians and epistemologists, who have presented arguments that the degrees of belief of an ideally rational agent should conform to the mathematics of probability theory—requiring that one's degree of belief in *not-P* be 1 minus one's degree of belief in *P*, and that if (one is sure that) *P*,*Q* can't both be true together then one's degree of belief in *P*-or-*Q* be the sum of one's degrees of belief in *P* and in *Q*. Philosophers have come to call degrees of belief that conform to the mathematics of probability theory *credences*.

The demands of rationality can be exacting, and while many aspire to meet them, few succeed. But applying a quantum model does not require ideal rationality, and one can understand what this involves even if one cannot muster the minimal rationality needed correctly to apply a quantum model. This is best explained by considering what it is to *accept* quantum theory. To accept quantum theory is to commit oneself to setting credence in each significant canonical magnitude claim equal to the probability specified by a legitimate application of the Born rule based on the best available quantum model, in the absence of more direct access to the truth value of the claim. But why should we accept quantum theory?

The strongest reason to accept quantum theory is provided by the success of its applications in predicting and explaining physical phenomena of a statistical nature. If we set credences in accordance with the Born rule we are led to expect and can come to understand the patterns displayed by these statistics. Each individual event contributing to a statistical pattern may be described by a canonical magnitude claim, stated in language distinct from that used to present any quantum model. In that sense magnitude claims are not quantum mechanical statements. But the evidence base for quantum theory is events so described, and any application of quantum theory implicitly assumes that some magnitude claims are true while others are false. In particular, any legitimate application of the Born rule to a measurement scenario presupposes that a well-conducted measurement has exactly one outcome: the silver atom's z-spin will be recorded as "up" (equal to $+\hbar/2$) or "down" (equal to $-\hbar/2$)—not both and not neither. The primary function of a quantum state assignment to a silver atom in applying a quantum model here is to issue good advice about how strongly to expect it to be recorded as up as opposed to down.

8.3 Truth, Content, and Objectivity

I think the notion of truth involved here is best interpreted in a deflationary way. $A_s \in \Delta$ is true if and only if the value of A on s is restricted to Δ ("The silver atom has z-spin $+\hbar/2$" is true if and only if the silver atom has z-spin $+\hbar/2$). If that is all that is meant by correspondence, then this is correspondence truth. But such correspondence is too thin to bear any explanatory weight and is not grounded in some physical or metaphysical relation of reference or denotation obtaining between language and world. There are many respects in which quantum theory is best understood along pragmatist lines, but it would be a mistake to adopt some Jamesian account of truth as "what is good in the way of belief", or "the expedient in the way of our thinking". If calling something true is more than a linguistic convenience (as in "Everything he said is true"), a pragmatist should seek a further function of truth as a norm of assertoric dialogue, as has Huw Price ([1988], [2011]).

It is not a classical pragmatist account of truth but a contemporary pragmatist approach to content that provides the key to understanding the significance of magnitude claims. For a pragmatist, a statement derives its content from the way it is used, making use prior to content. Brandom ([2000], p. 4) contrasts this with a representationalist approach that

> would identify the content typically expressed by declarative sentences and possessed by beliefs with sets of possible worlds, or with truth conditions otherwise specified. At some point it must then explain how associating such content with sentences and beliefs contibutes to our understanding of how it is proper to use such sentences in making claims, and to deploy beliefs in reasoning and guiding action. The pragmatist direction, by contrast, seeks to explain how the use of linguistic expressions, or the functional role of intentional states [beliefs], confers conceptual content on them.

I call this an approach to content rather than an account of content because it is concerned not to provide an analytic framework in which to represent the content of statements but to explain how a statement acquires whatever content it has. For Brandom, content is acquired inferentially:

> According to the inferentialist account of concept use, in making a claim one is implicitly endorsing a set of inferences, which articulate its conceptual content. ([2000], p. 19)

A magnitude claim acquires whatever content it has by its place in a web of inferences: I'll give examples in Chapter 12. One who makes that claim is implicitly committed to defending it by offering other claims from which it may be inferred, and one who accepts that claim is entitled to make other inferences from it. However reliable, it is important that not all these inferences be deductively valid: those that confer content are what Sellars called *material inferences*. It is also important that the inferential web that confers content on a claim connect ultimately to experience and to action: not all content-conferring inferences link claims to other claims. A claim may be defended by appeal to experience ("see for yourself"), and accepting a claim may entitle one to act accordingly on the basis of a practical inference.

As we saw in Chapter 5, the material inferences one can reliably make from a magnitude claim about a physical system depend on the system's environment. Since the claim derives much of its empirical content from such (deductively invalid) inferences, a claim about a system may have a rich significance in one environment but be effectively devoid of significance in a different environment. This helps to clarify the situation of (say) a neutron passing through an interferometer. A magnitude claim restricting its position to one path rather than another through the interferometer lacks empirical significance in the absence of environmental interactions suitable to decohere its position-space wave function. So one who accepts quantum theory will neither take the neutron to follow one path rather than another nor apply the Born rule to set credences in each possibility. The Born rule is legitimately applicable *only* to empirically significant canonical magnitude claims.

Or take the case of the silver atom passing through a Stern–Gerlach magnet. Application of an appropriate quantum model of its interaction with its environment prior to detection indicates no significant decoherence of either its position wave function $|\psi\rangle$ or its spin quantum state: passage through the magnet is tracked simply by transforming state (8.1) into state (8.3). One who accepts quantum theory will assign no nontrivial significance to a statement about the magnitude of the atom's spin or position prior to detection: and the Born rule may not be legitimately applied to such a statement.

However, application of an appropriate model of the atom's interaction with a detector will show that statements about the position of a system associated with the detector (either the atom itself or some part of the detecting system) have acquired a rich enough significance to legitimize application of the Born rule to them. The z-position of this system may be recorded. If it is within a small range of positive values, this is reported as a measurement of z-spin with outcome "up": if it is within a small range of negative values, this is reported as a measurement of z-spin with outcome "down". The Born rule may now be legitimately applied, directly to statements restricting this system's position to one of these ranges, and so indirectly also to these two possible outcomes. Prior to examining the relevant system's z-position, one who accepts quantum theory will assign equal credence of $1/2$ to each possible outcome. Only if the atom's interaction with the detector has stably decohered its z-spin may s/he also significantly claim that it then has the corresponding value of $\pm \hbar/2$.

After this clarification, we can return to the question of the representational role of the measure μ in a quantum model $\Theta = \langle \mathcal{H}, T, \psi, \mathcal{A}, \mu \rangle$. Recall that this is a mathematical object that assigns, at each $t \in T$, a number in the interval $[0,1]$ to each subspace $\mathcal{K}$ of $\mathcal{H}$. When discussing van Fraassen's modal interpretation I noted that each dynamical variable A on a quantum system is uniquely represented by an operator $\hat{A} \in \mathcal{A}$ acting on Hilbert space $\mathcal{H}$, and that this sets up a many–one correspondence between each magnitude claim $\mathbf{A}_s \in \Delta$ and a subspace $\mathcal{M}^A_\Delta$ of $\mathcal{H}$. Van Fraassen took the Born rule to yield probabilities only for the outcomes of measurements and offered a precise characterization of measurement by means of a particular quantum model of an interaction between the system in question and a second "apparatus" physical system. But it is important to stress that applications of the Born rule are not restricted to measurement outcomes: the Born rule may be applied to yield probabilities for any significant canonical magnitude claims.

Now we know what makes a magnitude claim significant, we can appreciate that while the measure μ is not itself a probability measure it can be applied to yield probabilities (only) of significant canonical magnitude claims according to the Born rule prescription

$$\Pr{}^{|\psi\rangle}(A_s \in \Delta) = \mu_\psi(\mathcal{M}^A_\Delta) = \langle\psi|\hat{P}^A(\Delta)|\psi\rangle.$$

Not only does μ not represent any element of physical reality, but even the probabilities it yields in a legitimate application of the Born rule do not do so—their function is to offer advice to one who accepts quantum theory on how to set credences for significant canonical magnitude claims.

If $A_s \in \Delta$ is a significant magnitude claim, then the probabilistic statement $\Pr(A_s \in \Delta) = p$ is true if and only if the probability that $A_s \in \Delta$ is p: so at least in this sense the probabilistic statement represents this as being the case. But the statements $A_s \in \Delta$ and $\Pr(A_s \in \Delta) = p$ have different functions. The primary function of the probabilistic statement is not to represent some physical entity, magnitude, or state of affairs, but to offer advice to an actual or hypothetical situated agent on how to set credences for the embedded magnitude claim. At least from the perspective of an analysis of science, the primary function of the magnitude claim is to represent a physical state of affairs.[13]

There is another important difference between a probabilistic statement $\Pr(A_s \in \Delta) = p$ and the embedded claim $A_s \in \Delta$. We noted already in Chapter 4 that a probabilistic statement based on application of quantum theory must be understood as relativized to the physical situation of an actual or hypothetical agent making the application. If these situations differ enough, $\Pr(A_s \in \Delta) = p$ relative to one physical agent-situation while $\Pr(A_s \in \Delta) = p' \neq p$ relative to another physical agent-situation. We saw also that this possibility is realized in some EPR–Bell measurement scenarios.[14] But the truth value of a magnitude claim $A_s \in \Delta$ is not so relativized—if Alice and Bob were to assign different truth values to a specific significant magnitude claim about a physical system in given circumstances, at most one would be right no matter what their respective physical situations. This does not make magnitude claims

---

[13] A radical pragmatist might deny that *any* statement has the primary function of representing the world. Price ([2013], p. 24) argues against what he calls Representationalism, the view that "the function of statements is to 'represent' worldly states of affairs, and true statements succeed in doing so". But in order to address the goals of predicting and explaining natural phenomena, science must be capable of representing those phenomena in language: in the case of quantum theory, this means assigning magnitude claims a primary representational role.

[14] Notably, scenarios in which Alice's and Bob's measurements on each prepared system occur at space-like separation—a notion I'll explain in Chapter 10.

more objective than probability statements about them. It is objectively true that a motorist in the United States probably drives on the right, even though this is so only relative to the (upright!) orientation of the motorists themselves.

I said that the universe does not contain physical entities or magnitudes represented by quantum states or wave functions, and nor is it a role of a quantum state to assign truth values to statements about the values of dynamical variables, whether or not these are classical. But a statement assigning a quantum state to a system in a given situation may be objectively true or false in just the same way as a statement of the form $\Pr(A_s \in \Delta) = p$. Both kinds of statement must be understood as relativized to the physical situation of an actual or hypothetical agent, and both kinds of statement can be understood as weakly representational. When so relativized, the statement that s has quantum state $|\psi\rangle$ is true if and only if s has quantum state $|\psi\rangle$, and in that sense represents s as having quantum state $|\psi\rangle$.

It is clear why quantum state assignments and probabilistic statements are similar in these respects once one understands their primary functions in an application of quantum theory. The primary function of a quantum state assignment is to supply probabilities for significant magnitude claims about physical systems, and the primary function of the resulting probabilistic statements is to provide objectively good advice to a physically situated agent about how strongly to believe these claims. So each kind of statement inherits both its objectivity and its relativity to the physical situation of a hypothetical agent from its primary function.

There are at least three respects in which quantum state assignments and probabilistic statements about significant magnitude claims are objective:

1. There is widely shared agreement on them within the scientific community.
2. A norm is operative within that community requiring resolution of any residual disagreements.
3. This norm is not arbitrary but derives directly from the scientific aims of prediction, control, and explanation of natural phenomena.

Flouting this norm would leave one unable to account for the patterns of statistical correlation among events described by true magnitude claims to which quantum theory is ultimately responsible.

I said that to accept quantum theory is to commit oneself to setting credence in each significant canonical magnitude claim equal to the probability specified by a legitimate application of the Born rule based on the best available quantum model, in the absence of more direct access to the truth value of the claim. I've shown how a quantum model may be used to generate probabilistic claims of the form $\Pr(A \in \Delta) = p$, but only when the right kind of decoherence renders the magnitude claim $A \in \Delta$ significant enough to permit a legitimate application of the Born rule. Deciding on the best available quantum model requires the expertise of the practicing physicist, which can be acquired only through extensive education and long experience. But there are situations in which direct access to the truth value of a magnitude claim can trump credences based on even the most expert application of quantum theory.

It is clear that this must happen sometimes since the strongest evidence for quantum theory is gleaned from experimental observation of the statistical patterns it so successfully predicts. It is only because scientists have access to the truth values of magnitude claims about each event forming part of such a pattern that is not derived through application of quantum theory that these statistical patterns can count as evidence for quantum theory. The scientist assigns probabilities to a range of magnitude claims about these events by applying a quantum model, then observes the frequency with which these claims are true. Even if the scientist's observations leave him or her somewhat uncertain about the truth value of an individual magnitude claim, these observed frequencies must be taken as data arrived at independently of quantum theory for it to count as evidence for that theory.

But there are circumstances in which quantum models may be conveniently applied to guide an agent's beliefs about magnitude claims about which s/he could, with difficulty, obtain more direct and detailed information. Suppose, for example, that Alice and Bob are performing experiments like those described in Chapter 4 on several million photon pairs, with the difference that Alice performs her polarization measurement on a photon in each pair a microsecond after Bob. Each of Alice and Bob records the result of every polarization measurement s/he makes as well as the precise time at which s/he makes it (so they can afterwards combine and coordinate their results into records of joint outcomes on individual pairs). Suppose also that Alice and Bob perform their individual experiments in different places with no means of communicating their settings or outcomes until they meet afterwards to compare notes. Immediately after recording the setting and outcome of an individual polarization measurement, Alice is unaware of Bob's setting and outcome for his photon in that pair. But she can apply quantum theory to guide her credences as to Bob's outcome, conditional on each of Bob's possible settings. In the situation as described, this will be the best she can do.

But Alice could do better if Bob had a way of rapidly communicating his setting and outcome on each of his photons to her immediately after recording it. This could provide Alice with reliable direct access to the truth values of magnitude claims about Bob's polarization measurements. If her physical situation included such reliable direct access, then she should not base her credences in these magnitude claims about Bob's photon-polarization measurement outcomes on the probabilities she derives by applying quantum theory to her own outcome for the corresponding photon in each pair: Bob's message reporting his actual outcome trumps quantum theory's advice.

I began this chapter by claiming to locate the truly radical nature of quantum theory not in what it says about the physical world but in its novel, indirect relation to the world. Unlike theories of classical physics, the novel mathematical structures in quantum theory's models are not applied directly to offer representations of physical situations. The function of these novel elements (Hilbert space, vectors, or other representatives of quantum states, operators, measures) is not to represent novel physical structures, but to provide objectively good advice to the one applying a quantum model concerning the significance and credibility of statements about properties of physical systems.

These magnitudes and systems are assumables not beables of quantum theory, since they are not represented by elements of its models. The advice is tailored to the physical situation of an actual or hypothetical agent. It is good insofar as it enables such an agent to predict and explain statistical patterns in physical events that can be described by making true statements about magnitudes on physical systems. To accept quantum theory is to commit to following the advice offered by applying its models. By doing so one is able to fulfill the basic scientific goals of predicting, controlling, and explaining natural phenomena.

While this certainly involves making representational statements about the physical world, the form of these statements is not novel to quantum theory—they are statements about entities and magnitudes acknowledged by the rest of physics, not statements about physical entities and magnitudes newly represented in models of quantum theory. Acceptance of quantum theory involves taking a novel attitude toward the content of these statements and the existence of these things. A magnitude claim about a physical entity may be significant when it is in one environment but not when it is in a different environment: and quantum theory can itself be applied to gauge its significance in different environments. We shall see later (in Chapter 12) that quantum field theory can be similarly applied to gauge when one may make significant claims about particles, fields, or other physical entities. We can understand how a statement's significance varies with environmental context once we recognize that any statement gets its content from the inferential relationships within which it is located.

Finally, it is important to stress that quantum theory is not about agents, their beliefs, or their language. A precise formulation of quantum theory will not speak of such things in its models any more than it will speak of agents' measuring, observing, or preparing activities. If quantum theory is about anything it is about the mathematical structures that figure in its models. Quantum theory, like all scientific theories, was developed by (human) agents for the use of agents (not necessarily human: while insisting that any agent be physically situated, I direct further inquiry on the constitution of agents to cognitive scientists). Trivially, only an agent can apply a theory for whatever purpose. So any account of a predictive, explanatory, or other application of quantum theory naturally involves talk of agents.

9

Probability and Explanation

The explanatory power of quantum theory is without parallel in the history of physics. From Schrödinger's explanation of the energy levels of the hydrogen atom, explanatory applications of quantum theory have now extended to phenomena as disparate as the stability of ordinary atomic matter; the structure of the periodic table of elements; radioactive decay; the energy levels of quark–antiquark systems; the superfluidity of helium; the interaction of electrons with light; ferromagnetism; the behavior of transistors, lasers, and neutron stars; correlations exhibited by photon pairs separated by many kilometers; Bose condensation; and even the quantum fluctuations that may account for the large-scale distribution of matter in the very early universe.

A distinctive feature of such explanations is that quantum theory does not itself represent any new physical entities or magnitudes as playing a role in bringing about these phenomena. We are able to use quantum models to show that other-wise surprising phenomena were to be expected and to say what they depend on. By repeatedly using the same types of model elements when applying similar kinds of models, we have been able to unify our understanding of otherwise diverse phenomena.

The Born rule plays a key role here: it figures, explicitly or implicitly, in all explanatory applications of quantum theory. Quantum theory contributes to our explanatory projects by providing us with a general set of techniques for calculating Born probabilities that tell us what we should expect in familiar as well as unfamiliar situations, and what these probabilities depend on. So to appreciate how quantum theory helps us explain these phenomena we need some account of probability in quantum theory. The pragmatist account I will offer is much more widely applicable and is not novel to quantum theory: here quantum theory reveals what philosophers should have known without its prompting. This contrasts with novel views of probability associated with alternative interpretations of quantum theory including some considered in the Interlude.

A natural phenomenon such as those I mentioned in the first paragraph is not an individual event but a reproducible regularity in the physical world. Using quantum theory, we have been able to explain a great variety of these. In an explanatory application of quantum theory, the Born rule yields probabilities. So we use quantum theory to explain reproducible regularities as probabilistic phenomena, including exceptionless regularities as special cases. Physics acknowledged probabilistic phenomena even before the advent of quantum theory, and classical physics was applied to their explanation. But what is a probabilistic phenomenon, and how can it be explained?

Experimenters often plot a smooth curve through data points to indicate the relationship between values of the plotted magnitudes. For example, if the data points represent the extension of a spring of a variety of materials when different small weights are suspended from it, this curve is typically a straight line. Robert Hooke noted this regularity in 1660, which has come to be known as Hooke's law. The phenomenon may be explained by the fact that small displacements of the spring's constituent atoms, molecules, or ions from their normal positions are proportional to the force causing the displacement.

Data points from actual experiments manifest Hooke's law: they do not constitute it. It is a fact about the data from many properly conducted experiments that they may be well modeled by Hooke's law. When we give an explanation of Hooke's law we explain this fact, not actual data obtained in experiments on springs. This example is typical: when a phenomenon is explained, the *explanandum* is not the actual data that manifest it, but the fact that those data may be well modeled by a generalization that abstracts from the complex details of the processes that produce data in particular circumstances, and idealizes and extrapolates from such data.

If the association between data specifying values of magnitudes is merely statistical, one cannot plot a single smooth curve to indicate any fixed relationship between them. Instead one can often give a probabilistic model. Such data are often the result of multiple complex processes whose details we can neither predict nor control. A roulette wheel provides a convenient example. Casinos depend both on the unpredictability of the outcome of any given spin and on a probabilistic model of the outcome statistics of large numbers of spins that assigns equal probability to the ball landing in each sector of the wheel. A richer example is provided by a biased coin. In this case the outcomes of many tosses will typically display statistics that reflect the bias of the coin, and a probabilistic model will accommodate this fact by taking the *probability* of each possible outcome of a toss to be related to the magnitude of the coin's bias. It is because the outcomes of spins of a roulette wheel and tosses of a biased coin are best modeled probabilistically that I call them probabilistic phenomena.

Here is a famous example of a probabilistic phenomenon originally taken to be accounted for on the basis of classical physics, but now explained using quantum theory. In his seminal investigations, Ernest Rutherford identified the less penetrating α-rays emitted by radioactive materials as charged helium ions. Initial experiments showed these α-particles could be deflected when passing through air or thin sheets of material. Rutherford suggested as a research project for Marsden, the student of his colleague Geiger, that he find out whether any α-particles could be deflected through large angles. The results astonished Rutherford: when a beam of α-particles is incident on a sheet of gold foil less than a thousandth of a millimeter thick, some α-particles are turned through an angle of more than $90°$.

It was quite the most incredible event that has ever happened to me in my life. It was almost as incredible as if you fired a 15 inch shell at a piece of tissue paper and it came back and hit you.

([1938], p. 68)

Rutherford is credited with the discovery of the atomic nucleus because he was able to account for this phenomenon by assuming that almost all the mass of an atom is concentrated in a tiny central, charged region. Since α-particles are also charged, this heavy, charged nucleus is able to exert a strong electrostatic force on any relatively light α-particle that comes close enough, producing a large change in its momentum and substantially altering its trajectory. Rutherford [1911] backed up this qualitative story with a detailed classical analysis applying a mathematical model closely analogous to those Newton had used to explain the hyperbolic orbits of comets in the solar system: in each case this posited an inverse-square central force—gravitational between sun and comet, electric between atomic nucleus and α-particle. This analysis led him to predict a number of quantitative features of the phenomenon, all of which were subsequently confirmed in further experiments by Geiger and Marsden [1913]. It was this verification of Rutherford's explanation of the phenomenon of α-particle deflection by thin sheets of metal that provided the powerful initial evidence for the existence of the atomic nucleus.

Rutherford's explanation illustrates several important points. Note first that what it explained was not a particular event but a general phenomenon—the way α-particles are deflected by thin sheets of metal. This phenomenon is a reliably reproducible regularity manifested in a wide variety of different circumstances, using different materials and different apparatus. It is not observable by unaided human senses, though by applying our senses to instruments we can collect data manifesting the phenomenon. That is what Geiger and Marsden did when painstakingly using a microscope to note the location of each of the thousands of flashes on a zinc sulphide screen they took to mark the angle of deflection of an α-particle. More precisely, what it explained (the *explanandum*) was not just that some α-particles are deflected through large angles when incident on a thin metal foil. Rutherford explained four quantitative facts about this phenomenon: that the probability of deflection through angle ϕ is proportional to the thickness of the scattering foil and to the square of the nuclear charge, and inversely proportional to the fourth power of $\sin \phi/2$ and of the velocity (and consequently momentum) of the α-particles.

Explaining these otherwise independent facts is one aspect of the power of Rutherford's explanation of the phenomenon. Another aspect is its power to unify this with other phenomena. α-particle scattering could be used to determine the nuclear charge of the scattering metal, which proved to be the same as the atomic number of the element of which it was composed—the number associated with its position in the periodic table of elements. Together with the assumption that α-particle decay subtracted two units from this number while β-particle decay added one unit, this led to an improved understanding of how changes of nuclear structure induced by radioactivity corresponded to variation in the chemical properties of the elements involved, and to the idea of atomic isotopes. Perhaps more significantly, Rutherford's explanation effectively unified the phenomenon of α-particle scattering by thin sheets of metal with other mechanical phenomena (celestial and terrestrial) and other electrical phenomena (such as the electrical force exerted by a very long charged wire). This unification was a consequence of his use of models of both classical mechanics and classical electrostatics.

Rutherford's basic model was of a single α-particle deflected by a single fixed point charge (a gold nucleus). But a gold foil contains a vast number of atoms distributed in some way throughout its volume, and an incident beam contains a vast number of α-particles, all of which he took to have slightly different, approximately parallel, trajectories. Even neglecting multiple scattering, to apply the basic model he needed to make assumptions about how the α-particle trajectories lined up with the positions of the gold nuclei. If, for example, every α-particle were headed straight for a gold nucleus, they would all be reflected back along their tracks. But he could safely neglect that possibility, reasoning that he could treat both nucleus and α-particle as point charges.

In a section of his [1911] paper entitled "Probability of single deflection through any angle", Rutherford assumed the (relative) probability of an α-particle being singly deflected because it enters some gold atom within a distance r of its central nucleus is equal to $\pi r^2 nd$, where d is the thickness of the foil and n is the number of gold atoms per unit volume, and went on to calculate "the fraction of the total number of particles which are deviated through an angle greater than ϕ". He clearly did not distinguish here between probability and relative frequency. Each of a huge number of α-particles was assumed to enter the foil along a particular trajectory, with no statistical correlation between their incoming trajectories and the locations of the gold nuclei in the foil. This was the kind of statistical assumption commonly made in classical physics when treating the motion of large numbers of molecules in the kinetic theory of gases. If Rutherford's explanation is cast in terms of relative frequencies, it may be given without mentioning probabilities. It would then apply to all and only instances of the phenomenon for which the relevant frequencies satisfy the necessary statistical assumption.

Rutherford gave his explanation of the phenomenon of α-particle deflection by thin sheets of metal some fifteen years before Heisenberg's and Schrödinger's seminal contributions to what we now call quantum theory. How we now explain that phenomenon using quantum theory is in some ways so different that from our present perspective the success of Rutherford's account seems almost fortuitous. By focusing on these differences we can get insight into what makes quantum theory revolutionary. But by abstracting from these differences we can also begin to appreciate the parallels between quantum and classical explanations of phenomena. We can come to see that explanations using quantum theory also derive their power from their ability to explain otherwise independent facts about a phenomenon and to unify otherwise diverse phenomena. And in each case we can come to see how a good explanation shows just what these phenomena depend on.

To explain what is known as Rutherford scattering today one can begin by solving the Schrödinger equation of non-relativistic quantum mechanics, subject to suitable boundary conditions. This presupposes that an incoming beam of α-particles may by assigned a suitable quantum state—in the simplest model its state vector will be a plane wave associated with α-particle momentum p and energy E:

$$\psi(x,t) = \exp\frac{i}{\hbar}(px - Et). \tag{9.1}$$

In the (assumed) absence of environmental interactions, assignment of state (9.1) licenses no significant statements about the trajectory of any incoming α-particle. The electrostatic interaction between nucleus and α-particles is modeled by an operator with the same dependence on the charge on and distance from the nucleus as the electric potential function Rutherford assumed in his classical explanation: but nothing in the Schrödinger equation corresponds to an associated electric force. Far away from the nucleus, a solution to this equation may be expressed as the sum of the incoming state (9.1) and a superposition of outgoing spherical waves centered on the nucleus. These depend on distance from the nucleus and angle ϕ from the direction of the incoming beam. Application of the Born rule yields the probability (density) of detecting an α-particle at angle ϕ after scattering by a metal foil of thickness d as a function of ϕ, d, p, and the nuclear charge. Perhaps surprisingly, this probability exhibits exactly the same functional dependence Rutherford predicted in his *classical* explanation of the phenomenon.

By using quantum theory one can explain essentially the same facts about the phenomenon that Rutherford took himself to have explained classically. By using this or related quantum models one can explain similar facts about otherwise diverse phenomena in atomic, nuclear, and subnuclear physics. The quantum explanation of Rutherford scattering is a paradigm for a vast range of applications of quantum scattering theory that have proved predictively accurate and explanatorily powerful as well as serving as a tool for further exploration of nuclear and subnuclear structure. These applications have illustrated the power of quantum theory to unify a wide range of phenomena, including many unknown to or inexplicable by physics in Rutherford's day. Moreover, quantum models based on the Schrödinger equation may be applied much more widely than to scattering problems. Some of the earliest applications were to explain the stability and detailed structural features of *bound* states of atoms and molecules. The unifying power of quantum mechanics now exceeds even that of classical mechanics.

But the quantum explanation of Rutherford scattering differs strikingly from Rutherford's own classical explanation. It makes essential use of probability without mentioning α-particle trajectories: these are not probabilities for deflection through angles in a certain range, but for being registered at a detector set to detect an α-particle within that range of angles. While Rutherford's use of probabilities was eliminable in favor of actual relative frequencies, the Born probabilities issuing from a legitimate application of a quantum model are *sui generis*. Moreover, the Born probability rule may be legitimately applied only to magnitude claims rendered significant by suitable environmental interactions, in this case involving a suitable α-particle detector.

By calling a Born probability for α-particle detection *sui generis* I mean two things:

(1) It is not to be understood as providing a partial remedy for ignorance of initial conditions (such as its incoming trajectory) that determine in each case whether a particular α-particle will be detected at a certain angle of deflection; and
(2) It is not definable as the relative frequency of detection at that angle in actual instances of the phenomenon.

In the [1926] paper proposing his probabilistic interpretation of the quantum state, Born made it clear that he would have endorsed point (1). The fact that this interpretation came to be called statistical suggests that many physicists would follow Rutherford in rejecting point (2). I'll discuss the pitfalls of statistical, or frequency, interpretations of probability in the next section.

Rutherford scattering is a probabilistic phenomenon. Geiger and Marsden had no way to predict or control the location of the flash on their zinc sulphide screen they took as detection of an individual α-particle. But after amassing statistics of thousands of such flashes they were able to note statistical regularities that could be modeled probabilistically. It turned out that in a probabilistic model that best fitted their data, the probability of detection at angle ϕ from the incoming beam is proportional to the thickness of the scattering foil, the square of the nuclear charge, and inversely proportional to the fourth power of $\sin \phi/2$ and of the velocity of the α-particles. It is these facts about this probabilistic phenomenon that Rutherford took himself to explain using classical physics, but that are today explained using quantum theory. But what *is* probability, and how *is* it related to statistics?

9.1 Probability

The basic mathematical principles of probability are simple and relatively uncontroversial. In fact they are so simple that they may be applied to things that have nothing to do with probability, such as piling sand in a sandbox! Suppose you wish to pile 1 unit of sand (1 pound, or 1 kilogram) in a suitably sized sandbox. There are lots of ways to do it: you don't have to end up with a flat pile. But each area of the box will end up covered by an amount of sand between 0 and 1 units, if two areas do not overlap then the sand covering their total area will be the sum of the amounts covering the individual areas, and the total area will be covered by 1 unit of sand. We can write these conditions as follows:

For any area A, $\quad 0 \leq \mu(A) \leq 1$

If A, B are areas with $A \cap B = \emptyset$ then $\mu(A \cup B) = \mu(A) + \mu(B)$

If the total area is S then $\mu(S) = 1$

where $\mu(A)$ is the amount of sand piled on area A. These principles define μ as a finitely additive measure with norm 1.[1] Mathematically, probability is a finitely additive measure with norm 1 on the set of subsets of some set.[2]

The sand measure is obviously not a probability measure (though it might arise from a probabilistic sand-piling process in which the amount of sand that covers each area depends on the probability for each grain that it will end up there). To say what probability is we need to distinguish it from other measures obeying the same mathematical principles: we need to give an interpretation of probability.

Many scientists are tempted to approach this task by asking to what in the physical world the term 'probability' refers. Noting that at least in some cases we can measure probabilities by compiling statistics from repeated trials and calculating relative frequencies of different outcomes, physical scientists and others have tried to define probability in terms of such relative frequencies. Unfortunately, proposed definitions have been shown to suffer from various problems, the most serious of which is circularity.

To say that the probability of a coin landing heads uppermost is (say) .49 does not mean that it will land heads 49 times out of every 100 tosses, since it is consistent to suppose that it lands n times for n any number between 0 and 100. The most one can say is that it will very probably land heads about 49 times. But that is no help in explaining what probability means because it uses the very notion that was to be defined. Nor does it help to appeal to hypothetical infinite sequences of trials in which the probability is 1 that the relative frequency of heads differs from .49 by less than any preassigned amount. We never have infinite sequences of identical repetitions of "the same" trial, and a probability of 1 is still a probability.

Perhaps in part because of the failure of frequency interpretations to latch on to a physical referent for the term 'probability', some have looked for a referent elsewhere. The social scientist Bruno de Finetti emphatically declared there was no such thing—that probability, like phlogiston, does not exist. But he also said:

probability [...] means degree of belief (as actually held by someone, on the ground of his whole knowledge, experience, information) regarding the truth of a sentence, or event E (a fully specified 'single' event or sentence, whose truth or falsity is, for whatever reason, unknown to the person). ([1968], p. 48)

---

[1] We can extend this to a countably additive measure by further requiring that if I is the set of all numbers $\{1, 2, 3, \ldots\}$ then if A_i ($i \in I$) satisfy $A_i \cap A_j = \emptyset$ for $i \neq j$ then $\mu(\cup_i A_i) = \Sigma_i \mu(A_i)$.

[2] If it is required to be countably additive then the subsets must form what is called a σ-algebra. That will always be true for applications of probability to quantum theory.

The concept of probability is tied just as closely to degree of belief as to frequency, but it is equally problematic to *define* probability as (actual) degree of belief. How can a person's degrees of belief be determined? Asked to specify a number between 0 and 1 as their degree of belief in some arbitrary proposition, most people would not know how to reply. A degree of belief may sometimes have more clear-cut behavioral manifestations in the odds at which someone is willing to bet on an event. But people are often willing to accept a set of bets which they are guaranteed to lose, or at least cannot win, when taken collectively. In such a case a person's degrees of belief cannot consistently be represented by any single probability measure.

Instead of appealing to a person's *actual* belief state one can seek to impose formal requirements of rationality on a person's preferences among options as expressed either verbally, or by actual choice behavior.[3] One can prove that expressed preferences meeting such formal requirements (of coherence in a precisely defined sense) can always be modeled by assigning numerical degrees of belief in different possible outcomes of choices satisfying the mathematical principles of probability theory. The rationality requirements are formal rather than substantive: one can still meet these requirements with a degree of belief .9 that there will be an earthquake tomorrow and a degree of belief .01 that it will rain, though not if one is also quite certain that both will happen tomorrow. Because degrees of belief abstracted from coherent preferences may vary from person to person, the view that probability just means degree of belief abstracted from coherent preferences is called personalist or subjectivist.[4]

Some have taken the use of probability in quantum theory to exemplify a third view of probability as a physical propensity—a disposition of a physical system to exhibit a certain kind of behavior in specific circumstances, not always but with a definite regular frequency.[5] Radioactive substances seem to exhibit such behavior in a wide variety of circumstances: the probability is 1/2 that a nucleus of tritium will have decayed after a period of time $T_{1/2}$ called its half-life (12.3 years).

The mathematical concept of probability was first applied in the context of games of chance, where one speaks of such things as the probability of getting a double six in a roll of two dice. If the world were deterministic, as suggested by the success of classical physics, then the outcome of a particular throw of the dice would be fixed by the prior physical state of the universe, so only ignorance of that state would excuse our assigning it a probability other than 0 or 1. But if the quantum world

---

[3] Frank Ramsey [1926] and L. J. Savage [1972] were prominent figures advocating such an approach.

[4] This appoach has also been called Bayesian because of its use of a simple mathematical theorem of probability (attributed to Thomas Bayes) to represent learning from experience and so to show how the beliefs of different persons can converge after taking account of common evidence. The personalist view of probability has been applied in quantum theory by a school of self-styled quantum Bayesians (or QBists) (see e.g. Fuchs, Mermin, and Schack [2014]).

[5] "In proposing the propensity interpretation I propose to look upon probability statements as statements about some measure of a property (a physical property, comparable to symmetry or asymmetry) of the whole experimental arrangement; a measure, more precisely, of a virtual frequency...." Karl Popper, ([1967], pp. 32–3).

were indeterministic, then the outcome of a particular throw of the dice would *not* be fixed by the prior physical state of the universe, but only rendered more or less likely. Then the prior condition of (the physically relevant portion of) the universe would provide an objective tendency to produce one outcome rather than another, and our use of probability would track that objective tendency instead of just making the best use of limited information about the physical conditions that determine what actually happens.

We already saw that the mathematics of probability is amenable to quite different interpretations. The philosopher David Lewis isolated two different concepts of probability and investigated the relations between them. He referred to the personalist subjectivist concept of formally rational degree of belief as *credence*, and an apparently more objective concept exemplified by radioactive decay as *chance*. Lewis thought these concepts were connected by a principle that he took to state all that we know about chance. Because of its importance he called it the Principal Principle. Here is how he described an instance of the principle in words:

If a rational believer knew that the chance of [an event *e*] was 50%, then almost no matter what he might or might not know as well, he would believe to degree 50% that [*e*] was going to occur. *Almost* no matter, because if he had reliable news from the future about whether [*e*] would occur, then of course that news would legitimately affect his credence.   (Lewis [1994], pp. 227–8)

The principle connects an agent's credence in *e* at time *t* to the chance of *e* at *t*. It implicitly defines chance by its role in determining rational credence in a way that renders all other information about what has happened leading up to *t* redundant. This appears to fit the motivating example of radioactive decay, in which the only thing that should matter when someone is deciding whether to believe a tritium nucleus will decay over the next 12.3 years is the probability it will decay, namely 1/2. It is natural to suppose that there is something about the physical situation of the nucleus and its surroundings that gives rise to its physical propensity to decay, and this may be called the objective chance of decay.

But is chance (as Popper said) a physical property of the whole experimental arrangement involved in its manifestation? Or when Lewis took his Principal Principle to capture all we know about chance, was this not an admission of ignorance but a claim that the Principle exhausts the content of the concept? For a pragmatist, a concept acquires its content from its use. The Principal Principle highlights the central application of the concept of chance, in guiding the strength of an agent's belief concerning matters the agent is (in general) not in a position to be certain about (certainty would require chance 1). But by itself it does not address the question of how an agent can come to know, or at least have some evidence for, the chance of *e*.

This is where frequencies can help, when they are available. If in doubt as to the chance of an asymmetric polyhedron landing with a particular face uppermost, one can toss it repeatedly and note the relative frequency of that type of event in a long sequence of trials. While this does not *define* the chance of its landing with that face uppermost on the very next trial, it does provide evidence of that chance. More precisely, it justifiably increases one's credence that the chance is close to the observed relative frequency.[6]

Whether or not one has this frequency data, one may be justified in applying a theory that specifies the probability of an event of type E in circumstances C, where e is of type E and the circumstances in which one wishes to know its probability are of type C. The justification for accepting such a theory may itself depend on frequency data collected in different applications of the theory, as well as its successful use in explaining otherwise puzzling phenomena.

But what is it to accept a theory like this? One straightforward answer is: to accept a theory is to believe that it is true. But another answer is more illuminating here: To accept a theory whose applications involve its general probabilities is to treat it as an expert by aligning one's credences to the specific probabilities implied by these general probabilities in each application. More precisely, it is to assign credence p to event e equal to the probability p of a type E to which e belongs in circumstances C, where C, E are the most specific types recognized by the theory that one is in a position to assign to e and its circumstances. This is perfectly compatible with the straightforward answer. To believe a probabilistic theory is true is to believe that it is right about the probabilities of events of various types. There is no difference between believing a theory is right about the probabilities of events and adopting degrees of belief in the occurrence of those events equal to the probabilities the theory prescribes. These are just alternative descriptions of the same cognitive state.

I take this last claim to be the key to how to think about quantum probabilities. A quantum model includes a mathematical object such as a measure over the subspaces of a Hilbert space. When the model is correctly applied, this measure yields probabilities for significant magnitude claims in accordance with the Born rule. These probability statements are objective in each of the three senses distinguished in Chapter 8. When true, such a statement correctly describes the probabilities it concerns. But these probabilities are not physical magnitudes, and statements about them are not magnitude claims. Their constitutive function is not to describe or represent physical reality but to advise a hypothetical agent on how strongly to believe the magnitude claims embedded in them. In physics it is those magnitude claims that have the function of describing or representing reality. But Born probability statements do not float free of physical reality. In the next section we'll see an example of how they may be tested against statistical data supplied by true magnitude claims. Such a test involves seeing how well Born probabilities match those of the probabilistic phenomenon manifested by data collected in actual experiments.

Not every probability statement based on a legitimate application of the Born rule specifies the chance or chances of the events it concerns. Lewis took the chance of an event to render redundant all other information about what happened earlier than the time at which it is specified. The Born rule may be legitimately applied to yield

---

[6] Assuming the unknown chance p is the same on every trial, the chance will be very low that the relative frequency in a long sequence of trials will deviate significantly from p. Applying the Principal Principle, one should assign this a correspondingly low credence.

probability statements concerning events when information about what happened earlier is *not* redundant. We saw an example of this in the last chapter, where Alice recorded her photon's polarization a microsecond after Bob but at that time was unaware of Bob's setting and outcome for his photon in that pair. If Alice applies quantum theory to guide her credences as to Bob's outcome, conditional on each of Bob's possible settings, she is not aligning her credences to the *chance* of that outcome, which is then either 0 or 1, depending on what Bob has in fact recorded.

From the pragmatist perspective, the chance of an event represents the ideal probability to which only an optimally informed agent could hope to align his/her credences. Probabilities that fall short of this ideal still have the same function of guiding credence. Chance is not to be distinguished physically or metaphysically by its position at the limit of a spectrum of probabilities that may be correctly assigned to an individual event. But it is important to stress that acceptance of quantum theory requires one often to place that limit elsewhere than at probability 0 or 1.

We can use the probabilistic phenomenon of Rutherford scattering to illustrate these ideas about probability and chance. The data provided by experiments like those of Geiger and Marsden manifest this general phenomenon while providing evidence for the four general quantitative facts about it specified in the previous section. These are all facts about how probabilities depend on the values of physical magnitudes. If Rutherford's classical explanation of the phenomenon were correct, none of these probabilities would yield the *chance* of an individual α-particle's being detected in a particular small range of angles of deflection: in each case that chance would be 0 or 1, depending on its actual initial trajectory and the actual locations of the gold nuclei (among other things). The situation is different according to the quantum explanation. One who accepts an explanation based on application of the Born rule to a quantum state like (9.1) takes this to give the chance of any individual α-particle's being detected in each small range of angles of deflection, and none of these chances would equal 0 or 1. In that sense quantum theory portrays this phenomenon as not merely probabilistic but indeterministic.

9.2 Explanation

To appreciate what is distinctive about explanations using quantum theory, it may help to have in mind a famous case in which regularities were explained using classical physics—Newton's explanation of Kepler's laws of planetary motion. The regular behavior displayed by planets as they orbit the sun is a phenomenon, and what we call Kepler's laws are facts about it. The explanation of this phenomenon is often presented as a deduction of Kepler's laws from Newton's, in conformity to the influential deductive-nomological (DN) model of scientific explanation proposed by Hempel in the middle of the last century (see his [1965]). Hempel's account of scientific explanation was a natural development of the Carnapian syntactic conception of the structure of a scientific theory I briefly discussed in Chapter 8: indeed, Carnap [1966] stressed the importance of laws in providing scientific explanations. A more adequate semantic conception takes this explanation rather to involve incorporating Keplerian models of planetary motion into Newtonian models. This makes no explicit mention of laws. But it naturally generalizes to permit incorporation of the same type of Keplerian model into the same type of Newtonian model, without regard to any particular physical systems to which these may be applied. The generalization could be applied to any undiscovered planets, either in our solar system or orbiting a distant star.

A state-space model of a Keplerian planet p specifies how a vector from the sun to the planet changes in length and direction throughout its orbit (assuming the sun remains stationary and both sun and planet are idealized as point objects. See the box in Chapter 8 for details). This vector traces out a periodic elliptical orbit with period T_p, average planet–sun distance a_p, and eccentricity e_p in such a way that it sweeps out equal areas in equal times. State-space models of different planets are related so that

$$\frac{T_p^2}{T_q^2} = \frac{a_p^3}{a_q^3}.$$

A state-space model of a (similarly idealized) Newtonian planet also includes its mass m_p and angular momentum L_p as well as the gravitational force on the planet due to the sun F_p. Each Keplerian model may be incorporated into a corresponding Newtonian model provided that T_p, a_p, e_p are appropriately related to m_p, L_p, and F_p: that is how Newtonian physics explains Kepler's first and second "laws". Moreover, neglecting interplanetary gravitational forces, Keplerian models of all planets together may be incorporated in this way into a *single* Newtonian model: that is how Newtonian physics explains Kepler's third "law". Box 9.1 shows how.

Box 9.1 Kepler's Laws Explained by Incorporation into Newtonian Models

Recall the theory of Kepler's "laws" on the semantic approach outlined in the box in Chapter 8. A model of the theory takes the form $\Theta = \langle a, e, T, r, \theta, t \rangle$ where e is a real number in the interval $[0, 1]$ and a, T are positive real numbers. It is a model of this Keplerian theory K if and only if two conditions hold:

$$r = \frac{a(1 - e^2)}{1 - e \cos \theta} \tag{9.2}$$

$$\tfrac{1}{2} r^2 \, d\theta/dt = \frac{\pi (1 - e^2)^{\tfrac{1}{2}} a^2}{T}.$$

We saw that these models can be used to state theoretical hypotheses such as

H_p: The orbit of planet p is faithfully represented by a model of K where a represents the semi-major axis of an ellipse of eccentricity e with the sun at one focus and T represents the time it takes to go once around this ellipse.

(continued)

Box 9.1 Continued

H_{All}: The orbit of every planet p is faithfully represented by a model of K where a_p represents the semi-major axis of an ellipse of eccentricity e_p with the sun at one focus and T_p represents the time it takes to go once around this ellipse.

H_3: If the orbits of two planets p, q are faithfully represented by models Θ_p, Θ_q then $\frac{T_p^2}{a_p^3} = \frac{T_q^2}{a_q^3}$.

Here H_{All} expresses the content of Kepler's first and second "laws", while H_3 expresses Kepler's third "law".

A Newtonian model of the motion of a single planet has the form of a structure $\Phi = \langle a^*, e^*, r^*, \theta^*, t^*, m, \mathbf{L}, \mathbf{F} \rangle$ whose first five elements are of the same mathematical form as $\langle a, e, r, \theta, t \rangle$; m is a positive number, $\mathbf{L}$ is a vector in three-dimensional space, and $\mathbf{F}(r^*, \theta^*)$ is a function into such vectors. A structure of this form is a model of Newtonian physics if and only if

$$\mathbf{F} = m\frac{d^2\mathbf{r}^*}{dt^{*2}}$$

$$\mathbf{F} = -G\frac{mM}{r^{*2}}\hat{\mathbf{r}}^*$$

$$\mathbf{L} = m(\mathbf{r}^* \times \frac{d\mathbf{r}^*}{dt^*}) \text{ is constant,}$$

where G is the gravitational constant and M is a positive number ($M \gg m$) used to represent the mass of the sun when the model is applied: $\mathbf{r}^* = r^*\hat{\mathbf{r}}^*$ is a vector from a fixed point o (representing the location of the sun in this application) to the point (r^*, θ^*).

To say that a model $\Theta = \langle a, e, T, r, \theta, t \rangle$ may be incorporated into a model $\Phi = \langle a^*, e^*, r^*, \theta^*, t^*, m, \mathbf{L}, \mathbf{F} \rangle$ is to say that there is a certain *embedding* of Θ into Φ: that is, a function f mapping elements $\langle a, e, r, \theta, t \rangle$ of Θ into corresponding elements of the substructure $\langle a^*, e^*, r^*, \theta^*, t^* \rangle$ of Φ such that

$$\langle a^*, e^* \rangle = \langle a, e \rangle$$

$\langle a^*, e^*, r^*, \theta^*, t^* \rangle$ satisfy equations (9.2) with the obvious substitutions

$\mathbf{L}$ is perpendicular to the plane of the ellipse (r^*, θ^*), where

$$\frac{|\mathbf{L}|}{m} = \frac{2\pi a^{*2}(1 - e^{*2})^{1/2}}{T^*}$$

$$\frac{\mathbf{F}}{m} = -\frac{4\pi^2 a^{*3}}{T^{*2}r^{*2}}\hat{\mathbf{r}}^*$$

where $\hat{\mathbf{r}}^*$ is a unit vector. The existence of such an embedding means that Θ is isomorphic to a substructure of Φ.

We can now express the fact that Newton's theory explains Kepler's laws by saying that every model of K may be embedded in this way as a substructure of a model of Newton's theory, and that there is such an embedding of the models Θ_p of K for *all* planets p at once into a single extended model of Newton's theory of the form $\langle a_1^*, e_1^*, r_1^*, \theta_1^*, t_1^*, m_1, L_1, F_1; a_2^*, e_2^*, r_2^*, \theta_2^*, t_2^*, m_2, L_2, F_2; \ldots \rangle$ with

$$\frac{a_i^{*3}}{T_i^{*2}} = \frac{a_j^{*3}}{T_j^{*2}} = \frac{GM}{4\pi^2} \text{ for all } i, j.$$

This extended model may now be used to formulate a theoretical hypothesis expressing the claim that the planets move on Keplerian ellipses in accordance with Newton's laws of motion, each acted on by the same inverse square gravitational force directed towards the sun.

This shows the sense in which Kepler's "laws" were to be expected. They were to be expected by anyone knowing the structure of models of Newton's theory of mechanics and gravitation; the material nature, approximate relative masses, and momentary states of the sun and planets; and the absence of non-gravitational planetary forces in the solar system. Note that no laws were explicitly involved in these explanations, either in presenting Keplerian or Newtonian models, in statements applying them to the planets, or in showing how to locate Keplerian models within Newtonian models. Nor did the explanations involve any causal statements. But the incorporation exhibits the dependence of Kepler's "laws" on Newton's theory. This dependence is asymmetric insofar as Keplerian models can be incorporated into Newtonian models but not vice versa.

Woodward ([2003], p. 219) has remarked on a common feature of causal and non-causal explanations:

[T]he common element in many forms of explanation, both causal and non-causal, is that they must answer what-if-things-had-been-different questions [*w*-questions].

An answer to a *w*-question about a phenomenon says what it (counterfactually) depends on, and how. When a phenomenon is explained by incorporating a representation of it into a theoretical model of a particular type, the theory also makes available many alternative models that may similarly be applied to answer a variety of *w*-questions. Applications of very similar Newtonian models show that the Newtonian orbit of a planet that moved much faster would not approximate an ellipse but a parabola or hyperbola, like a comet. In this way the explanation shows how Kepler's "laws" depend on the particular circumstances obtaining in the solar system. But Newtonian mechanics also makes available models including additional force terms. Used to represent purely hypothetical non-gravitational influences on planetary motion, these enable the theory to say how their orbits would have deviated from Kepler's "laws" in the presence of additional frictional or magnetic forces.

While Newton's theory of gravitation supplied only models representing gravity as an inverse square central force, Newtonian mechanics has a larger class of models representing how planets would have moved if gravitational force had not attracted massive particles along a line joining them, or had varied with the inverse cube of their separation. While strictly outside the bounds of Newtonian theory, one can even contemplate its extension to include models representing force as proportional to the square of acceleration, or space as four instead of three dimensional. Models of this extended theory could be used to represent planetary motions even in such radically counterfactual circumstances. Answering *w*-questions is a way of expressing confidence not only in a theory but also that the right model has been chosen in a particular application, while exhibiting the flexibility to modify the model (or even the theory) if it should turn out to be inapplicable.

It is Newtonian models' extra structure (compared to Keplerian models, say) that enables them to incorporate models of a wide variety of phenomena (including Rutherford scattering and the oscillations of a body suspended from a spring) and so to be used in explanations of all those phenomena. This is how Newtonian theory unifies a host of otherwise unrelated phenomena. A Newtonian model aids understanding by allowing us to think of features of the solar system, including the masses and motions of its constituents and the forces between them, in terms of the individual element of the model by which each is represented. This at least permits, and perhaps encourages, a realist attitude toward physical entities or magnitudes represented by that extra structure in the theory's models, such as masses and gravitational forces. For although representation does not require the reality of that which is represented, it at least provides a launching pad for an inference to the existence of what is represented in the best explanation of the phenomena.

An explanation is causal only if at least some answers formulated as counterfactuals expressing dependency relations may be understood causally. Now a phenomenon like Rutherford scattering is not a singular event but a general regularity, whose instances are reliably reproducible in a wide variety of different circumstances. The recent discovery of thousands of exoplanets (by the suggestively named satellite Kepler!) has indicated how widely instances of orbital planetary motion are reproduced in our galaxy. One way to explain a general phenomenon is separately to explain each of its actual instances, and to generalize by abstracting from these individual instances. This approach is characteristic of a causal explanation of a phenomenon that locates events of certain types in its instances, traces the particular causal connections among them, and infers corresponding general causal relations among these event types. But this is not the only way to explain a phenomenon.

When one explains by incorporating a model of a phenomenon into those supplied by a more general theory, generality is achieved when the phenomenon is initially modeled. A model of the phenomenon may be used to represent any similar instance, actual or merely hypothetical. Such an explanation is not causal. The model of the explanatory theory does not cause the model of the phenomenon: in physics, the models and their elements are typically mathematical and so lack causal powers. Nor are causal relations revealed when we turn our attention to the use of models in explanation. A fact about the phenomenon explained by alluding to a theoretical model is not just a fact about its actual instances: it concerns the general phenomenon itself, abstracted from its actual instances. But causal explanations, whether particular or general, concern only actual instances of a phenomenon.

Quantum theory helps us to see that many otherwise puzzling phenomena were to be expected, and enables us to say what each phenomenon depends on and so to answer a variety of w-questions about it. This will exhibit novel senses of expectation and dependence as compared with theoretical explanations provided by classical physics.

In Newton's explanation of Kepler's "laws", Rutherford's explanation of α-particle deflection by thin sheets of metal, or other examples of theoretical explanation using classical physics, one begins with a statement describing or otherwise representing the phenomenon to be explained. The phenomenon is explained by reference to one or more models provided by the explanatory theory. Such models not only enable one to *re*present the phenomenon by incorporating its representation into a model of the explanatory theory, but introduce additional theoretical elements that may be taken to represent novel aspects of the world underlying the phenomenon. That is how Newton explained Kepler's "laws" by appeal to theoretical models that included not only mathematical representations of the planetary trajectories whose features were to be explained, but also mass and force terms representing additional basic magnitudes. It is also how Rutherford explained the phenomenon of α-particle deflection by thin sheets of metal.

To use a physical theory, whether classical or quantum, to explain a phenomenon, one must describe or otherwise represent systems that manifest the phenomenon as well as the circumstances in which it obtains; but that representation need not be novel to the theory one uses to explain it. Indeed, to be acknowledged as available for potential explanation, the phenomenon must be specifiable independently of the theory that is to be used to explain it.

This original representation may be provided by some other theory, or in a language or representational system not associated with any recognized theory. It is common in physics to 'nest' successive representations of the phenomenon to be explained, sometimes starting with a rough description such as 'high-temperature superconductivity in iron pnictides', then modeling the physical systems involved within the framework provided by successive theories in progressively more abstract and idealized terms until one arrives at a representation to which a theory can be applied to (try to) explain the phenomenon. In the case of classical physics, the explanatory theory can then be used to provide an explanation by incorporating a representation of the phenomenon into its models. Detailed features of entities figuring in the phenomenon are then denoted by corresponding elements of a theoretical model. This theoretical representation will often improve the initial specification of the phenomenon. It may correct this, or enrich it by bringing out features that show the phenomenon's relation to other phenomena.

So representation plays a twofold role in theoretical explanations within classical physics. Its primary role is realized by elements used to represent the phenomenon to be explained. But theoretical models typically also include elements that (at least purport to) represent novel physical entities or magnitudes. The theoretical elements introduced in this secondary representational role then figure essentially in showing how a model of the *explanandum* phenomenon may be incorporated into a theoretical model. As we shall see, novel elements in quantum theoretic models do not play this same secondary representational role.

The first step in using quantum theory to help explain a phenomenon is to say what the phenomenon is. The phenomenon may initially be described in ordinary or scientific language, perhaps employing some mathematical representation. To bring the resources of quantum theory to bear one must represent it as involving properties of physical systems of a specific type or types in their environment. The phenomenon to be explained must now be redescribed by statements about the probabilities of magnitude claims or statements whose truth is determined by them, while the circumstances in which it obtains may be described in other non-quantum terms. Cartwright [1983] called this first stage of theory 'entry', giving a prepared description of the *explanandum* phenomenon and the conditions under which it occurs. The prepared description of a statistical regularity is by means of a probabilistic model, making it what I called a probabilistic phenomenon. But note that at this stage we have not yet entered the domain of quantum theory, as the prepared description is not of quantum states or Hilbert space measures, and includes no talk of operators.

When the mathematical modeling apparatus provided by quantum theory is now deployed, the prepared description of the phenomenon is not shown to be a deterministic or stochastic consequence of dynamical laws or principles of quantum theory. Explaining the *explanandum* phenomenon involves using the Born rule to derive the probabilistic statements in its prepared description. But no quantum model includes magnitude claims or their probabilities: these figure only in an application of a quantum model. This makes the explanatory application of quantum theory a more indirect business than simply incorporating a representation of the *explanandum* phenomenon into its models.

The Hilbert space measure in a quantum model can be applied to generate Born probabilities, and this use sets up a natural correspondence between the measure it assigns to a subspace and the Born probability of each magnitude claim associated with that subspace.[7] But only certain values of this Hilbert space measure may then be taken to represent a probability, and there is nothing in the model that specifies which may and which may not. Born probabilities are not novel magnitudes introduced by models

---

[7] Associated with subspace $\mathcal{M}$ are all magnitude claims of the form $\mathbf{A} \in \Delta$, where the operator that projects onto $\mathcal{M}$ is an element $\hat{P}^A(\Delta)$ from the spectral measure of the operator $\hat{A}$ corresponding to magnitude A.

of quantum theory, as models of classical mechanics introduced novel magnitudes of mass and force.

For a Hilbert space measure to incorporate a probability distribution over a set of statements concerning magnitude A about a physical system s, several conditions would have to be satisfied:

(1) Each statement of the form $A_s \in \Delta$ is in fact either true or false.
(2) $A_s \in \Delta$ is true if and only if $A_s \notin \mathbb{R} - \Delta$ is false.
(3) For any countable collection of disjoint sets $\{\Delta_i\}$ $A_s \in \bigcup_i \Delta_i$ is true if and only if $A_s \in \Delta_i$ is true, for some Δ_i.

Now there is a set of magnitudes modeled by operators in a Hilbert space of dimension greater than two for which these conditions cannot all be satisfied at once.[8] So for hardly any physical system is it possible consistently to incorporate all Born probability statements at once into any quantum model.

This formal impossibility is reflected in the restrictions on the significance of magnitude claims. The prepared description of the *explanandum* phenomenon imposes restrictions on the significance of magnitude claims about physical systems that display the phenomenon. Since the Born rule may be legitimately applied only to significant magnitude claims, it would be illegitimate to try to incorporate probabilities of insignificant magnitude claims into a quantum model. It is the statements in the prepared description of the *explanandum* phenomenon describing the conditions under which it occurs that determine which magnitude claims may be assigned Born probabilities here. The Born rule is not a law of quantum theory: it constrains neither mathematical elements in quantum models nor any physical magnitudes these might have represented. The Born rule governs *applications* of quantum models, and so emerges piecemeal in its legitimate applications.

Finally, Born probabilities are not novel quantum magnitudes insofar as applications of quantum theory are to phenomena whose description is already probabilistic. I noted that an *explanandum* phenomenon must be specifiable independently of the theory that is to be used to explain it. When quantum theory is applied to explain a phenomenon, the prepared description of that phenomenon must already involve probability statements. Quantum theory is used to explain *probabilistic* phenomena. We saw that this is true of Rutherford scattering of α-particles by the nucleus: it is also true of the phenomenon of spontaneous emission of radiation by atoms. In each case, the description of the phenomenon was given in probabilistic terms before modern quantum theory was even formulated, let alone used to explain it.

We use quantum theory to explain a phenomenon by showing how the probabilistic statements included in its prepared description are a consequence of legitimate applications of quantum models. The *explanandum* phenomenon is revealed as just

---

[8] Appendix B describes one such set formed from polarization components of three photons modeled in a Hilbert space of dimension 8. Proofs of this kind go back to Gleason [1957] and have been extended to more complex forms of quantum theory than those considered here.

one aspect of a broader pattern of probabilistic phenomena. But this involves no reinterpretation of the kind of probability involved. All these probabilities are objective, and they all function to provide good advice to an agent about how to set credences. If we set credences in accordance with a legitimate application of the Born rule, we are led to expect and can come to understand the regular patterns displayed by statistics of events that figure in *explanandum* phenomena. One who accepts quantum theory commits to setting credences in accord with legitimate applications of its models, just as one who accepts any probabilistic hypothesis commits to its recommended credences. Using the Born rule to explain a phenomenon by showing how it fits into a much broader pattern of phenomena may justifiably increase one's confidence when setting credences in accordance with a probabilistic statement about it. But this doesn't change the physical or metaphysical status of this probability.

Some people think that quantum theory is radical because it shows the world is "chancy"—that, rather than determining what must follow from what has gone before, its laws govern an ultimately stochastic process that gives rise to just one of many possible futures.[9] But the Born rule does not govern a stochastic process or represent its transition probabilities. The probabilities it delivers are not novel physical magnitudes introduced by quantum theory to characterize stochastic evolution as classical physics introduced mass and electric charge when characterizing supposedly deterministic evolution: they are the same kind of probabilities that figure in the description of the probabilistic phenomena they can be used to explain and predict. Though objective and subject to statistical tests (including some I'll soon describe), the objectivity of Born probabilities derives not from some novel representational role but from the norms associated with their assignment. Quantum theory tells an agent that an *explanandum* phenomenon was to be expected, since the probabilities specified in its prepared description are just those yielded by legitimate applications of the Born rule. It tells her what probabilistic phenomenon she should have expected in a variety of alternative conditions. It also helps her to see what the phenomenon depends on and how.

It is by exhibiting such dependence that the explanation goes beyond showing that the *explanandum* phenomenon was to be expected, by codifying the objective basis for this expectation and so saying *why* it was to be expected. Born probabilities derive from a legitimate application of the Born rule to a quantum state. Chapter 5 section 3 explained how the objectively correct quantum state assignment is backed by true significant magnitude claims. This will help to show how application of quantum theory indicates what the *explanandum* phenomenon objectively depends on, thereby showing why it was to be expected. I'll illustrate this account of the explanatory use of

---

[9] A process that has often been taken to be represented by the physical collapse of the quantum state. As we saw in Chapter 6, von Neumann even named this "process 1". But state vectors or other mathematical objects in quantum models are not used to represent novel physical magnitudes: their function is not descriptive but prescriptive. So reassignment of a quantum state to take account of additional information reflects no physical process.

quantum theory by reference to examples of interference rather like that exhibited in the two-slit experiment described in Chapter 2.

Groups working with Markus Arndt have performed a number of important multiple-slit, single-particle interference experiments with large molecules. In one recent experiment, Juffmann et al. [2009] prepared a beam of fullerene molecules with well-defined velocity, passed them through a multiple-slit interferometer in a high vacuum, and collected them on a carefully prepared silicon surface. They then moved the silicon about a meter into a second high-vacuum chamber and scanned the surface with a scanning tunneling electron microscope (STM) capable of imaging individual atoms on the surface of the silicon. After running the microscope over a square area of approximately 2 mm^2, they were able to produce an image of some one to two thousand fullerene molecules forming an interference pattern. They reported that the surface binding of the fullerenes was so strong that they could not observe any clustering of the molecules, even over two weeks. Clearly, they felt no compunction in attributing very well-defined, stable positions to the molecules on the silicon surface and even recommended developing this experiment into a technique for controlled deposition for nanotechnological applications.

Quantum theory helps us explain the probabilistic phenomenon manifested by the pattern in which the molecules were deposited on the silicon surface in this experiment. In rough outline, here is the structure of the explanation. Conditions in the oven, velocity selector, and experimental geometry warrant assignment to the incident molecules of a particular initial quantum state (as a first approximation, a plane wave function like (9.1)). The geometry of the interferometer supplies boundary conditions on this wave function that enable one to use the Schrödinger equation to calculate its values at the silicon surface. Applying the Born rule to this wave function at the silicon surface yields a probability density over magnitude claims, each attributing a different position on the surface to a fullerene molecule. This probability distribution features characteristic interference fringes of calculable relative magnitudes whose maxima and minima are spatially separated by calculable amounts. The relative frequencies of molecules in each region of the screen clearly manifest exactly these fringes in this instance of the probabilistic phenomenon. The regular pattern of fullerene molecules deposited on the silicon surface—a pattern that is revealed in the subsequent image produced by the STM—is thereby explained as an instance of this probabilistic phenomenon.

By applying a model of quantum theory here we are able to see what the probabilistic phenomenon manifested by the pattern depends on and to answer many w-questions about it. In a variant of this phenomenon with a different position of the slits or of the silicon surface the probability distribution would have featured fringes with different separations, in accordance with application of the Born rule to the modified quantum state associated with the different geometry. If the quantum state assigned to the fullerenes had been a mixture rather than a superposition of pure states, each corresponding to passage through a single slit, the probability distribution would have featured no interference. This could be checked by recording the pattern displayed under varied experimental conditions. Such variations in the set-up would also be described in non-quantum terms, for example by claiming that gas pressure in the interferometer was raised significantly higher than 10 mbar, or that the slits were subjected to intense illumination during the experiment. The phenomenon depends also on the mass of the fullerene molecules; by applying a quantum model here we can say how it would differ if they were replaced with uniformly heavier fullerenes, or with a mixture of fullerenes with different masses.

Notice that although the Born rule played a critical role in this explanation, there was no mention of measurement or observation, at least not until the (assumed) final stage at which someone looked at the STM screen on which the image of the deposited molecules was displayed. The primary *explanandum* is not that person's subjective visual experience, but the probabilistic phenomenon manifested by the objective pattern in which the fullerene molecules were deposited on the silicon surface. Although the deposition pattern is clearly unobservable by unaided human senses, any agent aware that a reliably working STM has produced this image is both licensed and warranted in claiming it is an image of an objective pattern formed in the experiment. The license comes from the massive environmental decoherence of the molecules' wave function produced by binding to the silicon surface. The warrant comes from the (assumed) fact that the STM is working well and has been reliably operated.

Notice also that the pattern in which the molecules were deposited on the silicon surface in this experiment was not explained causally but as an instance manifesting a probabilistic phenomenon. The *explanandum* facts about that phenomenon concerned how its probability distribution depended on the values of magnitudes such as the mass and velocity of the fullerenes and the dimensions of the interferometer. These magnitudes cannot be said to cause that probability distribution, and nor can the Born probabilities that explain it. Since probabilities are neither physical events nor aspects of physical events they cannot enter into causal relations. But one can certainly cause an *event* by raising its probability, as we'll see in the next chapter. So it is appropriate to ask whether quantum theory may be used to give a causal explanation of the events that give rise to an instance manifesting a probabilistic phenomenon, such as the incidence of individual fullerenes at particular locations on the silicon surface in the experiment of Juffmann et al. [2009].

The answer to that question depends on what one requires of a causal explanation. To see what is at stake here it will be helpful to return to the example of Rutherford scattering. By recasting Rutherford's account of this phenomenon as an exemplary causal explanation one can see how far quantum theory falls short of this causal ideal.

Rutherford initially appealed to probability in his explanation of α-particle deflection by thin sheets of metal, so he thought of this as a probabilistic phenomenon. But he went on to talk of numbers of collisions, and fractions of particles deflected through different angles. This suggests recasting his explanation in terms of relative frequencies rather than probabilities. So recast it offers an explanation of the dependence (on ϕ, d, p, and the nuclear charge) of the relative frequency with which α-particles are deflected through an angle close to ϕ in an actual episode of α-particle deflection by thin sheets of metal. The explanation would then rely on the assumption that the frequency with which α-particles come within a distance r of a metal atom's nucleus is in fact proportional to $\pi r^2 nd$. This may be a reasonable assumption, but it is not a consequence of classical mechanics and electrostatics.

What one can explain using just classical mechanics and electrostatics is the entire trajectory followed by an individual α-particle with initial momentum p directed toward a point at distance r from the center of a single metal atom's nucleus with given mass and nuclear charge. One can think of its encounter with the nucleus with this initial momentum as the cause of the particle's subsequent trajectory, with the effect that its final momentum corresponds to deflection through angle ϕ. In that case each individual α-particle trajectory constitutes a continuous causal process from initial emission to final deflection: and an entire episode of α-particle deflection by thin sheets of metal is simply the aggregate of these individual causal processes.

If Rutherford's classical explanation is recast in this way it makes no reference to probability, and it cannot explain Rutherford scattering as a probabilistic phenomenon that extends beyond its actual instances. But if its essential statistical assumption is true it can offer a satisfyingly causal explanation of each actual episode of Rutherford scattering. Can an explanation of Rutherford scattering using quantum theory be so recast?

The interaction of a single α-particle with the nucleus of a metal atom may be modeled in different ways by quantum theory. The simplest model in ordinary (non-relativistic) quantum mechanics represents the nucleus as located at a point that serves as the center of an electric potential represented in the Hamiltonian entering into the Schrödinger equation for a wave function assigned to the α-particle (representing its quantum state). At any time t when the α-particle is assumed to be still distant but headed toward the nucleus with momentum p and energy E, the wave function (9.1) may be assigned in the simplest model. This assignment is backed by non-quantum conditions describing the source of the particle (say, as a known decay product of a specific radioactive element located far away from the target nucleus). These backing conditions also warrant ascription of momentum p and energy E to the α-particle at such an early time t, and its having momentum p and energy E then may be considered a cause of its detection at the screen.

But the wave function assigned to the α-particle does not represent it as having a position at any time prior to detection: the absence of decoherence precludes significant claims about its position. The wave function for later times becomes a superposition of state (9.1) with outgoing spherical waves centered on the nucleus: the absence of decoherence now precludes significant claims about its momentum as well as position. By using quantum mechanics to model the interaction of an α-particle with the nucleus of a metal atom in this way, one forswears saying anything significant about its trajectory prior to detection. Quantum theory cannot be used to represent the interaction of a single α-particle with the nucleus of a metal atom as a continuous causal process.

One could avoid this conclusion if the solution to the Schrödinger equation as it evolves from initial wave function (9.1) itself represented such a process. But it is not the function of a quantum state to represent the behavior of some novel physical entity or magnitude. Von Neumann called Schrödinger evolution "process 2": but this is no more a physical process than his process 1—the notorious collapse of the wave function on measurement. Schrödinger himself at first tried to understand his eponymous equation as describing the behavior of a new physical field. But if quantum states function the way I explained in Part I, the wave function does not describe the world, but rather offers good advice on how to describe the world and what to believe about it as so described.

Even if one cannot use quantum theory to represent as a continuous causal process what goes on between initial cause (an α-particle's having momentum p and energy E) and final effect (detection of an α-particle at a particular location on the screen), there are good reasons to consider these events to be causally related. Bringing a radioactive source of α-particles up to a zinc sulphide screen is a reliable way to produce flashes on the screen! I'll say more about such reasons in the next chapter.

What Rutherford scattering teaches about causal explanation is readily applied to the experiment of Juffmann et al. [2009]. Any explanation of either as a probabilistic phenomenon proceeds at a general level, abstracted from whatever actual instances of the phenomenon there happen to be. The quantum explanation involves assignment of an appropriate initial quantum state and calculation of a final state using the Schrödinger equation. This is useful because a legitimate application of the Born rule to this final state implies a probability distribution. The explanation shows the phenomenon was to be expected insofar as this Born probability distribution matches the distribution characterizing the probabilistic phenomenon. It shows what this distribution depends on, notably including the quantum state and its backing conditions. That dependence is counterfactual, but not causal.

Moreover, one can causally explain key events in an instance of the phenomenon as effects of others. An individual flash on a zinc sulphide screen is caused by production of an α-particle in the decay of a radioactive atom; deposition of a fullerene molecule on the silicon surface is caused by emission of a fullerene molecule from a heated oven. By making statistical asssumptions about many similar types of cause events one can explain the statistical pattern in many similar types of effect events in an episode manifesting the probabilistic phenomenon. But this causal explanation will make no use of quantum models, and so is not really a quantum explanation. To get a quantum explanation of this statistical pattern one must advert to it as a concrete instance manifesting the general probabilistic phenomenon that quantum theory does help one to explain—but not causally.

So far I have ignored a complication involved in the use of quantum theory to explain the phenomenon of Rutherford scattering. What is explained is not exactly the probability distribution for an α-particle to be deflected through an angle close to ϕ: rather, it is the probability distribution for an α-particle to be *detected* at a location on the screen making an angle close to ϕ with the direction of its initial momentum. The distinction is important, since we cannot assume that the α-particle is ever *at* that location. One could try to use a model of decoherence to justify attributing it a determinate location at the screen, but the attempt may fail. Certainly, it would not be so straightforward as the application of a model of decoherence to justify attribution of a determinate location to the fullerenes adhered to the silicon screen in the experiment of Juffmann et al. [2009]. Perhaps decoherence would only license attribution of a determinate position to a grain of zinc sulphide that emits light after interacting with a *non-localized* α-particle?

Many probabilistic interference phenomena involving beams of light may be explained analogously to the explanation of the probabilistic phenomenon manifested in Juffmann's experiment with fullerenes. But here it is clear that a quantum model cannot be applied directly to yield Born probabilities for photon positions since decoherence does not license significant magnitude claims about photon positions at a detector. Recall this passage quoted in footnote 6 to Chapter 3:

we should not talk about a wave propagating through the double-slit setup or through a Mach–Zehnder interferometer; the quantum state is simply a tool to calculate probabilities. Probabilities of the photon being somewhere? No, we should be even more cautious and talk only about probabilities of a photon detector firing if it is placed somewhere.

(Zeilinger et al. [2005])

In applying a quantum model to explain two-slit and other interference patterns involving severely attenuated light beams, the Born rule can be legitimately applied to magnitude claims not about photons but about systems in the detector. So in this case the Born rule is legitimately applied to something other than the light beam which is the target of this application of a quantum model. The discrete response of the detector makes it natural to think of the beam as composed of photons. But this application of quantum theory directly yields advice not about photons but about systems in the detector. Photons are not beables of quantum theory: but they do provide a convenient way of speaking of certain quantum assumables.

A standard single-photon detector generates a "click" when interaction with detected light ejects a single electron from a metal surface, leading to a rapid avalanche of other electrons which produces a detectable current. The resulting decoherence renders magnitude claims about the energy of the ejected electron significant, including the true statement that this energy is sufficient to free it from the metal surface. But it may be taken thereby indirectly to justify a claim of the form R concerning a photon detection event:

R: The transverse position of the photon is detected between r and $r + \Delta r$.

R is not a canonical magnitude claim. But it acquires content from its place in a web of such justified inferences, even though this does not include an inference from R to the statement that the transverse position of the detected photon was between r and $r + \Delta r$ when it was detected.[10] So it is legitimate to extend the application of the Born rule to statements of the form R concerning the detection (but not the actual position) of photons, and to take such applications to explain the interference pattern.

This is relevant to another experiment that has been taken to confirm an important general phenomenon: In the interference between more than two paths, the interference terms are the sum of the interference terms associated with these paths taken two at a time. The Born rule provides an explanation of this general probabilistic phenomenon: here is a particular example based on the Born probabilities for the case of three-path interference. Label the paths A, B, C, the probabilities Pr (with path subscript) and the interference terms I (with path subscript). Then

$$\mathrm{Pr}_A = \langle A|A \rangle = ||A||^2, \text{etc.}$$
$$\mathrm{Pr}_{AB} = \langle A+B|A+B \rangle = ||A||^2 + ||B||^2 + I_{AB}, \text{etc.}$$
$$\mathrm{Pr}_{ABC} = \langle A+B+C|A+B+C \rangle = ||A||^2 + ||B||^2 + ||C||^2 + I_{AB} + I_{BC} + I_{CA}$$

Hence (I) $I_{ABC} = I_{AB} + I_{BC} + I_{CA}$.

Sinha et al. [2010] matched experimental statistics against an instance of (I) in a variety of experiments involving the interference of light at up to three slits. One of these involved single photons, detected in coincidence with a second "herald" photon from an entangled pair by avalanche photodiodes. The experimental interference pattern is generated by moving a multimode optical fiber uniformly across a plane intercepting light from the slits and counting the relative number of photons detected in each small region. The recorded statistics conform well to those expected on the basis of three-path instance (I) of the general phenomenon, when one takes this application of the Born rule to yield probabilities for statements of the form R.

This experiment displays regularities at two levels. Each sub-experiment with a particular slit configuration manifests a probabilistic phenomenon that depends both on that configuration and on other fixed elements of the experimental set-up. In addition, the results of the sub-experiments manifest the probabilistic fact (I). This probabilistic fact is implied by the Born rule, together with the assumption that the fixed elements of the experimental arrangement back the same initial quantum state assignment in each sub-experiment.

Assuming the Born rule is a law of nature governing physical systems, one could take fact (I) as a law that nomically depends on the Born rule. In that case one could regard its explanation from the Born rule as fitting Hempel's deductive statistical model of one species of scientific explanation within the genus described by his

---

[10] I'll say more in Chapter 12 about how a claim derives what content it has from the inferences to which it commits one, together with the inferences from which one is entitled to infer the claim.

covering law model.[11] But the Born rule is not a law of quantum theory: it emerges only piecemeal in its applications. When applying models of quantum theory to a variety of different physical systems, one can use different instances of the Born rule to explain probabilistic phenomena each manifests in basically the same way. This exhibits a wider pattern to which they all conform. Here quantum theory helps us to explain not by showing what the probabilistic phenomenon causally depends on, but by exhibiting the pattern which its instances manifest as itself fitting into the still broader pattern of phenomena that quantum theory helps us to explain through use of the Born rule. Here we see a clear example of how the unifying power of quantum theory contributes to its explanatory credentials.

One of the greatest triumphs of quantum theory is the explanation of the stability of ordinary atomic matter. Its inability to account for this most familiar of phenomena was perhaps the most obvious failing of classical physics. Why is the atom stable against collapse of electrons into the nucleus, given the electrical attraction between atomic electrons and nuclear protons? Classical physics predicted just such a collapse, with an accelerating electron rapidly radiating away an unbounded amount of potential energy as it spiralled into the nucleus.

As an exceptionless regularity, the stability of ordinary atomic matter may be considered a special (probability 1) case of a probabilistic phenomenon.[12] Attempts to use quantum theory to explain this phenomenon typically proceed by applying a quantum model to prove that the average, or expectation value $\langle E \rangle$, of the internal energy of an atom has a finite lower bound E_{min}. In the simplest treatment, this expectation value is calculated by applying the Born rule to an arbitrary wave function that is a solution to the Schrödinger equation for the nuclear protons and atomic electrons subject only to their mutual electrical interactions. For the case of the hydrogen atom, the lower bound of 13.6 eV is attained by the ground state wave function. But why does the existence of such a lower bound establish the stability of the atom?

Quantum theory helps one to explain an atom's stability by showing the probability for it to collapse is 0. If this probability were greater than 0 then $\langle E \rangle$ would have no (finite) lower bound. A magnitude claim attributing an internal energy in a certain range to an atom is significant enough to permit application of the Born rule only when the atomic wave function has undergone enough and the right kind of environmental decoherence. But application of the Born rule in such circumstances will always assign probability 1 to the claim that it has a finite internal energy, and so probability 0 for it to collapse with the release of unbounded energy.

This explains the stability of atoms, because it also indicates what this probabilistic phenomenon depends on. It depends counterfactually but not causally on whatever

---

[11] See Hempel [1965].
[12] I follow custom in setting aside instability involving the atomic nucleus, including radioactive decay and spontaneous nuclear fission.

quantum state may be assigned to the atom. Considered as processes, the continued existence of each individual atom depends causally on that of its constituents. Atomic constitution is specified by non-quantum claims about the number, charges, and masses of its constituents. Any application of a model of non-relativistic quantum mechanics to a system so constituted will assign a pure or mixed quantum state that is a solution to the Schrödinger equation, since in this form of quantum theory these are the only quantum states that are taken to be backed by true magnitude claims about its constituents. But a legitimate application of the Born rule to such a state will always yield probability 0 for a magnitude claim assigning the atom an internal energy less than any finite amount.

The assertion that all explanations of physical phenomena using quantum theory involve application of the Born rule may strike some as implausibly broad. For example, the explanation of the discrete energy levels of hydrogen and other atoms was one of quantum theory's earliest and most striking successes. This may seem to be independent of the Born rule, but in fact it is not. A general solution to the Schrödinger equation for an isolated hydrogen atom may be expressed as a superposition of wave functions $\{\varphi_i\}$:

$$\psi = \Sigma_i c_i \varphi_i. \tag{9.3}$$

A legitimate application of the Born rule to each component wave function φ_i in this superposition would yield probability 1 for magnitude claim $E = E_i$ about the energy of the atom, and probability 0 for $E \neq E_i$. Since $\Sigma_i |c_i|^2 = 1$, it follows that a legitimate application of the Born rule to ψ yields probability 0 for a magnitude claim $E \notin \cup_i E_i$ saying that the atom's energy is not equal to any E_i associated with component φ_i in this superposition of ψ. So a legitimate application of the Born rule to *any* state assigned to a hydrogen atom will always yield probability 1 to its energy having a value equal to some E_i. This explains the energy levels of the hydrogen atom as a special (probability 1) case of a probabilistic phenomenon.

10
Causation and Locality

10.1 Introduction

While quantum theory is generally acknowledged to help us explain a huge variety of natural phenomena, there is one set of phenomena that many people believe remains locally inexplicable—that is, they cannot be explained without instantaneous action at a distance. Surprisingly, quantum theory itself drew our attention to these phenomena when Bell [1964] used it to predict them even before they were observed. These are patterns of non-localized correlations like those I first discussed in Chapter 4. That discussion was oversimplified, both physically and philosophically. Removing the oversimplifications will prove to be no mere academic exercise. On the contrary, by carefully examining the way we can use quantum theory to explain the puzzling correlations we can learn a lot about the concepts of chance, causation, and explanation and how they work together in science.

A principle of local causality played a crucial role in Bell's argument that

certain particular correlations, realizable according to quantum mechanics, are locally inexplicable. They cannot be explained, that is to say, without action at a distance. ([2004], p. 152)

His statement of this principle presupposed familiarity with basic but elementary features of spatiotemporal relations in the theory of relativity. So I'll begin this chapter by explaining these features for readers unfamiliar with them. This will also make it easier to show how to adapt Lewis's Principal Principle to apply to chances in relativistic space-time. We'll see how general Born probabilities may then yield more than one chance of the same event at the same time. Causation is related to chance by the intuitive principle that a cause raises the chance of its effect. Applied to Bell's correlations this principle appears to show that a distant outcome may be an immediate cause of an outcome nearby. But the intuitive principle requires qualification if causation is to retain its conceptual ties to possible interventions. So while non-localized sets of essentially simultaneous events do have a common cause, they are not causally related to one another. Since we can use quantum theory to say why they were to be expected and what they depend on, it follows that we can use quantum theory to explain patterns of non-localized correlations, including Bell's particular correlations, with no instantaneous action at a distance.

10.2 Relativistic Spatiotemporal Relations

When Newton formulated his theories of mechanics and gravitation he adopted familiar concepts of space and time, commenting:

I do not define time, space, place, and motion, as being well known to all.

In everyday life we commonly assume that some objects are moving while others are at rest, that two momentary events either occur at the same time or one occurs some definite amount of time earlier than the other, and that any localized events occur a definite distance apart. But our confidence in some absolute distinction between motion and rest is easily shaken after we learn that the Earth that provides our customary standard of rest rotates and orbits a star in a rotating galaxy that will collide with another in about 4 billion years in an expanding universe.

Though he did not know it, Newton's theories can be formulated and applied with no absolute distinction between rest and uniform motion. Without it the distance between two momentary events is not uniquely defined unless they occur simultaneously: to specify how far apart they occur one must appeal to some uniformly moving object (one accelerated by no forces), arbitrarily designated as being at rest. Such an object is said to move inertially, or to determine an inertial frame with respect to which distances as well as time intervals may be referred.

Intervals of Newton's absolute time required no such reference. But the insight that led Einstein to his special theory of relativity was that the only way to take account of a key fact about light *did* require time intervals as well as spatial distances to be referred to an inertial frame. The key fact is the invariance of one particular speed between different inertial frames—the enormous speed at which light travels in a vacuum (3×10^8 meters/second). According to the familiar concepts of space and time adopted by Newton,[1] if a light ray coincides with each of two momentary events e, f that occur at different points of space separated by a distance d and a time interval t in inertial frame F, then they are separated by the same duration t but a *different* distance d' in an inertial frame F' moving with respect to F. If this were so the speed of light referred to F' would have to differ from its speed referred to F, contrary to the invariance of the speed of light between inertial frames. To implement the invariance of the speed of light, not only distances but durations between such events must be required to differ when referred to inertial frames associated with different states of uniform motion.

If a light ray in a vacuum coincides with each of momentary events e, f that occur at different points of space separated by a distance d and a time interval t in inertial frame F, then $d = ct$, where c is the constant speed of light in a vacuum. Since c is such an enormous magnitude, it's often convenient to represent durations and distances on scales in which $c = 1$, such as years and light years or seconds and light seconds. The invariance of the speed of light may then be expressed by the equation

$$t^2 - d^2 = t'^2 - d'^2$$

[1] Even after we have abandoned his absolute distinction between rest and uniform motion.

in which d', t' represent the distance and duration of the interval between e, f in inertial frame F'.

If e_1, e_2 are arbitrary momentary events that occur at different points of space separated by a distance d and a time interval t in inertial frame F, the quantity $I \equiv t^2 - d^2$ is called the *space-time interval* between e_1, e_2. The invariance of the speed of light implies that the space-time interval between any two momentary point-events that coincide with a light ray in a vacuum is always zero, even if those events are separated in space and time. But in the theory of relativity the space-time interval I is required to be invariant for *arbitrary* space-time points p, q whether or not these are possible locations of two momentary point-events that coincide with a light ray in a vacuum.[2] In any inertial frame, the space-time interval I_{pq} between any two points p, q is given by

$$I_{pq} = (t_p - t_q)^2 - (d_p - d_q)^2,$$

where p,q are said to be *time-like, space-like*, or *null separated* depending on whether $I_{pq} > 0, I_{pq} < 0$, or $I_{pq} = 0$.

It is often easiest to represent such relations in space-time diagrams. The use of two-dimensional figures to represent three-dimensional spatial relations is common (see figures 2.1 and 2.7). Representing four space-time dimensions in a two-dimensional figure need be no more difficult, either because only one or two spatial dimensions are singled out for special interest or because what is being represented is sufficiently symmetric. In either case, the fact that no more than three dimensions need to be represented makes it possible to employ conventional diagrammatic techniques. A space-time diagram conventionally represents duration t from an initial time in some inertial frame by a vertical axis, with a horizontal axis representing distance d from a spatial origin along one spatial axis; a second spatial axis may be added if necessary using standard perspective. A point in the diagram represents a space-time point—a possible location for a momentary point-event. A vertical line represents a possible history of a point object at rest in some particular inertial frame, while a horizontal line represents an instant of time in that frame.

Choosing natural units of time and distance in which the speed of light is 1, a straight line inclined at 45° to the horizontal represents the possible history of a light ray in a vacuum. A momentary flash of light emitted at the origin O of such a diagram would expand spatially as a sphere: its *history* could then be represented in the diagram as the top half of a cone with vertex at O and axis vertical. The inevitable omission of a representation of the third spatial dimension doesn't matter because light travels at the same speed in all directions. For obvious reasons, the possible history of light diverging from an emission event or converging to an absorption event at O is called the *light cone*

---

[2] This statement is correct for the Minkowski space-time of the *special* theory of relativity. Other Lorentzian space-times of the *general* theory of relativity have the same light-cone structure only in the tangent space to the manifold. Modifying the present discussion of relativistic spatiotemporal relations to encompass such relativistic space-times would require minor, distracting technical tweaks of the argument in this chapter to reach the same philosophical conclusions.

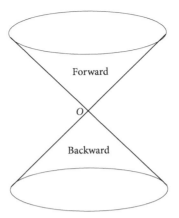

Figure 10.1 The light cone at O

at O (even though this is strictly a three-dimensional *hyper*cone in space-time)—see Figure 10.1. Any history of a massive point object that momentarily coincides with O would be presented by a continuous curve through O, always confined to the interior of this light cone and with a tangent always at less than 45° to the vertical (so its velocity is always less than that of light). When used to represent the history of an idealized physical agent, such a curve is called the agent's *world line*.

Every point p on the light cone is null separated from O ($I_{Op} = 0$) even though it is represented by a point of the diagram at some distance from O. Because of the invariance of the null interval, the light cone at a space-time point is represented in the same way in every inertial frame. The distinction between the forward (part of the) light cone and the backward light cone is also invariant: if a point is represented as being in the forward (backward) light cone in one frame, it will be represented as being in the forward (backward) light cone in every other frame. Points in the light cone at O are time-like separated from O; those in the forward light cone are (invariantly) *later*, while those in the backward light cone are (invariantly) *earlier*. Points outside its light cone are space-like separated from O. A point-event occurring outside the light cone at O has no invariant temporal order relation to O: it is earlier in one inertial frame, simultaneous in another, and later in a third. Relations of temporal order as well as duration between space-like separated points are not invariant, just as relations of spatial order as well as distance between time-like separated points are not invariant.

10.3 Relativistic Chance

Recall Lewis's claim that the Principal Principle connecting it to degree of belief expresses all we know about chance. The basic idea was that at time t a rational agent would set credence in the occurrence of an event e equal to the chance at t of e, in the absence of reliable news from the future about whether e would occur.

To accommodate it to the novel space-time structure of relativity, this idea needs to be modified. In the absence of any absolute temporal structure, what are we to understand by t or news from the future?

We naturally think of an agent as deciding what to believe and how to act at a moment of time. But it is not realistic to suppose that either deliberation, belief formation, or action is literally instantaneous, so that it makes sense to ask at what precise nanosecond (10^{-9} seconds) any of these occurred. The time t is an idealization of the moment at which the Principle is to apply, as is the position $\mathbf{r}$ of the one to whom it is applied. The analogous idealization in relativity is to apply the Principle to a hypothetical agent at a space-time point p.

What are we to understand by the future, reliable knowledge of which would vitiate application of the Principle? The intuition behind Lewis's restriction on its application was that while there are natural means for an agent to acquire knowledge of any past event by consulting records and making observations, there can be no records or observations of the future. This intuition has a foundation in Newtonian physics, which imposes no upper limit on the speed of physical processes: no matter how far away something occurred, there is no natural barrier to finding out about it now through some such process. Acceptance of relativity requires revision of this assumption.

If no physical process propagates faster than the speed of light, an agent at p has no direct epistemic access to events outside his/her backward light cone. So in this context the relativistic analog of the future is the region of space-time outside the backward light cone at p. This includes not only those events in (or on) the future light cone of p, but also events space-like separated from p, even though many of these occurred *earlier* with respect to the inertial frame defined by the agent's state of motion at p.

So in the space-time of relativity, the conceptual connection between chance (Ch) and the degree of belief (Cr) it prescribes is captured in this formulation of the Principle that implicitly defines chance:[3]

The chance of e at p, conditional on any information I_p about the contents of p's past light cone, satisfies: $Cr_p(e/I_p) =_{df} Ch_p(e)$.

Since two agents in the same state of motion but in different places have different "pasts" (i.e. different backward light cones), it follows that they may consistently assign an event different chances at what they take to be the same time, as illustrated in Figure 10.2, where $Ch_p(e) \neq Ch_q(e)$.

---

[3] See Ismael [2008]. I have slightly altered her notation to avoid conflict with my own. Here 'e' ambiguously denotes both an event and the proposition that it occurs. Cr stands for credence: an agent's degree of belief in a proposition, represented on a scale from 0 to 1 and required to conform to the standard axioms of probability.

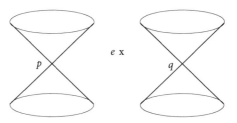

Figure 10.2 The chances of *e*

10.4 Chances from Quantum Probabilities

We saw in the previous chapter that one way to arrive at the chance of an event is to apply the general probabilities of a scientific theory to events of that kind in those circumstances. In a physical situation of a type that backs assignment of a specific quantum state, a licensed application of quantum theory's Born rule supplies general probabilities for certain kinds of events. Some, but not all, such applications yield the *chances* of particular individual events. Chapter 4 described several such applications of quantum theory to a type of situation in which results of distant linear polarization measurements are recorded at locations **1,2** in a brief interval of time T. I can now give a more precise description of that type of situation in the language of relativity theory.

Suppose Alice and Bob are each at rest with respect to an inertial frame in which they carry out their experiments at widely separated locations. Each records the outcomes of linear polarization measurements on a sequence of systems of a type to which each has correctly assigned state $|\eta\rangle$, based on their common knowledge of physical conditions in the overlap of the backward light cones of the recording events. After synchronizing their clocks, Alice and Bob both record the outcome of a linear polarization measurement in the same very short time interval T in their frame. They do this repeatedly, but keep records only of those instances for which both of them have recorded something.[4] In a change of notation, I now let **1** designate a small space-time region including Alice's recording event on a given occasion, while **2** similarly designates Bob's recording region. Assume regions **1,2** are space-like separated.[5] These spatiotemporal relations are depicted in Figure 10.3.

Here t_1, t_2, t_3 are instants of time with respect to Alice's and Bob's inertial frame F and all of them lie within F-interval T. p, q, r are space-time points simultaneous in F while p' is earlier than p, q, r in F. Indeed p' is earlier than q with respect to *every* frame, but since it is space-like separated from each of p, r it bears no invariant temporal order relation to either of those points.

---

[4] It will simplify things to assume from now on that this covers all instances—Alice's and Bob's detectors are perfectly efficient so they each have a record of every system.

[5] Two space-time regions are space-like separated if and only if every space-time point in one is space-like separated from every space-time point in the other.

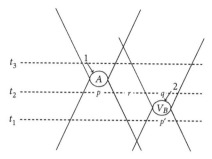

Figure 10.3 Alice's and Bob's space-time diagram

The diagram shows an instance in which Bob's record is of type V_B, indicating vertical polarization with respect to his polarizer axis. Alice's record is labeled by the random variable A, which may take on either value V_A, H_A depending on her recording e_A on a particular occasion (event v_A, h_A respectively). Alice and Bob may each apply quantum theory to arrive at the chances of events v_A, h_A on each occasion. One naturally associates point p with Alice, points q, p' with Bob, and point r with neither. But in fact quantum theory may be applied by anyone to determine the chances of v_A, h_A relative to any of these points. The chances of each of v_A, h_A may well differ when relativized to one space-time point rather than another: but they are not determined by the actual epistemic or physical state of any agent (though we should all strive to assign them correctly and some may succeed in doing so!).

We saw in Chapter 4 that Alice and Bob are licensed to apply the Born rule to yield the probability for Alice's record to be of types V_A, H_A, either unconditionally or conditional on Bob's record being of type V_B. These probabilities are as follows:

$$Pr_a^{|\eta\rangle}(A) = \frac{1}{2} \ [= Pr_b^{|\eta\rangle}(B)]$$

$$Pr_{a,b}^{|\eta\rangle}(V_A|V_B) = \frac{Pr_{a,b}^{|\eta\rangle}(V_A \& V_B)}{Pr_b^{|\eta\rangle}(V_B)} = \frac{1/2 \cos^2 \angle ab}{1/2} = \cos^2 \angle ab$$

Similarly, $Pr_{a,b}^{|\eta\rangle}(H_A|V_B) = \sin^2 \angle ab$.

Assume that the polarizers were set to angles $a°, b°$ at events in the overlap of the backward light cones of 1,2. Then $a, |\eta\rangle$ together provide the most complete specification of the circumstances acknowledged as relevant in assigning a Born probability to an event of type A at p, so $Ch_p(e_A) = Pr_a^{|\eta\rangle}(A) = 1/2$: similarly, $Ch_{p'}(e_B) = Pr_b^{|\eta\rangle}(B) = 1/2$. But in assigning a Born probability to an event of types V_A, H_A at q, a more complete specification of circumstances is relevant, since Bob's recording was made in region 2, which lies in the backward light cone of q.

The additional information this makes accessible at q may be taken into account by conditionalizing the Born probability of an event of either of these two types on Bob's setting b and record V_B, in which case $Ch_q(v_A) = Pr_{a,b}^{|\eta\rangle}(V_A|V_B) = \cos^2 \angle ab$,

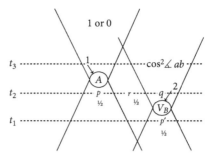

Figure 10.4 The chances of v_A

$Ch_q(h_A) = Pr_{a,b}^{|\eta\rangle}(H_A|V_B) = \sin^2 \angle ab$. Alternatively, it may be accounted for by using the state assignment $|V_B\rangle$ to the remaining system L relative to q in applying the Born rule as follows

$$Pr_a^{|V_B>}(A) = |\langle A|V_B\rangle|^2.$$

This gives the same chance for v_A and for h_A relative to q: but it illustrates the fact that assignment of a quantum state can be a way of specifying the circumstances relevant to assignment of a Born probability by incorporating the record of a measurement outcome accessible from the space-time point relative to which the assignment is made. Figure 10.4 shows the chances of v_A relative to points in different regions of the space-time depicted in Figure 10.3. In the special case where the axes of Alice's and Bob's polarizers are perfectly aligned ($a° = b°$), $\cos^2 \angle ab = 1$.

10.5 Chance, Causation, and Intervention

Causes make things happen by increasing the chance that they will. This is as true of natural processes as it is of intentional actions. In both cases the chance is sometimes increased so much that we discount the possibility that the cause will not have the expected effect: barring a freak intervention, the fall or dive of a heavy object into a pool of water will be followed by a displacement of its surface. But only in a deterministic world could the occurrence of an event be necessitated by a cause corresponding to the world's complete state at an earlier time.

A cause raises the chance of its effect if that chance would have been lower in its absence. So causation and chance are counterfactually related. But c does not cause e in every case of distinct events c, e for which the chance of e would have been lower in the absence of c: c, e may be joint effects of a common cause, or e may cause c. To eliminate such candidates, David Lewis ([1986], [2004]) required that the contemplated counterfactual scenario not involve "back-tracking"—counterfactual variations in c's past.[6] He developed an elaborate set of conditions governing the

---

[6] Except for a very brief "transition period" he took to be necessary to avoid physical discontinuities.

evaluation of the necessary counterfactuals to ensure that their application squared with most of our causal judgments concerning scenarios of the kind we can imagine coming across in everyday life. Such an approach gives no useful guidance in the unfamiliar circumstances Alice and Bob face when trying to determine what causal relations obtain between events involved in an instance manifesting a probabilistic phenomenon that application of quantum theory leads one to expect. This needn't mean neglecting the counterfactual relation between causation and chance. To see what it comes to we should adopt the pragmatist approach and ask what purpose is served by a concept of causation that requires such a relation.

Why do we have our concept of causation? Because, as situated agents, we need to assess the consequences of our alternative decisions. More specifically, a situated agent needs to estimate the chances of different outcomes, conditional on alternative choices. In Huw Price's [2012] terms, "causal dependence should be regarded as an analyst-expert about the conditional credences required by an evidential decision maker". Just as the chance of A tells you everything you need to know to figure out how strongly to believe A, the causal dependence of A on B tells you everything you need to know about A and B when contemplating doing or preventing B (assuming you are not indifferent about A).

Of course not every case in which A causally depends on B provides any actual agent with an option of doing or preventing B: the impact of an asteroid may have caused the extinction of the dinosaurs without being in the power of any agent. It is rather that when applying causal concepts we take the perspective of a *hypothetical* situated agent, able to decide whether, and how, to intervene and alter the course of natural processes. This explains the counterfactual relation between causation and chance. In evaluating a candidate for a relation of causal dependence between e and c, the appropriate counterfactual scenarios to consider are those in which a hypothetical agent opted to bring about or prevent c. If c, e are distinct actual events and the chance of e would have been greater if c rather than something else had been the result of an intervention, then e causally depends on c.

We can now evaluate causal dependence relations between key events involved in the scenarios depicted in Figure 10.4. Besides events in regions 1,2 these include events a, b at which the axes of Alice's and Bob's polarizers (respectively) were set: assume for now these events occurred in the overlap of the backward light cones of 1,2. They also include one or more events $\{o_i\}$ in that overlap truly describable by magnitude claims which together back assignment of state $|\eta\rangle$.

Consider a case in which Alice's and Bob's polarizers were perfectly aligned. If the outcome in region 2 had been of type H_B instead of V_B, then $Ch_q(v_A)$ would have been 0 instead of 1. Suppose that the polarization axis for the measurement in region 2 had been rotated through 60° from this position: then $Ch_q(v_A)$ would have been 1/4 or 3/4, depending on the outcome in region 2. Or suppose that no polarization measurement had been performed in region 2: then $Ch_q(v_A)$ would have been 1/2. This shows that $Ch_q(v_A)$ depends counterfactually on the polarization measurement in 2 and also on its outcome. Don't such counterfactual variations in $Ch_q(v_A)$ amount to *causal* dependence between space-like separated events? There are several reasons why they do not.

The first reason is that while $Ch_q(v_A)$ would be different in each of these counterfactual scenarios, in none of them would $Ch_p(v_A)$ differ from 1/2, so the "local" chance of v_A is insensitive to all such counterfactual variations in what happens in 2. If one wishes to infer causal from counterfactual dependence of "the" chance of a result in 1 on what happens in 2, then at most one of two relevant candidates for "the" chance displays such counterfactual dependence. For those who think of chance as itself a kind of indeterministic cause—a localized physical propensity whose actualization may produce an effect—$Ch_p(v_A)$ seems better qualified for the title of "the" chance of v_A than $Ch_q(v_A)$.

The importance of intervention in specifying the counterfactual relation between causation and chance provides the second reason. Consider the situation of a hypothetical agent Bob at p' deciding whether to act by affecting what happens in 2 to try to get outcome v_A in 1. Bob can choose not to measure anything, or he can choose to measure polarization with respect to any axis b. If he were to measure nothing, $Ch_q(v_A)$ would be 1/2. If he were to measure polarization with respect to the same axis as Alice, then $Ch_q(v_A)$ would be either 0 or 1, with an equal chance (at his momentary location p') of either outcome. Since he can neither know nor affect which of these chances it will be, he must base his decision on his best estimate of $Ch_q(v_A)$ in accordance with Ismael's [2008] Ignorance Principle:

Where you're not sure about the chances, form a mixture of the chances assigned by different theories of chance with weights determined by your relative confidence in those theories.

Following this principle, Bob should assign $Ch_q(v_A)$ the estimated value $1/2.0 + 1/2.1 = 1/2$, and base his decision on that. Since measuring polarization with respect to the same axis as Alice would not raise his estimated chance of securing outcome v_A in 1, he should eliminate this option *whether or not he could execute it*. His estimated value of $Ch_q(v_A)$ were he to measure polarization with respect to an axis rotated 60° from Alice's is also 1/2 ($1/2.1/4 + 1/2.3/4 = 1/2$). Similarly for any other angle.[7]

But what if Bob had simply arranged for the measurement in 2 to have had the different *outcome* h_B? Then $Ch_q(v_A)$ would have been 0 instead of 1. No-one who accepts quantum mechanics can countenance this counterfactual scenario. The Born rule implies that $Pr_b^{|\eta\rangle}(H_B) = 1/2$, and anyone who accepts quantum mechanics accepts the implication that $Ch_{p'}(h_B) = 1/2$. So anyone who accepts quantum mechanics will have credence $Cr_{p'}(h_B/I_{p'}) = 1/2$ no matter what he takes to happen in the backward light cone of p' (as specified by $I_{p'}$). If he accepts quantum mechanics, Bob

---

[7] This recapitulates part of the content of proofs that Alice and Bob cannot use these correlations to exchange signals faster than light. Bell ([2004], pp. 237–8) shows why manipulation of external fields at p' or in 2 would also fail to alter Bob's estimated value of $Ch_q(v_A)$.

will conclude that there is nothing it makes sense to contemplate doing to alter his estimate of $Ch_{p'}(h_B)$, and so there is no conceivable counterfactual scenario in which one in Bob's position arranges for the measurement in 2 to have had the different outcome h_B. In general, there is causal dependence between events in 1 and 2 only if it makes sense to speak of an intervention in one of these regions that would affect a hypothetical agent's estimated chance of what happens in the other. Anyone who accepts quantum theory should deny that makes sense in this case.

Perhaps the most basic reason why counterfactual dependencies, of the chance(s) of v_A on space-like separated events in or near region 2, are no sign of causal dependence is that chances are not localized physical magnitudes and are incapable of entering into causal relations. That Bell thought they behaved like physical magnitudes is suggested by the [1975] paper in which he introduced local causality as a natural generalization of local determinism:[8]

In Maxwell's theory, the fields in any space-time region 1 are determined by those in any space region V, at some time t, which fully closes the backward light cone of 1. Because the region V is limited, localized, we will say the theory exhibits *local determinism*. We would like to form some no[ta]tion of *local causality* in theories which are not deterministic, in which the correlations prescribed by the theory, for the beables, are weaker.   (Bell [2004], p. 53 emphasis in original)

It seems Bell thought the chances prescribed by a theory that is not deterministic were analogous to the fields of Maxwell's electromagnetism, so that while local determinism specified the localized magnitudes (e.g. fields), local causality should specify the localized *chances* of such magnitudes, where those chances (like local beables) are themselves localized physical magnitudes.

Others have joined Bell in this view of chances as localized physical magnitudes. But quantum theory teaches us that chances are *not* localized physical propensities whose actualization may produce an effect. Maudlin says what he means by calling probabilities objective:

[T]here could be probabilities that arise from fundamental physics, probabilities that attach to actual or possible events in virtue solely of their physical description and independent of the existence of cognizers. These are what I mean by *objective probabilities*.
              (Beisbart and Hartmann, eds. [2011], p. 294 emphasis in original)

Although quantum chances do attach to actual or possible events, they are not objective in this sense. As we saw, the chance of outcome v_A does not attach to it in virtue solely of its physical description: the *chances* of v_A attach also in virtue of its space-time relations to different space-time locations. Each such location offers the epistemic perspective of a situated agent, even in a world with no such agents. The existence of these chances is independent of the existence of cognizers. But it is only because we are not merely cognizers but physically situated agents that we

---

[8] Remember, Bell introduced the neologism 'beable' in contrast to 'observable' to denote (what a theory takes to be) an element of reality—a physical magnitude that takes on a value whether or not it is observed.

have needed to develop a concept of chance tailored to our needs as informationally deprived agents. Quantum chance admirably meets those needs: an omniscient God could describe and understand the physical world without it. While they are neither physical entities nor physical magnitudes, quantum chances are objective in a different sense. They supply an objective prescription for the ideal credences of a hypothetical agent in any space-time location. Anyone who accepts quantum theory is committed to following that prescription.

I used the idea of intervention to argue against any causal dependence between events in 1 and 2: anyone who accepts quantum theory accepts that it makes no sense to speak of an intervention in one of these regions that would affect a hypothetical agent's estimated chance of what happens in the other. So even though the outcome v_B in 2 backs the assignment $|V_B\rangle$ to system L at q, the outcome in 1 does not depend causally on v_B: for similar reasons, neither does the outcome in 2 depend on that in 1. The same idea can now be used to show that both these outcomes *do* depend causally on at least one of the events in the overlap of the backward light cones of 1 and 2 that warranted assignment of state $|\eta\rangle$—an event $o \in \{o_i\}$ truly described by magnitude claims that backed this assignment.

Let r be a point outside the future light cones of v_A, v_B but within the future light cone of event o. Let $v_A \uplus v_B$ be the event of the joint occurrence of v_A, v_B. This is an event of a type to which the Born rule is applicable: the application yields its chance $Ch_r(v_A \uplus v_B) = Pr^{|\eta\rangle}_{a,b}(V_A \& V_B) = 1/2 \cos^2 \angle ab$. We already saw that $Ch_r(v_A) = Pr^{|\eta\rangle}_{a,b}(V_A) = 1/2 = Pr^{|\eta\rangle}_{a,b}(V_B) = Ch_r(v_B)$. The event o affects all these chances: had a different event o' occurred backing the assignment of a different state (e.g. $|H_A\rangle|V_B\rangle$), or no event backing any state assignment, then any or all of these chances could have been different. Since it makes sense to speak of an agent bringing about the event o rather than o' by intervening at s in its past light cone, we have

$$Ch_r(v_A \uplus v_B | do - o) \neq Ch_r(v_A \uplus v_B | do - o')$$
$$Ch_r(v_A | do - o) \neq Ch_r(v_A | do - o')$$
$$Ch_r(v_B | do - o) \neq Ch_r(v_B | do - o')$$

where $do - o$ means o is the result of an intervention without which o would not have occurred. It follows that $v_A, v_B, v_A \uplus v_B$ are each causally dependent on o: o is a common cause of v_A, v_B even though the probabilities of events of these types do not factorize. The same reasoning applies to each registered photon pair on any occasion at any settings a, b. So the second requirement on explanation is met: the separate recording events, as well as the event of their joint occurrence, depend *causally* on one or more of the events that serve to back assignment of state $|\eta\rangle$ in this scenario.

Since quantum state assignments are relative to the physical situation of a hypothetical agent applying the theory, $|\eta\rangle$ may also be the correct state to assign relative to a space-time point t time-like later than both 1, 2 only because $|\eta\rangle$ is backed by magnitude claims concerning event t' in the backward light cone of t that is also time-like later than 1, 2. In such a case a hypothetical agent Tom located at t should expect recording events in 1, 2 to manifest the probabilistic correlations specified by $|\eta\rangle$. But Tom could not use $|\eta\rangle$ to *explain* those correlations, since the chances of the recording events and their joint occurrence will all be either 0 or 1 at t, irrespective of $|\eta\rangle$.[9]

By rejecting any possibility of an intervention expressed by $do - v_B$ or $do - h_B$, anyone accepting quantum theory should deny that $Ch_{p(q)}(v_A|do-v_B) \neq Ch_{p(q)}(v_A|do-h_B)$ is true or even meaningful. Nevertheless $Ch_q(v_A|v_B) \neq Ch_q(v_A|h_B)$: in this sense v_A depends counterfactually but not causally on v_B. Does such counterfactual dependence provide reason enough to conclude that v_B is part of the explanation of v_A? An obvious objection is that because of the symmetry of the situation with 1 and 2 space-like separated there is an equally strong reason to conclude that v_A is part of the explanation of v_B, contrary to the fundamentally asymmetric nature of the explanation relation. But one can see that this objection is not decisive by paying attention to the contrasting epistemic perspectives associated with the different physical situations of hypothetical agents Alice* and Bob* with world lines confined to interiors of the light cones of 1, 2 respectively.

As his world line enters the future light cone of 2, Bob* comes into position to know the outcome at 2 while still being physically unable to observe the outcome at 1. His epistemic situation is then analogous to that of a hypothetical agent Chris in a world with a Newtonian absolute time, in a position to know the outcome of past events but physically unable to observe any future event. Many have been tempted to elevate the epistemic asymmetry of Chris's situation into a global metaphysical asymmetry in which the future is open while the past is fixed and settled. It is then a short step to a metaphysical view of explanation as a productive relation in which the fixed past gives rise to the (otherwise) open future, either deterministically or stochastically.[10]

Such a move from epistemology to metaphysics should always be treated with deep suspicion. But in this case it is clearly inappropriate in a relativistic space-time since the "open futures" of agents like Alice* and Bob* cannot be unified into *the* open future. This prompts a retreat to a metaphysically drained view of explanation as rooted in cognitive concerns of a physically situated agent, motivated by the need to unify, extend, and efficiently deploy the limited information to which it has access.

For many purposes it is appropriate to regard the entire scientific community as a (spatially) distributed agent, and to think of the provision of scientific explanations as aiding *our* collective epistemic and practical goals. This is appropriate insofar as localized agents share an epistemic perspective, with access to the same information about what has happened. But Alice* and Bob* do not have access to the same information at time t_2 or t_3 since they are then space-like separated. So it is entirely

---

[9] This scenario is realized in so-called delayed choice entanglement-swapping experiments (see Peres [2000], Ma [2012]). Two polarization-entangled pairs of photons are prepared independently, and the linear polarization of one photon from each pair is measured. Depending on the outcome of a joint measurement on the remaining pair of photons, Tom should assign one of four different entangled states *to the pair that has already been detected*. This is not a case of retrocausation, since that state does not specify their prior physical condition. Nor can it be used to explain the already observed correlations, since application of the Born rule to this state does *not* yield the chances of events that manifest them.

[10] See, for example, Maudlin ([2007], pp. 173–8).

appropriate for Bob* to use v_B to explain v_A and for Alice* to use v_A to explain v_B. This does not make explanation a subjective matter, for two reasons. There is an objective physical difference between the situations of Alice* and Bob* underlying the asymmetry of their epistemic perspectives: and by adopting either perspective in thought (as I have encouraged you to do, dear reader) anyone can come to appreciate how each explanation can help make Bell's correlations seem less puzzling.

Each event figuring in Bell's particular correlations can be truly described by a canonical magnitude claim. We may choose to describe some, but not all, such events as an outcome of a quantum measurement on a system: the chances of many of those events depend counterfactually on the particular entangled state assigned at t_1—if that state had been different, so would these probabilities. But this dependence is not causal. In quantum theory, neither states nor probabilities are the sorts of things that can bear causal relations: in Bell's terminology, they are not beables.

When relativized to the physical situation of an actual or hypothetical agent, a quantum state assignment is objectively true or false—which depends on the state of the world. More specifically, a quantum state assignment is made true by the true magnitude claims that back it. One true magnitude claim backing the assignment of $|V_B\rangle$ to system L relative to q reports the outcome of Bob's polarization measurement in region 2: but there are others, since this would not have been the correct assignment had the correct state assignment at p' been $|H_A\rangle|V_B\rangle$. We also need to ask for the backing of the entangled state $|\eta\rangle$.

There are many ways of preparing state $|\eta\rangle$, and this might also be the right state to assign to some naturally occurring photon pairs that needed no preparation. In each case there is a characterization in terms of some set of true magnitude claims describing the systems and events involved: these back the state assignment $|\eta\rangle$. It may be difficult or even impossible to give this characterization in a particular case, but that is just an epistemic problem which need not be solved even by experimenters' skilled in preparing or otherwise assigning this state. $|\eta\rangle$ will be correctly assigned relative to p' only if some set of true magnitude claims backing that assignment is accessible from p': events making them true must lie in the backward light cone of p'.

A quantum state counterfactually depends on the true magnitude claims that back it in somewhat the same way that a dispositional property depends on its categorical basis. The state $|\eta\rangle$ may be backed by alternative sets of true magnitude claims just as a person may owe his immunity to smallpox to any of a variety of categorical properties. If Walt owes his smallpox immunity to antibodies, his possession of antibodies does not cause his immunity: it is what his immunity consists in. No more is the state $|\eta\rangle$ caused by its backing magnitude claims: a statement assigning state $|\eta\rangle$ is true only if backed by some true magnitude claims of the right kind. A quantum state is counterfactually dependent on whatever magnitude claims back it because backing is a kind of determination or supervenience relation, not because it is a causal relation.

In this view, a quantum state causally depends neither on the physical situation of the (hypothetical or actual) agent assigning it nor on any of its backing magnitude claims. The correct state $|V_B\rangle$ to be assigned to L relative to q is not causally dependent on anything about Bob's physical situation, even if he happens to be located at q; it is not causally dependent on the outcome of Bob's polarization measurement in region 2; and it is not causally dependent on how Bob sets his polarizer in region 2. But a quantum state assignment is not just a function of the subjective epistemic state of any agent: If Bob or anyone else were to assign a state other than $|V_B\rangle$ to L relative to q, he or she would be making a mistake.

Application of the Born rule to derive the probabilistic correlations manifested by Alice's and Bob's recordings shows these were to be expected. Their recordings depend counterfactually but not causally on the quantum state $|\eta\rangle$. They also depend counterfactually on that state's backing conditions, as described by true magnitude claims: some of those counterfactual dependencies are causal. The status of the quantum state disqualifies it from participation in causal relations, but true magnitude claims may be taken to describe what I called assumables—beables recognized but not represented by quantum theory. If assignment of state $|\eta\rangle$ is backed by true magnitude claims describing an event (or events) $\{o_i\}$ occurring in the overlap of the backward light cones of space-like separated regions 1, 2 then events in those regions recording linear polarizations are causally dependent on at least one such event o. By meeting both minimal requirements on explanation, the application of quantum theory enables us to explain Bell's correlations.

10.6 Locality and Local Causality

Bell ([2004], p. 152) claimed that these correlations could not be explained without action at a distance. Was he right? Is this explanation local? Several senses of locality are relevant here. We may begin with Bell's intuitive principle, quoted in Chapter 4:

(IP) The direct causes (and effects) of events are near by, and even the indirect causes (and effects) are no further away than permitted by the velocity of light.

He used this to motivate the following principle of Local Causality:

(LC) A theory is said to be locally causal if the probabilities attached to values of local beables in a space-time region 1 are unaltered by specification of values of local beables in a space-like separated region 2, when what happens in the backward light cone of 1 is already sufficiently specified, for example by a full specification of all local beables in a space-time region 3. (Bell [2004], pp. 239–40)

Figure 10.5 is a space-time diagram of the situation he had in mind.

What did Bell mean by a local beable? Here are two quotes in which he tries to explain:

[I]t could be hoped that some increase in precision might be possible by concentration on the beables, which can be described "in classical terms", because they are there. The beables must

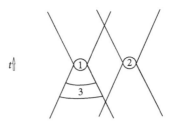

Figure 10.5 Local causality

include the settings of switches and knobs on experimental equipment, the currents in coils, and the readings of instruments. (Bell [2004], p. 52)

We will be particularly concerned with local beables, those which . . . can be assigned to some bounded space-time region. . . . It is in terms of local beables that we can hope to formulate some notion of local causality. (p. 53)

Bell ([2004], pp. 240–1) argued that ordinary quantum mechanics is not locally causal, using the special case of the scenario depicted in Figure 10.4 in which Alice's and Bob's polarizer axes are perfectly aligned, so $a = b$ and $\cos^2 \angle ab = 1$, as in Bohm's version of the EPR argument. He begins:

Each of the counters considered separately has on each repetition of the experiment a 50% chance of saying 'yes'.

Each of the chances $Ch_p(v_A), Ch_{p'}(v_B)$ is $1/2$, as Bell says: but $Ch_q(v_A) = 1$. After noting that quantum theory here requires a perfect correlation between the outcomes in 1, 2, he continues:

So specification of the result on one side permits a 100% confident prediction of the previously totally uncertain result on the other side. Now in ordinary quantum mechanics there just *is* nothing but the wavefunction for calculating probabilities. There is then no question of making the result on one side redundant on the other by more fully specifying events in some space-time region 3. We have a violation of local causality.

It is true that (with $b = a$ here) $Pr^{|\eta\rangle}_{a,b}(V_A|V_B) = 1$, $Pr^{|\eta\rangle}_{a,b}(H_A|V_B) = 0$ (since $Pr^{|\eta\rangle}_{a,b}(V_A \& V_B) = 1/2$, $Pr^{|\eta\rangle}_{a,b}(H_A \& V_B) = 0$) while $Pr^{|\eta\rangle}_a(A) = Pr^{|\eta\rangle}_b(B) = 1/2$. But does that constitute a violation of local causality? The factorizability condition

$$Pr_{a,b}(A \& B) = Pr_a(A) \times Pr_b(B) \qquad \text{(Factorizability)}$$

is straightforwardly applicable to a theory whose general probabilities yield a *unique* chance for each possible outcome in 1 prior to its occurrence. In the case of quantum theory, however, the general Born rule probabilities yield *multiple* chances for each possible outcome in 1, each at the same time (in the laboratory frame): $Ch_p(v_A) = 1/2$, but $Ch_q(v_A) = 1$ (assuming the outcome in 2 is of type V_B). When (LC) speaks of "*the* probabilities attached to values of local beables in a space-time region 1 being unaltered by specification of $[V_B]$ in a space-like separated region 2" (my italics), which probabilities are these?

Since the connection to (IP)'s motivating talk of 'cause' and 'effect' is provided by the thought that a cause alters the chance of its effect, (LC) is motivated only if applied to the *chances* of v_A, h_A in region 1. But $Ch_p(v_A)$ is *not* altered by specification of V_B in space-like separated region 2: its value depends only on what happens in the backward light cone of 1, in conformity to its role in prescribing $Cr_p(v_A)$. Of course $Ch_q(v_A)$ does depend on the outcome in 2. If it did not, it could not fulfill its constitutive role of prescribing $Cr_q(v_A/I_q)$ no matter what information I_q provides about the contents of q's past light cone. It follows that $Ch_q(v_A)$ is not altered but merely *specified* by specification of the result in 2.

Only for a hypothetical agent whose world line has entered the future light cone of 2 at q is it true that specification of the result in 2 permits a 100% confident prediction of the previously totally uncertain result on the other side. A hypothetical agent at p is not in a position to make a 100% confident prediction: for such an agent the result in 1 remains totally uncertain: what happens in 2 makes no difference to what s/he should believe, since region 2 is outside the backward light cone of p. That is why it is $Ch_p(v_A)$, not $Ch_q(v_A)$, that says what is certain at p. Newtonian absolute time fostered the illusion of the occurrence of future events becoming certain for everyone at the same time—when they occur if not sooner. Relativity requires certainty, like chance, to be relativized to space-time points—idealized locations of hypothetical knowers.

So does ordinary quantum mechanics violate local causality? If "the probabilities" (LC) speaks of are $Pr_a^{|\eta\rangle}(V_A)$, $Pr_a^{|\eta\rangle}(H_A)$, and the condition that these be unaltered is understood to be that $Pr_a^{|\eta\rangle}(V_A) = Pr_{a,b}^{|\eta\rangle}(V_A|B)$, $Pr_a^{|\eta\rangle}(H_A) = Pr_{a,b}^{|\eta\rangle}(H_A|B)$, then ordinary quantum mechanics violates (LC). But if this is all (LC) means, then it is not motivated by (IP) and its violation does not imply that the quantum world is non-local in that there are superluminal causal relations between distant events. For (LC) to be motivated by causal considerations such as (IP), "the probabilities" (LC) speaks of must be understood to be chances, including $Ch_p(v_A)$ and $Ch_q(v_A)$. But neither of these would be altered by the specification of the outcome $[V_B]$ in a space-like separated region, so local causality would then not be violated. Although it remains unclear exactly how Local Causality is supposed to be applied to quantum mechanics, one way of applying it is unmotivated by (IP), while if it is applied in another way quantum mechanics does *not* violate this local causality condition.

I argued in the previous section that the explanation of these correlations using quantum theory involves no superluminal causal dependence. As stated, the condition of Local Causality is not straightforwardly applicable to the quantum explanation since it presupposes the uniqueness of the probability to which it refers. (Factorizability) is violated, but Bell ([2004], p. 243) preferred to see (Factorizability)[11] as not a

---

[11] Bell actually expressed this preference about (Conditional Factorizability) (see §4.3), but he had already noted that for quantum theory (Conditional Factorizability) reduces to (Factorizability).

formulation but a consequence of 'local causality': I have argued that it is not. To retain its connection to (IP), a version of Local Causality should speak of chances rather than general probabilities. A version that equates the unconditional chance of v_A to its chance conditional on v_B holds, no matter how these chances are relativized to the same space-time point.[12] But a version that is clearly motivated by the intuitive principle (IP) would rather equate the unconditional chance of v_A to its chance conditional on *a counterfactual intervention that produced* v_B. However, this version is inapplicable since acceptance of quantum theory renders talk of interventions producing v_B not merely counterfactual but senseless.

The explanation one can give by applying quantum theory appeals to *chances* that are localized, insofar as they are assigned at space-time points that may be thought to offer the momentary perspective of a hypothetical idealized agent whose credences they would guide. But these chances are not quantum beables and they are not physical propensities capable of manifestation at those locations (or anywhere else). Apart from the settings of the laboratory equipment, the only causes figuring in the explanation are localized where the physical systems are located whose magnitudes back the assignment of state $|\eta\rangle$. That chances are not propensities becomes clear when one drops the assumption that a, b occur in the overlap of the backward light cones of 1 and 2, as depicted in Figure 10.6.

If a, b are set at the last moment, the chance of $v_A \uplus v_B$ that figures in its explanation may be located at a point r *later* in the laboratory frame than v_A, v_B. If chance were a physical propensity it should act *before* its manifestation. But chances aren't propensities (proximate causes of localized events); they are a localized agent's objective guide to credence about epistemically inaccessible events.

I will conclude by noting one sense in which the explanation one can give using quantum theory is *not* local as it stands. Though it is a (non-factorizable) cause of events of types A, B in regions 1, 2 respectively, the event o is not connected to its effects

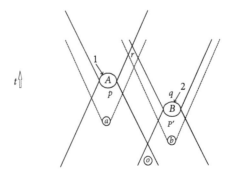

Figure 10.6 A late explanatory chance

[12] The chance of v_A at p does not depend on conditions in space-like separated region 2. The chance of v_A at p conditional on v_B must be clearly distinguished from $Ch_p(v_A|v_B)$, which is just the ratio of two chances: $Ch_p(v_A \uplus v_B)/Ch_p(v_B)$.

by any spatiotemporally continuous causal process described by quantum theory. This puts the explanation in tension with the *first* conjunct of Bell's ([2004], p. 239, emphasis added) intuitive locality principle (IP):

The direct causes (and effects) of events are near by, and even the indirect causes (and effects) are no further away than permitted by the velocity of light.

o is separated from both recording events in regions 1, 2 in time, and from at least one in space. If o is not merely a cause but an *unmediated* cause of these events then it violates the first conjunct of (IP) because it is *not* nearby. But if one adopts the present view of quantum theory, the theory has no resources to describe any causes mediating between o and these recording events. So while their quantum explanation is not explicitly inconsistent with the first conjunct of (IP), mediating causes could be found only by constructing a new theory.

11
Observation and Objectivity

11.1 Introduction

We normally think of observation as a passive process in which our sense organs inform us about the physical world without significantly affecting it. A pot actually boils just as fast whether or not it is watched. There are cases when observation does have a minor influence on its object: putting your foot in the bathwater to see how hot it is will make the water very slightly cooler or warmer unless foot and water are both at the same temperature. Classical physics had the resources to deal with such effects of observation. A model of the physical process of observation would enable one to estimate if not calculate its effect on what was observed. This possibility removed any temptation to deny that observation is simply a way of finding out about a physical world of entities and properties that exist whether or not they are observed, even when observation does disturb them.

But observation has often been allotted a more active role in quantum theory. Recall Jordan's view (quoted in Chapter 1) that observations not only disturb what is to be measured, they produce it; and Mermin's statement that the moon is demonstrably not there when nobody looks. If observation produces rather than revealing what we observe, this casts doubt on the objective existence of unobserved entities and properties. In Chapter 1 we saw d'Espagnat and Heisenberg apparently giving voice to such skepticism. In this and the next chapter I respond to this threat to objectivity. I use the view of quantum theory developed in Part I to show how quantum theory *does* inform us about a physical world that objectively exists, whether or not we observe it.

However, accepting quantum theory turns out to involve acknowledging minor but surprising limitations on the content of all our descriptions of this world. This poses a challenge to the philosophical view that a statement's content can be understood just by saying what in the world would make it true. A richer account of how statements acquire content has the resources to respond to this challenge, as the next chapter will show.

Out of context, Chapter 1's quotes from Jordan, Mermin, Heisenberg, and d'Espagnat could be misinterpreted simply as expressions of philosophical idealism. But in the context supplied by quantum phenomena and the theory physicists have successfully applied to them they are better construed as reporting a conclusion of scientific investigation—that observations of the quantum domain do not generally reveal entities and properties that were there anyway. Clearly observation alone cannot establish that conclusion. But it receives support from many theoretical considerations explained in Part I. Interference phenomena like those described in Chapter 2 intimate but do not establish this conclusion. Chapter 4 adduced arguments against Einstein's suggestion that one could understand the introduction of probability in quantum theory as necessary only because no observations could reveal all the independently existing properties of a quantum system at once. And Appendix B spells out details of two powerful arguments against such naive realism.

However, one can accept that observations of the quantum domain do not *generally* reveal properties that were there anyway, while maintaining that there are some observations of some systems that *do* reveal many of their properties—enough to provide a rich, objective description of the physical world around us. This will not be a description in the language of quantum theory itself: though objective, quantum states and Born probabilities have other functions, as we have seen. It will be a description in a language based on magnitude claims made available independently of quantum theory. This description should be capable of underwriting the objectivity of our everyday descriptions of the world. Minimally, to secure an objective evidential base for quantum theory, it should make possible the objective description of the operation of apparatus involved in quantum measurements, including their outcomes.

The paradox of Wigner's friend threatens even this minimal requirement. Eugene Wigner ([1967], p. 172) presented his friend paradox to motivate the view that consciousness has a special role to play in quantum theory—by collapsing the quantum state. In a more recent discussion, d'Espagnat [2005] argued that a consistent treatment of the Wigner's friend scenario may be given (without collapse) if the descriptive content underlying quantum theory is restricted to probabilistic predictions flowing from the Born rule, but only if these concern conscious experiences of observers.

But fortunately we can understand quantum theory without forsaking the objective description of nature. Quantum theory should be understood to provide situated agents with resources for predicting and explaining what happens in the physical world, not conscious observations of it. We can meet this threat to objectivity by judicious appeal to decoherence. Even in the Wigner's friend scenario, differently situated agents agree on the objective content of statements about the values of physical magnitudes. In more realistic circumstances Quantum Darwinism also permits differently situated agents equal observational access to evaluate their truth. Quantum theory is a resource for situated *agents* (rather than observers) with no requirement that these be conscious.[1] Quantum theory has nothing to say about consciousness or the conscious experiences of observers. When quantum theory is understood this way, Wigner's friend scenarios may be treated consistently and without ambiguity.

---

[1] As far as we know, conscious humans are the only agents currently able to avail themselves of this resource. But this view admits the possibility that non-human, or even non-conscious, situated agents may come to use quantum theory.

The chapter proceeds as follows. Section 2 analyzes some distinct notions of objectivity relevant to observation in quantum theory. Section 3 recalls the paradox of Wigner's friend. Section 4 shows how to resolve the paradox while maintaining the objectivity of description in quantum theory. Section 5 addresses a more general worry about objectivity of description. Section 6 relates the foregoing treatment of objectivity to a recent suggestion (Ollivier et al. [2004]) that objective properties emerge from subjective quantum states through Quantum Darwinism. It argues that while Quantum Darwinism can help explain the intersubjective verifiability of observations, it does not establish but rather presupposes that observational records and what they record can be objectively described. A pragmatist approach to quantum theory maintains the objectivity of a statement about the values of physical magnitudes, but it recognizes that the statement's content depends on the environmental context. The chapter concludes by showing how this dependence affects even our use of the familiar language of everyday affairs, including laboratory procedures, to assign objective properties—beables—to objects (Bell [2004]). It is only because ordinary objects are enormous on an atomic scale and continuously subject to the rich environment in which they are embedded that we have been able to ignore the consequent limitation on the contents of such everyday statements.

11.2 Objectivity

Objectivity (and its polar opposite, subjectivity) can mean many things. Dreams are a paradigm of subjectivity. The subject matter of a dream is not objective: a dream does not portray what really happens. The mode of presentation is not objective: this subject matter is accessible only to the dreamer. What is presented as well as its mode of presentation strongly depends on the specific as well as the general features of the dreamer—her individual physiology, psychology, and prior experiences, as well as her humanity (Dick [1968]). By contrast, a description or representation in physics has an objective subject matter if its content represents physical reality, and an objective mode of presentation to the extent that how this is represented does not depend on the specific and general features of the one whose representation it is.

The prominence of notions of observation or measurement in standard formulations of quantum theory raises concerns about the objectivity of descriptions in that theory. If the Born rule is understood to yield probabilities only for _results of observations/measurements_ then one can question the objectivity of these results. The orthodox view—that a quantum measurement cannot be understood generally as revealing the value of the measured magnitude—has now been amply supported by "no-go" theorems.[2] This challenges the objectivity of subject matter in a description of the result of an observation. One can respond to this challenge by proposing an analysis of observation/measurement as an objective physical process. But the assumption

---

[2] As discussed in Chapter 4: for further details see also Appendix B.

that quantum theory can itself represent this process leads to the notorious quantum measurement problem discussed in Chapter 6. The alternative of characterizing observation independently as a physical interaction with a macroscopic apparatus, and/or involving irreversible amplification involves the "shifty split" between quantum and classical descriptions justifiably criticized by Bell [2004].

With his keen appreciation of how difficult it is to provide a satisfactory physical characterization of observation/measurement in quantum theory, Wigner came to think that a measurement in quantum theory occurs only when a conscious observer becomes aware of a result. On this view, the Born rule yields probabilities only for alternative conscious experiences. It thereby threatens both the objectivity of the subject matter of description in quantum theory and the objectivity of its mode of presentation. Quantum theory, it seems, is then concerned to predict and (perhaps) explain "communicable human experience. In other words the set of all the impressions human minds may have and communicate to others" (d'Espagnat [2005]).[3]

Suppose one so restricts the scope of quantum theory. Then the descriptive statements to which the Born rule attaches probabilities fail to be objective in several respects:

1. Their subject matter is not a physical reality independent of our experiences.
2. Their mode of presentation depends on the individual consciousness.
3. This consciousness is generically *human*.[4]

One could seek to restore this third respect of objectivity by generalizing beyond humans to observers capable of any sufficiently sophisticated form of conscious experience. But since consciousness is not well understood (especially in non-humans), to do so would further obscure the content of these descriptive statements, which would still fail to be objective in either of the first two respects.

While the quantum theory of environmental decoherence does not by itself resolve the quantum measurement problem, many believe it may contribute to a resolution within some appropriate interpretative framework. Such a framework has been provided in Part I. From the present perspective, the key interpretative proposal is to remove any talk of observation or measurement from a formulation of the Born rule by taking this to generate a function with the *mathematical* features of a probability measure over statements that simply describe values of magnitudes rather than results of observing them. The "no-go" theorems block this approach if one further assumes that every statement over which these functions are defined always has a well-defined

---

[3] The Wigner's friend scenario casts doubt even on such communicability. D'Espagnat's paper seeks to relieve this doubt.

[4] Chapter 8 already explained respects in which quantum state assignments and probabilistic statements about significant magnitude claims are objective. In the course of this chapter I will distinguish further aspects of objectivity. The notion is many-faceted so it is best not to begin by defining objectivity in an attempt to encompass them all.

content, so an agent should believe it to a degree specified by its Born "probability"[5] to avoid refutation by statistics of actual observations. But suppose, on the other hand, that how much significance attaches to a statement about a system depends on the extent of environmental decoherence suffered by its quantum state in a model. Then the numbers yielded by the Born rule have the import of genuine probabilities only for significant statements: Born rule "probabilities" for statements lacking such significance do not correspond to (actual or hypothetical) frequencies, and should not guide an agent's degrees of belief in these insignificant statements.

On this view any reference to observation or measurement has been eliminated from the Born rule and basic principles of quantum theory, and so there is no reason to suppose that descriptive statements that arise in quantum theory are subjective in any of the respects (1)–(3) noted above. But there is a fourth aspect of objectivity to consider, as the following quote makes clear.

A view or form of thought is more objective than another if it relies less on the specifics of the individual's makeup and position in the world, or on the character of the particular type of creature that he is.   (Nagel [1986], p. 5)

If the significance for an agent of a descriptive statement about the value of a magnitude depends on how that agent is situated in the world, then that statement may lack a kind of objectivity. Differently positioned agents could understand the statement differently: they may come to a no-fault disagreement about whether it is true, or even meaningful. Quantum theory as presented in Part I faces such a challenge to the objectivity of our descriptive statements in quantum theory—a challenge that is highlighted by the Wigner's friend scenario, as we shall see.

11.3 Wigner's Friend

The "paradox" of Wigner's friend presents a challenge to the objectivity of physical description within quantum theory. To set up the "paradox", imagine a human experimenter (Wigner's friend, John[6]) who records in a device d the result of a quantum measurement of dynamical variable Q he performs on a system s inside his isolated laboratory.[7] Eugene and John both agree in assigning to s initially (at time t_i) the superposed quantum state

$$|\varphi\rangle = \sum_i c_i |\varphi_i\rangle, \text{ where } \hat{Q} |\varphi_i\rangle = q_i |\varphi_i\rangle.$$

---

[5] "Probability" rather than probability, because here the function has the mathematical character of a probability measure but the theorems show its values cannot play probability's requisite practical and/or epistemic role.

[6] Eugene Wigner and John von Neumann were both pupils in the same Budapest high school: Wigner was a year ahead.

[7] To call the laboratory 'isolated', is to require by fiat the absence of any decohering interactions with its external environment. So we are talking of a ridiculously impractical *Gedankenexperiment*, as Schrödinger explicitly said he was when describing his cat scenario. The point of doing so is to show how this raises a problem for one view of quantum theory and then to explain why this problem does not arise if quantum theory is understood in the way presented in Part I.

Here $\hat{Q}$ is the operator uniquely corresponding to Q, the $|\varphi_i\rangle$[8] are pairwise orthogonal, and no c_i equals 0. Let C be the statement that the value of a recording magnitude M on d is m at time t_f after s has interacted with d. John looks at d at t_f and makes statement C. On the basis of this observation, he assigns a collapsed quantum state to $s + d$.[9] Environmental interactions within the laboratory rapidly entangle this collapsed state with the state of everything else in the laboratory. But the total quantum state $|\psi_J\rangle$ of the enormously complex system composed of John, d, and everything else inside the laboratory will continue to reflect the collapse induced by John's measurement.

Meanwhile, Eugene, who has remained outside the laboratory, assigns a state $|\psi_E\rangle$ to the enormously complex system composed of John, d, and everything else inside the laboratory, based on all the information about the properties of systems to which he has access in this situation. For Eugene, John's measurement involves only linear, unitary interactions—between s and d, between d and John, and between all these systems and the rest of the laboratory. Accordingly, he evolves the state $|\psi_E\rangle$ unitarily, with no collapse from t_i to t_f. Since $|\psi_J(t_f)\rangle \neq |\psi_E(t_f)\rangle$, John's and Eugene's quantum states differ after John's measurement but before Eugene enters the laboratory. As these states are usually understood, $|\psi_J(t_f)\rangle$ represents a definite result of John's measurement (recorded by the state of d, John's memory, etc.) while $|\psi_E(t_f)\rangle$ excludes any such definite result and its traces. To retain its internal consistency, this view of quantum states must deny the objectivity of John's measurement result, since differently situated agents (John and Eugene) disagree about whether C is true at t_f.

On any view of quantum states, since $|\psi_E(t_f)\rangle$ satisfies

$$\hat{O}|\psi_E(t_f)\rangle = o|\psi_E(t_f)\rangle$$

where $\hat{O}$ is taken to correspond to some extremely complex dynamical variable O on the entire laboratory system, Eugene could in principle (though certainly not in practice) distinguish between $|\psi_J(t_f)\rangle$ and $|\psi_E(t_f)\rangle$ by recreating many exactly similar laboratory systems and measuring O on each system: $|\psi_E(t_f)\rangle$ would give the same result o on each, while $|\psi_J(t_f)\rangle$ would almost certainly yield a statistical spread of different results. This same procedure would also, in principle, enable Eugene to distinguish between $|\psi_E(t_f)\rangle$ and the mixed state $\rho_J(t_f) = \sum_i |c_i|^2 |\psi_J(t_f)\rangle_i \langle\psi_J(t_f)|_i$ Eugene may assign to reflect his ignorance of the result of John's measurement.[10]

---

[8] This means that $|\varphi\rangle$ is a nontrivial superposition of eigenvectors $|\varphi_i\rangle$ of $\hat{Q}$ with corresponding eigenvalues q_i: see appendix A, § A.2.

[9] He does so in accordance with a view of quantum states due to Dirac and von Neumann according to which a quantum state completely describes the dynamical properties of a system to which it is assigned. In this view Q has value q_i in state $|\varphi_i\rangle$ but no determinate value in state $|\varphi\rangle$. While himself subscribing to this orthodoxy, Wigner maintained that it is only John's consciousness that induces the collapse. I use the word 'collapse' to refer to a hypothetical physical change in the properties of a system supposedly represented by a reassignment of its quantum state.

[10] To keep things simple, here I assume John made an ideal measurement on the state $\sum_i c_i |\varphi_i\rangle$, thereby collapsing the state of $s + D$ onto some eigenstate $|\varphi_k\rangle|\chi_k\rangle$ with probability $|c_k|^2$.

Wigner's own way of resolving this paradox was to give consciousness (and only consciousness) the distinctive physical role of inducing collapse of the quantum state. For him, it was the interaction with John's consciousness that produced a discontinuous physical change inside the laboratory, resulting in the final state $|\psi_J(t_f)\rangle$ and not $|\psi_E(t_f)\rangle$. Such a change would be detectable in principle in the Wigner's friend scenario, though quite impossible to detect in practice. I will offer a different resolution involving no such collapse. This involves the different understanding of quantum states described in Part I (especially Chapters 4 and 5).

11.4 Paradox Resolved

Consider the situation of Wigner's friend John inside his isolated laboratory. On the view of quantum theory presented in Part I (see § 5.2), John is licensed by decoherence to apply the Born rule to the superposed state of s and advised by quantum theory to adjust his credences in statements about the value of recording magnitude M on d at time t_f so they match the corresponding Born probabilities. This is because the quantum state he consequently assigned to $s + d$ immediately before has, by t_f, been robustly decohered in d's "pointer basis" after interaction with the environment inside his laboratory. He is then warranted by his own direct observation of d in making the significant statement C that the value of M on d is m.

Meanwhile, Eugene, who has remained outside the laboratory, assigns state $|\psi_E\rangle$ to the enormously complex system composed of John, d, and everything else in the laboratory. By assumption, the state $|\psi_E\rangle$ has not been decohered through interaction between the laboratory and its external environment by t_f, so this state does not license Eugene to entertain any significant statement about the value of a magnitude on the laboratory or anything in it. In this situation, Eugene should not apply the Born rule to $|\psi_E\rangle$ and not adjust any of his credences accordingly. If and only if the laboratory ceases to be isolated so $|\psi_E\rangle$ decoheres (as he subsequently enters the laboratory, for example) would Eugene be in a position to apply the Born rule to significant statements about the laboratory or traces it leaves in an external recording device (such as his own brain).

Recall the difficulties presented by the Wigner's friend scenario noted in the previous section:

1. As these states are usually understood, $|\psi_J(t_f)\rangle$ represents a definite result of John's measurement (recorded by the state of d, John's memory, etc.) while $|\psi_E(t_f)\rangle$ excludes any such definite result and traces of it in John's laboratory.
2. On any view of quantum states, since some extremely complex dynamical variable O on the entire laboratory system has the determinate value o in state $|\psi_E(t_f)\rangle$ but not in state $|\psi_J(t_f)\rangle$, Eugene could in principle (though certainly not in practice) distinguish between $|\psi_E(t_f)\rangle$ and $|\psi_J(t_f)\rangle$ (or indeed $\rho_J(t_f)$) by recreating many exactly similar laboratory systems and measuring O on each.

These difficulties present a challenge to the objectivity of physical description here. (1) apparently implies that, by assigning different quantum states to the contents of John's laboratory at t_f, John and Eugene come to disagree about the truth value of the descriptive statement C (i.e. about whether C is true or false). If so, they can both be right only if that truth value is relative to the physical situation of the agent making it—in conflict with the fourth aspect of objectivity noted in section 2. But (2) shows that such relativization of truth value to agent situation still leaves it unclear how John and Eugene can consistently apply quantum theory here. Independent of the practicality of Eugene's discriminatory measurements, $|\psi_E(t_f)\rangle$ and $|\psi_J(t_f)\rangle$ (or $\rho_J(t_f)$) are distinct states, yielding incompatible Born probabilities concerning the possible values of certain magnitudes on the entire laboratory. Relativization of the laboratory's quantum state to the situations of Eugene, John, respectively leaves it ambiguous on which of these states quantum theory advises them to base their expectations about these possible values.

The first step in responding to this challenge is to reject the usual understanding of the states $|\psi_J(t_f)\rangle$, $|\psi_E(t_f)\rangle$. Neither state has the function of representing the physical properties of John's laboratory or anything in it. Since neither state represents anything bearing on the result of John's measurement (e.g. whether or not C is true at t_f), assignment of both states to John's laboratory at the same time could not lead John and Eugene to assign (apparently) conflicting truth values to C.

But John and Eugene each assume that John will perform a measurement on s by interacting s appropriately with d, and that the environment inside the laboratory will decohere the state of $s + d$ in d's "pointer basis" (without inducing any physical collapse in that state). So both John *and* Eugene are licensed to entertain statement C and (beforehand) to set their credences for C at t_f equal to the corresponding Born rule probability. This implies they agree that C has a truth value at t_f. But while Eugene remains outside the laboratory, only John is in a position to look at d at and after t_f and so determine what that truth value is. This secures the objectivity of the description C, in the sense that differently positioned agents (John and Eugene, in this case) agree that C has a truth value after t_f, and do not disagree about what that truth value is. At this stage, John, but not Eugene, is in a position to *know* that C is true rather than false. Eugene can choose whether or not to enter the laboratory to try to find out whether C is true.

Suppose Eugene decides to see for himself the outcome of John's observation by entering the laboratory just after t_f. This will decohere $|\psi_E(t_f)\rangle$, permitting Eugene to apply the Born rule to this state to adjust his degrees of belief in various significant statements about the contents of the laboratory, including C, John's record of the outcome in his notebook, John's verbal report, etc., as well as correlations between such statements. The Born probabilities of these statements (joint as well as single) based on the state $|\psi_E(t_f)\rangle$ will lead him to the following confident expectation: whatever may be the outcome of John's observation, the laboratory will contain multiple mutually supporting records of it. Nothing about $|\psi_E(t_f)\rangle$ will tell him what John's outcome actually was. But he can easily find that out by asking John and observing any of the other multiple, correlated records inside the laboratory. The important point is that $|\psi_E(t_f)\rangle$ does not *exclude* an outcome at t_f, so by entering the laboratory Eugene simply finds out what happened—he does not *make* it happen.

If Eugene were instead (able) to remain outside the laboratory but interact with it so as to measure the value of O, he should use $|\psi_E(t_f)\rangle$ to calculate a Born probability of 1 that a suitable recording device, applied to the laboratory and decohered in its "pointer basis" by external environmental interactions, will record a value of O equal to o. Even though Eugene assumes that John records the outcome of his measurement at t_f, by t_f there has been no physical interaction between Eugene and anything inside the laboratory that could serve to inform Eugene of that outcome. So Eugene cannot base his expectation of the outcome of a measurement of O on $|\psi_J(t_f)\rangle$, and would be mistaken if he were to base that expectation on $\rho_J(t_f)$. There is no ambiguity about to what state Eugene should apply the Born rule when setting his credences concerning the outcome of his O measurement.

On what quantum state should John base his credence as to the outcome of an external O measurement? The outcome of John's measurement should prompt him to update the quantum state he assigns to certain subsystems of his laboratory so as to take account of the new information he has acquired. For example, since he takes C not only to have a high content at and for some time after t_f but also to be true then (based on his own observation of d), his state assignment to d should reflect this. So he should not only believe C but also assign d a quantum state that assigns Born probability 1 to C—the state $|\chi_m\rangle$ (or more realistically a mixture of states, each of which assigns Born probability 1 to C). This is how he squares the inferences to which C entitles him—that repeated observations of d will confirm the continued truth of C—with credences based on application of the Born rule to the quantum state he assigns to d.

He may go on to assign quantum state $|\varphi_m\rangle|\chi_m\rangle$ to $s+d$ if he has justified confidence that the interaction has not disturbed s, since his expectations based on that state would then be borne out, even concerning measurements in his laboratory of correlations between dynamical variables on s and d. That will be so only because decoherence by their environment within the laboratory has delocalized phase relations between the states of $s + d$. But if he were to consider the entire contents of his laboratory, he would have to acknowledge that the outcome of an external O measurement by Eugene on this vast composite system would be sensitive to phase relations between its subsystems that have *not* been decohered by any interactions with the environment outside the laboratory. So he would be mistaken to base his credence as to the outcome of an external O measurement on assignment of state $|\psi_J(t_f)\rangle$ to the entire system these compose. He can hypothetically take the perspective of Eugene, but if he does so he must then assign the same state $|\psi_E(t_f)\rangle$ that he agrees is the correct state for someone in Eugene's situation. There is no ambiguity as to which quantum state should be used to calculate Born probabilities for the results of an external O measurement performed on the entire laboratory, in the sense that differently positioned agents (John and Eugene, in this case) agree on this state. Of course, only Eugene is then in a position to make such a measurement.

What will Eugene find if he enters the laboratory after measuring the value of O in this way? d will no longer have a record of the truth of C. Even if John still exists, he will not remember that C was verified as true at t_f. Because $\hat{O}$ fails to commute with $\hat{M}$, as well as each operator corresponding to an effective "pointer position" on every other system correlated with this through interactions within the laboratory, the external interaction required to record the value of O will have so disturbed the laboratory and its contents as to remove all traces that C was true at t_f. But one can't change the past: C was indeed true at t_f, as John then verified with his own eyes.

There is, in principle, an even more dramatic way to erase all traces of C. Since the entire laboratory and its contents constitute an isolated system, Eugene will take $|\psi_E\rangle$ to have evolved unitarily from its state $|\psi_E(t_i)\rangle$ prior to John's measurement on s to its state $|\psi_E(t_f)\rangle = \hat{U}_{if}|\psi_E(t_i)\rangle$. The state Eugene should ascribe to the contents of the laboratory at t_i should reflect his belief that his friend has not yet performed the planned measurement: this will assign Born probability 1 to statements about d, John, and other items in the laboratory that suffices to substantiate this full belief. Mathematically, there will exist a Hamiltonian that would induce the time-reversed evolution of $|\psi_E\rangle$ so that at a later time t_g (where $t_g - t_f = t_f - t_i$) it is restored to its value before John's measurement: $|\psi_E(t_g)\rangle = \hat{U}^\dagger_{fg}|\psi_E(t_f)\rangle = |\psi_E(t_i)\rangle$.[11]

If Eugene had the powers of a quantum demon, he could instantaneously replace the original Hamiltonian by this time-reversing Hamiltonian at t_f, thereby restoring $|\psi_E\rangle$ at t_g to its original value at t_i. Suppose that he does so, and postpones his entry into the laboratory until t_g. Since the quantum state of the entire laboratory is identical to what it was before John had made any measurement, Eugene must fully expect that if he then asks John about the result of his measurement, John will say he has not yet performed any measurement. He must further fully expect that his own examination at, and at any time after, t_g of D, John's notebook, and anything else inside the laboratory will reveal no record of any such measurement ever having been made. Eugene's action at t_f has, by t_g, erased all traces of John's measurement and its result: indeed, Eugene has succeeded in erasing all traces of everything that happened inside the laboratory between t_i and t_f. Once again, he has not changed the past. But there has been a wholesale loss of history, understood as reliable information about the past.

11.5 Objective Content Secured

I have shown why John inside the laboratory and Eugene outside do not assign inconsistent truth values to statement C. Moreover, each assigns the same rich content to C based on the decoherence of the state of $s + d$ induced by environmental interactions inside the laboratory. This certifies the objectivity of C's content in the Wigner's friend scenario. The certification depends on the fact that both John and

---

[11] $\hat{U}^\dagger_{fg}$, the adjoint of $\hat{U}_{fg}$, satisfies $\langle \psi, \hat{U}^\dagger_{fg}\zeta\rangle = \langle \hat{U}_{fg}\psi, \zeta\rangle$ for arbitrary ψ, ζ.

Eugene apply essentially the same model of decoherence to the same initial quantum state of $s + d$.

But while quantum states are objective they are also relational: the quantum state an agent should assign to a system generally depends on the (actual or hypothetical) physical situation of that agent. If differently situated agents assign different quantum states to a system, this raises the possibility that their models of its decoherence will so differ as to lead each to assign a different content to certain statements concerning it. Relativization of quantum state assignments threatens relativization of the content of magnitude claims like C. This would undermine their objectivity in the fourth respect pointed out in section 2.

Even though John and Eugene are differently situated in the Wigner's friend scenario, each can choose to adopt the perspective of the other's situation for the purpose of applying quantum theory. Eugene should assign the same initial state as John to s because each knows that this is the state on which John will perform his measurement. Eugene and John are each in the same state of ignorance as to the initial state of d and its (lack of) correlations with that of s, so there is no reason for them to assign different states either to d alone or to $s + d$. (Their conclusion—that interactions within the laboratory robustly decohere the state of $s + d$ and so endow C with rich content—is not sensitive to fine details of the initial state of $s + d$.)

One might object that Eugene's physical situation *requires* him to assign the state $|\psi_E\rangle$ to the contents of the laboratory, and that since the phase of $|\psi_E\rangle$ is not delocalized into its environment Eugene should not assign a rich content to C. An alternative objection would be that C has no unambiguous content for Eugene, since he has no principled reason to base his assessment of that content on the degree of decoherence of the "internal" state of $s + d$ rather than that of $|\psi_E\rangle$.

But in assessing the content of a magnitude claim such as C about a system (d in this case), an agent like Eugene should base his model of decoherence on quantum state assignments incorporating everything he is in a position to know about the physical situation of d, which may itself be represented in other magnitude claims. We have assumed that Eugene is in a position to know that John will initiate a certain interaction between s and d inside the laboratory. Eugene's physical situation does not require him to assign the state $|\psi_E\rangle$ to the contents of the laboratory when assessing the content of C: to do so would be to neglect information to which he has access in this situation that is relevant to assignment of quantum states to s and d.

There are circumstances in which agents are so differently situated that they correctly assign different quantum states to the same system, where each agent's assignment is based on all information to which the agent has access, given his/her physical situation.[12] If models of decoherence based on such different assignments were to result in these agents assigning different contents to the same magnitude claim,

---

[12] We encountered an example of this in Chapters 4 and 10, when Alice and Bob perform space-like separated polarization measurements on an entangled photon pair. Knowing his outcome, Bob can assign

that could threaten the objectivity of description in quantum theory. To assess the threat we need to ask how far different agents' state assignments may differ, and how this affects the models of decoherence they should employ in assessing the content of relevant magnitude claims.

The example in footnote 12 prompts the following restriction on different agents' state assignments: If Alice assigns a mixed state to a system while Bob assigns a pure state, then Bob's state vector lies in the support of Alice's density operator. (The support of a density operator is the smallest subspace containing all its eigenvectors with non-zero eigenvalues.) An argument for the following generalization of this restriction has been offered (Brun et al. [2002]): Assignments of different density operators by multiple agents are compatible if and only if the supports of all these operators have at least one vector in common. The argument assumes that mutually compatible state assignments each incorporate some subset of a valid body of currently relevant information about the system, all of which could, in principle, be known by a particularly well-informed agent.

That assumption can fail if Alice and Bob each perform non-disturbing space-like separated measurements of a different component of photon polarization on photons L,R respectively from an entangled pair. After recording polarization B_R for his photon R, Bob should assign polarization state $|B_L\rangle$ to Alice's photon L in the region of his forward light cone that does not overlap Alice's. If his recording of linear polarization were to proceed by a non-disturbing polarization measurement, then in that same region he should assign polarization state $|B_R\rangle$ to photon R, and state $|B_L\rangle |B_R\rangle$ to the pair. If Alice's coincident recording of the linear polarization A_L of photon L were to proceed by a non-disturbing polarization measurement, then in the region of her forward light cone that does not overlap Bob's she should assign polarization state $|A_L\rangle$ to photon L, $|A_R\rangle$ to R, and $|A_L\rangle |A_R\rangle$ to the pair. No vector lies in the supports of both density operator representations of states $|B_L\rangle |B_R\rangle$, $|A_L\rangle |A_R\rangle$, though both states do lie in the same four-dimensional polarization subspace.

Suppose differently situated agents assign quantum states to a system S that differ in some way consistent with the assumption of (Brun et al. [2002]). Assuming this is the only difference between the models of decoherence they use in assessing the content of magnitude claims about S, there are no grounds for thinking they will arrive at different assessments. By assumption, all their models share the generic features of applying the same Hamiltonians to the same initial state of S's environment with the same limitations on prior system/environment entanglement. Since a model with these generic features will decohere the phase of a generic pure state of S to the same extent in the same "basis",[13] that is how it will model the decoherence of every vector in

---

a pure polarization state to the photon entering Alice's detector. Since this information is not accessible to Alice, she correctly assigns that photon a mixed polarization state.

[13] The scare quotes mark the need to allow for models in which decoherence effectively "diagonalizes" s's density operator in an overcomplete basis of wave functions peaked around precise values of both position and momentum.

the subspace in which lies the vector common to the supports of all the states assigned by differently situated agents subject to the assumption of Brun *et al.* [2002]. It will also model the decoherence of vectors like $|B_L\rangle|B_R\rangle$, $|A_L\rangle|A_R\rangle$ that don't satisfy that assumption. It follows that the model of decoherence each agent uses to assess the content of magnitude claims about S will lead each agent to the same assessment of the contents of these claims. This means they will all agree on the content of these magnitude claims, thereby securing their objectivity.

11.6 Independent Verifiability

It is ironic that observation poses a threat to the objectivity of physical description in quantum theory, since observation is generally used to *settle* questions about objectivity in science and daily life. Doubting the objective presence of the dagger he saw, Macbeth tried to grasp it; to prove that Banquo's ghost occupied his own place at table, he beseeched his guests to see for themselves. Classical physics permits multiple, independent observations on a system to verify a claim about its state, since none of these need irremediably disturb that state.

But suppose one is wholly ignorant of the conditions that would back assignment of a quantum state to an individual system. No single observation on it can reliably disclose that state. Repeated observations are no better, since observing a system typically irreparably disturbs its quantum state. So even if a system's wholly unknown quantum state could be ascertained by a single observation, this finding could not be checked in further observations, either by the original agent or by others. So a wholly unknown quantum state of an individual system is not as objective as a corresponding classical state in yet another, epistemic, sense: it is not independently verifiable. Ollivier *et al.* [2004] put the point like this:

> The key feature distinguishing the classical realm from the quantum substrate is its objective existence.

They propose what they call an operational definition of objectivity for a property of a quantum system, according to which such a property is simultaneously accessible to many observers who are able to find out what it is without prior knowledge and who can arrive at a consensus about it without prior agreement. Their idea is that such a property is objective to the extent that multiple records of it exist in separate portions of the environment, so that "observers probing fractions of the environment can act as if the system had a state of its own—an objective state". They say that:

> The existence of an objective property requires the presence of its complete and redundant imprint in the environment as necessary and sufficient conditions.

While Ollivier *et al.* [2004] never say exactly what they mean by a property, it seems clear they would count a canonical magnitude claim locating a value in Δ of a magnitude on system S as a property assignment to S.

The state of a system in classical physics is specified by a point in phase space: this is equivalent to an assignment of a value to each dynamical variable on that system, that is, an assignment of a property locating the value of this magnitude in a unit set. The authors' stand-in for the objective state of a quantum system S interacting with its environment is one of the eigenvectors of the operator $\hat{M}$ on S corresponding to a dynamical variable M (a "preferred observable") selected by environmental decoherence (see Chapter 5 and Appendix C). A canonical magnitude claim locating a value of M on system S in Δ is taken to assign S an *objective* property solely on the grounds that a complete and redundant imprint is present in the environment.

There are other magnitudes N_r on S represented by operators $\hat{N}_r$, each of whose eigenvectors is close to an eigenvector of $\hat{M}$. A canonical magnitude claim locating a value of N_r in Δ on system S also counts as assigning S a reasonably objective property because of a complete and (slightly less) redundant imprint in the environment. Their idea seems to be that the proliferation of imprints of properties of a quantum system in its environment progressively objectifies its properties until these come to mimic properties that characterize a classical state. An eigenvector associated with the preferred observable stands in for an objective state by specifying what properties of the system are objectified by the environmental interactions to which it is subjected.

[A]mplification of a preferred observable happens almost as inevitably as decoherence, and leads to objective classical reality.

This is Quantum Darwinism: "the idea that the perceived *classical reality* is a consequence of the selective proliferation of information about the system". It is not an account of classical reality, or of the actual objective state of a quantum system. If I have understood it correctly, it is an account of how independent acts of observation on a system's environment can produce consensus on properties of a quantum system irrespective of whether or not that system has such properties. One is reminded of Wittgenstein's remark in his *Philosophical Investigations*:

As if someone were to buy several copies of the morning paper to assure himself that what it said was true.

Wittgenstein's avid reader is actually in better shape than multiple quantum observers. The morning paper may have correctly reported what happened. But if no magnitude claims about a quantum system are true then it lacks the properties observers attribute to it, so whatever proliferates is in fact *misinformation*. It is important not to be misled by the causal language of imprints into thinking that objective properties of the environment are caused by objective properties of a decohered quantum system. As it stands, Quantum Darwinism (Ollivier *et al.* [2004]) fails to provide an adequate account of objectivity in the sense of the independent verifiability of magnitude claims because it does nothing to show how either those magnitude claims or statements about their environmental imprints can be objectively true, given the quantum states of system and environment. Groupthink does not amount to intersubjective verification. But the view of quantum theory presented in Part I can help the Quantum Darwinist take this crucial last step: and after s/he has taken it, s/he can be recruited as an ally in the common cause of securing the objectivity of quantum description. The contribution of Quantum Darwinism is then its account of how magnitude claims that are already objectively true can come to be widely known, even when they concern extraordinarily sensitive physical systems.

Environmental interactions modeled by decoherence endow certain magnitude claims with a significant content that requires even differently situated agents to seek agreement on their truth values. When a system is decohered by its environment, these will include claims about the value of its preferred observable. They will also include claims about magnitudes on subsystems of the environment that are correlated with the system's preferred observable. Since properties not only of the system but also of subsystems of its environment are in this sense objective, it makes sense to ask whether objective statements about a system can be independently verified by observing various portions of its environment. Quantum Darwinism may now offer illuminating answers to this question by providing quantum models of interactions between a quantum system and its multipartite environment. This is a way to show how differently situated agents can come to agree on the truth values of significant magnitude claims about a system without disturbing its state.

11.7 Conclusion: A Limit to Transcendental Objectivity

Scientific knowledge is a product of a scientific community. It is a presupposition of the scientific enterprise that no insurmountable barrier prevents independent investigators from communicating with each other and coming to agree on what they know about the world. That presupposition could fail if one scientist could not trust the sincere claims of another about the results of his or her investigations. And it could fail if statements about the world lacked the objective content needed to constitute scientific knowledge.

Reflection on the paradox of Wigner's friend has persuaded some people that quantum theory cannot be understood without careful attention to the role of conscious human experience (Wigner [1967, p. 172], d'Espagnat [2005]). But quantum theory as presented in Part I permits a consistent and unambiguous treatment of the paradox without reference to consciousness. Situated agents can use quantum theory to make objective statements about the values of magnitudes in the physical world, not just about observations of them. This has helped us to predict and explain an enormous variety of otherwise puzzling physical phenomena. A true statement about the value of a physical magnitude states an objective physical fact. Ordinarily, such facts are readily independently verifiable. But acceptance of quantum theory requires one to countenance the possibility that in extraordinary circumstances information about the value of a magnitude could be irretrievably erased, making this fact no longer verifiable.[14]

A magnitude claim ascribes a property to a physical system. The claim is objective in all four senses delineated in section 2. But the *content* of this claim is a function of the environmental context, and this may tempt one to say that that the property it ascribes is somehow not fully objective. Indeed there is an ideal of objectivity that is not met here, according to which a claim correctly ascribes an objective property to something just in case the world contains a corresponding object which intrinsically possesses that property when represented as it is in itself, irrespective of its environment and not just from the perspective of physically situated and cognitively limited agents. Nagel [1986] called this detached perspective the view from nowhere. A property or object that meets the ideal of being correctly represented in the view from nowhere may (with apologies to Kant) be called transcendentally objective.

A recent extension of the Wigner's friend "paradox" shows why acceptance of quantum theory requires rejection of the transcendental objectivity of properties ascribed in some important magnitude claims.[15] Suppose that Alice and Bob decide to conduct measurements of various polarization magnitudes, each on one of a pair of photons assigned entangled state $|\eta\rangle$ (see Chapter 3). Being lazy, they do not perform the measurements themselves, but delegate that task to their friends, Alice' and Bob' respectively, each of whom performs the required measurement in his or her otherwise completely isolated laboratory.

Suppose Alice' records the outcome of a linear polarization measurement with respect to axis a' while Bob' records the value of a linear polarization measurement with respect to axis b'. Assuming no collapse, each of Alice and Bob then correctly assigns the following quantum state Ψ to the combined system $L \oplus Al' \oplus R \oplus Bob'$, where (for example) Al' is a physical system consisting of Alice''s body and everything else in her laboratory prior to the entry of photon L whose polarization she measures:

$$\Psi = \hat{U}^A \hat{U}^B \frac{1}{\sqrt{2}} \left(|H^L\rangle |H^R\rangle + |V^L\rangle |V^R\rangle \right) \otimes |Al'\rangle \otimes |Bob'\rangle.$$

Here $\hat{U}^A$ is a unitary operator modeling the measurement interaction between photon L and Al' in Alice''s laboratory and $\hat{U}^B$ is a unitary operator modeling the measurement interaction between photon R and Bob' in Bob''s laboratory.

Now Alice and Bob bestir themselves and join the action. Alice "undoes" the entanglement between the states of L and Al' by turning on an interaction between them modeled by $\hat{U}^{A\dagger}$, while Bob "undoes" the entanglement between the states of R and Bob' by turning on an interaction between them modeled by $\hat{U}^{B\dagger}$, after which they

---

[14] This consequence is not peculiar to quantum theory. Classical physics also allows for the (in principle) possibility of "rewriting history" (e.g. by restoring the exact local situation prior to the melting of an ice cube in a glass of water).

[15] I first learned of this in a talk by Matthew Pusey at a workshop on Information-theoretic Interpretations of Quantum Mechanics held at Western University, Canada in June 2016.

each correctly assign $L \oplus R$ the same original state $|\eta\rangle$. Temporarily forgetting about their friends, Alice then records the outcome of a linear polarization measurement on L with respect to axis a while Bob records the value of a linear polarization measurement on R with respect to axis b. This entire sequence is repeated very many times, beginning each time with a pair of photons correctly assigned state $|\eta\rangle$. Of course what I have just described could never be carried out in practice, but this in no way affects the validity of the following argument.

Assuming transcendental objectivity of measurement records, on each occcasion Alice, Bob, and each of their friends actually made a record of the polarization of a photon (L or R respectively) with respect to an axis. But none of them is in a position to say what all those records were, since by "disentangling" the states of the friends and their laboratories from their respective photons, Alice and Bob irretrievably destroyed all traces of Alice''s and Bob''s measurement outcomes. None of the participants is in a position to take the view from nowhere. But from a detached perspective we can assume such records exist on every occasion, and consider how they may be statistically distributed over, say, a million repetitions of this sequence of operations.

For many choices of axes a, b, a', b' the universal applicability of quantum theory to this scenario implies that for state $|\eta\rangle$ the outcomes of pairs of joint measurements of polarization along (a, b), (a, b'), (a', b), (a', b') violate a Bell inequality (the CHSH inequality—see Chapter 4). But any statistical distribution of actual outcomes realized in many repetitions of the sequence of operations just described will almost certainly conform to that inequality.[16] So anyone who accepts quantum theory must deny the transcendental objectivity of records of the outcomes of linear polarization measurements on photons correctly assigned polarization state $|\eta\rangle$.

Does this challenge the presupposition of the scientific enterprise I stated at the beginning of this section? Not seriously. The conditions of the extended Wigner's friend scenario could never be realized in practice, and moreover exclude the possibility of communication to Alice and Bob of the knowledge acquired by their friends about the outcomes of their experiments. So the extended Wigner's friend scenario raises no problem of trust among sincere investigators. The (very distant) threat is only to an ideal according to which independently acquired scientific knowledge must always be universally sharable within the community. And, to repeat, this scenario in no way threatens the objectivity of the content of magnitude claims, though it does restrict that content.

Science is based on observed facts, and quantum theory is no exception. What makes quantum theory exceptional is what it teaches us about the nature of these facts. Bell ([2004], p. 41, emphasis in original) introduced the term 'beable' because:

---

[16] This follows from a theorem of Fine [1982] according which satisfaction of the CHSH inequalities is a necessary and sufficient condition for the existence of a joint distribution of four random variables, each of whose values represents a possible outcome of a corresponding polarization measurement.

it should be possible to say of a system not that such and such may be *observed* to be so but that such and such *be* so.

When using quantum theory to make a significant statement about the value of a magnitude on a physical system one is saying that it be so, often using:

> the familiar language of everyday affairs, including laboratory procedures, in which objective properties—*beables*—are assigned to objects.

Perhaps this makes this magnitude a beable in Bell's sense: he does say:

> the beables... can be described "in classical terms", because they are there. The beables must include the settings of switches and knobs on experimental equipment, the currents in coils, and the readings of instruments. ([2004], p. 52)

In Chapter 8 I called the listed magnitudes "assumables" rather than beables of quantum theory because they are not represented in that theory's models. But Bell thought that in a "serious, fundamental formulation" quantum theory would have its own beables:

> The beables of the theory are those elements which might correspond to elements of reality, to things which exist. (Bell [2004], p. 174)

These passages at least suggest that acceptance of quantum theory in no way modifies the content of statements about values of magnitudes—a content that is somehow established by a fixed representation relation between language and the world (actual, or merely possible if a beable *be* not!). One passage may be understood to commit Bell to what I have called the transcendental objectivity of claims about the readings of instruments such as those taken to record the outcomes of photon polarization measurements. But as we have just seen, acceptance of quantum theory precludes adherence to the transcendental objectivity of such records.

In any case, a pragmatist cannot accept a representational account of how content accrues to a statement (Brandom [1994], [2000]). To understand quantum theory one needs instead to adopt an alternative account of what gives a statement content. By modifying inferential relations involving magnitude claims quantum theory affects their content, rendering this contextual. By making the content of a magnitude claim about a system a function of the environment, acceptance of quantum theory cautions one against taking that claim to attribute an intrinsic property to an object that exists independently of environmental context, even while insisting on the objective content of the claim. This should make one think differently even about the content of everyday statements about ordinary things like the settings of switches and knobs on experimental equipment, the currents in coils, and the readings of instruments.

12

Meaning

12.1 Introduction

Quantum theory is a new kind of science because it warrants our acceptance of its fundamental status in twenty-first-century physics despite lacking any beables of its own. That is what makes it a truly radical theory—or so I claimed in the introductory part of this book. After describing the theory and explaining how its models are applied it is time to make good on this claim. To do so I need to talk about meaning—about how scientific terms and concepts acquire their content and how we use some of them to describe and represent the physical world when applying quantum theory.

Although representation and conceptual content are familiar topics in recent philosophy, currently prominent analyses of these topics are not helpful in accounting for the significance and representational content of key terms that figure in the application of quantum models. But there is a less popular philosophical approach to content that is just what is needed to understand the significance of talk of quantum states, Born probabilities, and magnitude claims: Brandom [1994], [2000] calls it pragmatist inferentialism. Building on the pragmatist idea that meaning derives from use, this takes public use of language to manifest patterns of appropriate inference both internal to language or thought and in external links to perception and action. Since it is statements, not terms, that bear one another inferential relations, statements are the primary vehicle of content. A term acquires whatever content it has from the way it contributes to the content of statements involving it.

Quantum theory is revolutionary because our non-representational use of quantum models has illustrated a new style of scientific theorizing. In explaining how quantum models are used, I have called upon the resources of pragmatist philosophy at a number of points: in stressing the priority of use over representation in quantum modeling, in accounting for the function of Born probabilities, and in saying what is and what is not causal about explanations of phenomena using quantum theory. These resources were independently available. Pragmatist approaches to theory-use, probability, and causation need no defense from quantum theory: they can be, and have been, supported by quite general philosophical arguments.

The same is true of a pragmatist inferentialist approach to conceptual content. As we shall see, application to quantum theory reveals its power by helping clear up long-standing worries about the theory's conceptual foundations. But the approach is quite general. According to pragmatist inferentialism, concepts of classical physics, of the rest of science, and of daily life *all* get their content from how they help determine the inferential role of statements in which they figure. Here too quantum theory reveals what philosophers should have known anyway. To emphasize how quantum theory reveals the power of this and other currently underappreciated strands of pragmatist philosophy this book might be better known as *The Quantum Revelation in Philosophy*.

All our concepts, including those of physics, derive their content from use in human communication. Even quite primitive non-human animals perceive their environment and act under the guidance of their perceptions of it. But representation in perception and action does not require the kind of rich conceptual content accessible only to those able to make statements in a common language. Humans have used concepts to navigate their physical as well as social environment as long as they have been able to communicate with language. The concepts of contemporary physics are a sophisticated development from those first deployed in the physics of Galileo, Descartes, and Newton in the seventeenth-century scientific revolution—concepts of space, time, motion, and material object. Humans, along with many organisms lacking a capacity for language, are able to represent space, time, motion, and body in perception. So these first physical concepts were naturally employed to describe what all normal human adults were capable of representing even without language.

But it took the genius of Galileo and Descartes to appreciate the power of converting descriptions of the physical world in terms of space, time, motion, and material object into mathematical representations of magnitudes and extending these representations beyond what was humanly perceptible to a vastly enlarged universe. Newton's mechanics then introduced the concept of force (including gravitational force), requiring a distinction between weight and mass: its development permitted introduction of further concepts such as energy (kinetic as well as potential), power, and momentum (angular as well as linear). His calculus permitted a much richer mathematical framework in which to manipulate representations of all these magnitudes. But Newtonian physics had begun to deploy concepts of new magnitudes not directly represented in perception.

How did concepts such as force and mass acquire their content, if not directly by association with perceptual representations? In each case, there remained a limited connection to perception. If mass is a measure of quantity of matter, then doubling the volume of, say, water will double its mass. But this connection does not extend to materials of different densities. Similarly, bodily exertion may in some cases provide a crude comparison of the sizes and directions of forces. Concepts of mass and force acquired their content as dynamical variables initially through their roles in Newton's theory. Operation of a balance directly compares the effect of the same force (of gravity) on different objects: it is use of Newton's laws of motion that licenses an inference to a statement comparing their masses. These same laws license inferences from masses and observation of their (accelerated) motion to the forces responsible for this motion. Use of Newton's law of gravitation similarly allows inferences to the masses of the Earth, other planets, and sun.

Later developments in classical physics provided additional theoretical connections with new dynamical variables such as electric current and magnetic field strength. The connections endowed such new concepts with content by providing inferential links to statements about mass, force, and other Newtonian variables: they simultaneously enriched the conceptual roles of the latter variables. Acceptance of relativity theory then required modifications in all these conceptual roles because of the theory's revision of our concepts of space, time, motion, and gravitation. It is possible to describe this as a process in which we came closer to the truth about the same magnitudes, and that our concepts of mass, force, etc. then better reflected the character of these beables. One doesn't have to adopt the operationist view that each change of theoretical framework changes what magnitudes 'mass' and 'force' denote, corresponding to the different measurement procedures permitted within each framework.

Quantum theory introduces some new concepts into physics and modifies the content associated with other concepts, including the dynamical variables of classical physics. In the next section I show how key concepts introduced with quantum theory acquire their content. Subsequent sections explore what modifications in the content of other concepts are required by acceptance of quantum theory.

12.2 Some Novel Quantum Concepts

The new concepts introduced with quantum theory have much looser links to magnitudes such as length, duration, momentum, and energy, but acceptance of quantum theory significantly modifies the conceptual roles of those magnitudes. The most important of these new concepts are observables, quantum states, and Born probabilities, each represented by some mathematical object. An observable is ordinarily represented by a linear (indeed self-adjoint) operator; a quantum state is ordinarily represented by a state vector, wave function, or density operator; a Born probability is represented by a number between 0 and 1 inclusive.

When commenting on how Dirac [1930] understood the concept of an observable in Chapter 2 (footnote 6), I remarked that it is now applied to any dynamical variable corresponding to a self-adjoint operator. This was an oversimplification. Following Heisenberg's insistence that a quantum theory should concern only those magnitudes that could be measured, Dirac sought to formulate the principles of quantum theory in terms of only measurable dynamical variables. This was what motivated his restricting the theory to *observables*. He took that restriction to exclude dynamical variables that may take on complex (not just real) numbers as values. The Born rule implies that there is no chance of getting anything other than a real value in a measurement of a dynamical variable represented by a self-adjoint operator. So Dirac would appear to have succeeded if a quantum observable is just a dynamical variable corresponding to a self-adjoint operator.[1]

Indeed it is now ordinarily assumed that every observable on a system corresponds to a self-adjoint operator on a Hilbert space used to assign it states. That assumption likely explains but does not excuse an ambiguity in the term 'observable', which may be taken to refer either to a dynamical variable or to the corresponding operator. When applied to the operator the concept of an observable is used in a novel way to refer to a mathematical element of a model of quantum theory. But the operator itself is not used to represent a physical entity or magnitude newly posited by quantum theory: a self-adjoint operator is not a beable. Instead the operator is used in a legitimate application of the Born rule to calculate the expectation value or probability distribution of the measurable dynamical variable to which it corresponds.

Many dynamical variables of classical physics are associated with self-adjoint operators in quantum models: these include (components of) position, momentum, and angular momentum; kinetic, potential, and total energy; and components of electric and magnetic field intensity at each point. But already in a finite-dimensional Hilbert space there are many self-adjoint operators that don't obviously uniquely correspond to any classical dynamical variable; and even the space in which translational states are assigned to a single particle is infinite-dimensional, with a huge (in fact infinite) set of self-adjoint operators.

If *every* self-adjoint operator were to correspond to an observable, then quantum theory would seem to introduce a host of novel physical magnitudes. But for various reasons one should deny that every self-adjoint operator corresponds to a dynamical variable.[2] Spin is often cited as a novel quantum magnitude. But spin is a form of angular momentum—a classical dynamical variable—and while quantum theory treats all angular momentum in a novel way (e.g. by quantizing its possible values), magnitude claims about spin components and total spin are not claims about some physical magnitude newly posited by quantum theory.

There are dynamical variables unknown to classical physics about whose magnitudes acceptance of quantum theory licenses one to make claims. Quantum theory may play a heuristic role in alerting us to the possible existence of novel "classical" beables—assumables—without describing or representing any itself.[3] That would not

---

[1] Dirac ([1930], pp. 36–7) proposed a further restriction, but following von Neumann's rigorous reformulation this has no longer seemed appropriate.

[2] If a system is subject to superselection rules then only operators that commute with all superselection operators can represent observables: so an observable on a system of indistinguishable particles must correspond to an operator that commutes with an operator implementing a permutation symmetry. If $\hat{\psi}$ is a Dirac field operator in quantum field theory with adjoint $\hat{\psi}^\dagger$, then $\hat{\psi} + \hat{\psi}^\dagger$ is a self-adjoint operator representing no observable.

[3] The Standard Model has certainly enabled us to make claims about magnitudes (strangeness and color) and entities (the top quark and the Higgs field) unkown to classical physics. But while its quantum gauge field theories license claims about them, these magnitudes and entities are neither part of nor represented by models of those theories.

undermine the thesis of this book—that quantum theory is revolutionary not because it describes or represents new beables, but because by applying this theory we have been able better to describe the physical world without such quantum beables.

One way to appreciate how loosely quantum states, observables-as-operators, and Born probabilities are related to dynamical variables including length, duration, momentum, and energy is to see that none of these new concepts denotes a magnitude with dimension: it cannot be represented on a scale of *units* like one based on the meters, seconds, and kilograms of the SI system. This corresponds to the fact that one cannot find the value of a state vector, operator, or Born probability simply by measuring values of dynamical variables on a single system and performing theoretical calculations.

Probability figured in physics before quantum theory. Physicists first used probability theory to model experimental data and assess its theoretical significance. Probability was then introduced into physical theory by statistical mechanics, though its role in that theory remained controversial: statistical mechanics was not regarded as a fundamental theory within classical physics—that role was reserved for Newtonian mechanics and Maxwell's theory of electricity and magnetism. Quantum theory's novelty was to give probability a key role in fundamental physics.

By performing many measurements of a dynamical variable on a large collection of similar systems one can amass statistics justifying an inference to its probability distribution. Repeating this for several variables, each corresponding to a suitable operator, one can infer the quantum state that yields their probability distributions via the Born rule. Applying the Born rule with an operator believed to correspond to a dynamical variable (and therefore to represent an observable), one can then calculate the expectation value [4] of that observable in this state. Extensive experimental statistics may warrant confident conclusions about a quantum state, the correspondence between dynamical variables and operators, and expectation values of observables in that state. But, unlike in classical physics, a single data point is almost wholly uninformative; and reliance on extensive statistics requires the assumption that each data point results from just the same type of measurement on an exactly similar system.

This is a clue to the novel functions of quantum states, observables-as-operators, and Born probabilities in quantum theory. These novel concepts don't acquire their content by representing beables—purported elements of physical reality denoted by corresponding magnitude terms. They acquire their content through the way they function in applications of quantum theory: in no case is that function to represent anything in the physical world.

The function of a quantum state (represented by a wave function, vector, or density operator) is not to represent properties of a physical system to which it is assigned, nor anyone's knowledge of its properties. What gives a quantum state its content is not what it represents but how it is used—as an *informational bridge*. I introduced the

---

[4] See Appendix A, § A.2.

notion of a quantum state's backing conditions in Chapter 9 and gave more examples in Chapter 10. Knowing about some physical situation (its backing conditions), an agent may assign a quantum state to form expectations about certain other possible physical situations (described by canonical magnitude claims of the form *The value of dynamical variable M on physical system s lies in* Δ). Quantum states are objective: only expectations based on correct state assignments are generally reliable. But a quantum state represents neither its backing conditions nor any canonical magnitude claim, though it helps one to understand how these are linked.

Since its main function is to provide information on how strongly to expect significant magnitude claims to be true given prevailing backing conditions, a quantum state assignment may be said to represent probabilistic relations between backing conditions and canonical magnitude claims. These probabilistic relations are objective: they would exist in a world without agents, as long as that world featured patterns of statistical regularity that were sufficiently stable to be modeled by Born probabilities of the values of dynamical variables specified by canonical magnitude claims. We can use quantum theory successfully because (or at least insofar as) they do exist in our world.

As physically situated, and so epistemically limited, a user of quantum theory can assign quantum states on the basis of what that agent is in a position to know, in order to form reasonable expectations about what that agent is not in a position to know. That is how quantum states function as informational bridges. A quantum state provides a sturdy informational bridge only if it would be the right state to assign for any agent in that physical, and therefore epistemic, situation—only if it represents the actual probabilistic relations between its accessible backing conditions and inaccessible magnitude claims.

The metaphor of a bridge illuminates the way concepts of quantum states, observables-as-operators, and Born probabilities get their content in quantum theory. A bridge is an artificial construction that rises above some natural obstacle, connecting solid ground on either side. A quantum state is built on the solid ground of the backing conditions and significant magnitude claims it links. In between, the absence of suitable environmental interactions presents a natural obstacle to other significant claims. Backing conditions and significant magnitude claims describe or represent physical reality: they are about beables—quantum assumables. A quantum state operates at a higher level of abstraction—it is not a beable. But its use to yield Born probabilities for significant magnitude claims whenever its backing conditions are met is the key to quantum theory's great predictive and explanatory success.

In Chapter 5 I explained circumstances in which quantum states may be assigned and the grounds for their assignment. We can now see how the inferences involved contribute to the content of a statement assigning a quantum state—a *quantum state assignment* (QSA).

Some of these inferences are formal—they concern the mathematical representation of quantum states. Distinct mathematical objects represent the same quantum state if and only if they yield the same measure μ on subspaces of the Hilbert space $\mathcal{H}$ of the system whose states they may represent: so $|\psi\rangle$, $-|\psi\rangle$, $e^{i\theta}|\psi\rangle$, and $\hat{\rho} = |\psi\rangle\langle\psi|$ all represent the same quantum state. But since all QSAs are relative to the situation of an actual or hypothetical agent, a system may be assigned state $\hat{\rho}_1$ relative to one agent-situation but $\hat{\rho}_2$ relative to a different agent-situation, even though $\hat{\rho}_1, \hat{\rho}_2$ yield incompatible Hilbert space measures μ_1, μ_2. Chapter 5 provided an example in which a system was assigned a pure state relative to one agent-situation but a mixed state relative to a different agent-situation. There are even circumstances (see §11.5) in which a system may be consistently assigned distinct pure states, each relative to a different agent-situation. But all these QSAs assign states in the same Hilbert space. They may all be thought to assign states to the same abstract system, defined solely by the dimensionality of this Hilbert space, as when an abstract two-dimensional Hilbert space system is called a qubit.

Other inferences are what Sellars [1953] called *material* inferences. These make a more substantial contribution to the content of a QSA. They importantly include inferences from backing conditions to a QSA they back.

Often physicists assign a quantum state to a system based on prior manipulations they describe as *preparing* that state. A polarization state may be prepared by passing light through a crystal of Iceland spar and selecting the ordinary ray that emerges. A specific excited internal quantum state of atoms in a beam may be prepared by passing them through a suitable sequence of laser and microwave pulses: additional laser beams can be used to modify their translational quantum states so these would yield negligible probabilities for velocities outside a narrow range. The conditions backing such state assignments may be specified by describing the devices used to prepare the state and the processes involved. These conditions may be described without invoking quantum theory. Even if they are not specified by making magnitude claims, it is reasonable to assume that the conditions' holding depends on the truth of unspecified, underlying magnitude claims (on which they may be said to supervene).

Many QSAs are independent of any manipulation. Quantum states are often not prepared but assigned to systems already there in the big world outside any laboratory. A macroscopic object such as a Stern–Gerlach magnet or a football may be assigned a quantum state sharply peaked in its center-of-mass position because environmental interactions are well modeled as leading to effectively instantaneous decoherence in (approximate) position basis: because of the object's relatively enormous mass, this state is still compatible with a very sharply defined momentum. An isolated atom in a cold enough environment anywhere in the universe may be assigned its lowest energy (or ground) quantum state.

One cannot define a quantum state in terms of the inferences involved in assigning it: More inferences are added with every new application of quantum theory. Moreover, these concern only the circumstances of application of the quantum state concept, whose content is also conferred by the consequences of such application. I have said that quantum states function not descriptively but prescriptively. Content-conferring inferences *from* a QSA do not lead directly to magnitude claims or descriptive statements whose truth they determine. These inferences fall into two classes which I will call probability inferences and significance inferences.

The main function of a QSA is as input to the Born rule. A legitimate application of the Born rule issues in probability assignments to canonical magnitude claims. These in turn imply statements about expectation values of dynamical variables as well as other features of their probability distributions such as variances and covariances. As I explained in Chapter 9, these probability statements are objective, though they don't describe propensities and need not correspond to frequencies or anyone's actual degrees of belief. They have the same function as any probability statement—the normative function of requiring an agent not in a position to know whether a statement is true to match credence in that statement to the probability it prescribes. A material inference from a QSA to a probability statement implied by a legitimate application of the Born rule helps confer content on that QSA: it is a probability inference. There are very many such inferences. Each application of quantum theory to a new kind of system adds to their number and variety. A quantum state cannot be defined by probability inferences, even together with inferences from its backing conditions.

Chapter 9 showed how we can use quantum theory to explain many otherwise surprising probabilistic phenomena. Each such explanation involves a legitimate application of the Born rule to a quantum state, and this in turn requires probability inferences from that state. So a QSA acquires content from every successful explanation using it in quantum theory, just as this exhibits the explanatory power of that state assignment. A single QSA may be used to explain a variety of different phenomena, thereby unifying them and certifying the truth of that QSA as required by the best explanation of all these phenomena. But this does nothing to show that a quantum state is a beable—that $|\psi\rangle$ or $\hat{\rho}$ represents a beable intended to correspond to an element of physical reality. The truth of a QSA does not consist in its correspondence to a state of physical reality: a quantum state is linked to physical reality more indirectly, through its role in prescribing credence via application of the Born rule and advising on the significance of magnitude claims, including certifying some as significant enough to license application of that rule.

So a QSA further gains content through inferences to the significance of magnitude claims. The significance of a magnitude claim depends on how a system is interacting with its environment. Quantum theory may be used to model that interaction by assigning a quantum state to system plus environment and applying the quantum theory of environmental decoherence as described in Chapter 5. If the model shows rapid and robust decoherence of the system's quantum state in a basis of states naturally associated with some dynamical variable,[5] then every magnitude claim assigning a real-number value to that variable will have a precise content: one such claim will be

---

[5] The association is set up by the self-adjoint operator uniquely corresponding to that dynamical variable. The states in question are pairwise orthogonal eigenvectors of that operator (see Appendix A).

true, the rest false. That is a situation in which one may legitimately apply the Born rule to assign probabilities to canonical claims about the value of this variable.

But the Born rule is frequently applied without justifying or even making use of any such model of decoherence. Then physicists must rely on tacit knowledge or rules of thumb to decide when and to what magnitude claims it is legitimate to apply the Born rule. This does not mean we lack an exact formulation of quantum theory: we can precisely describe its mathematical models, and the Born rule may be stated without using any ambiguous term like 'measurement'. Practical knowledge is involved in applying any physical theory: even what looks like a precise rule for applying a theory ultimately depends on human consensus on how that rule is to be applied. No new conceptual problem arises when applying quantum theory's Born rule.

But the significance inferences that contribute to the content of QSAs inevitably also affect the content of the magnitude claims whose significance they concern. In this way acceptance of quantum theory affects the content of all magnitude claims, though for most such claims (especially concerning ordinary objects above a microscopic scale) the effect may be safely neglected.

12.3 The Content of Magnitude Claims

Philosophers customarily regard a claim as meaningful if and only if it expresses a definite proposition when made in an adequately specified context: otherwise it is taken to be meaningless. An inferentialist approach to the content of an empirical claim accepts a role for context but replaces this "digital" view of content with an "analog" view. When a quantum state specifies the context for a magnitude claim, it does so by specifying the inferential power of particular magnitude claims in that context: inferential power comes in degrees, and so, therefore, does content. The quantum state exercises its core function only for a magnitude claim with sufficiently high content. That function is to advise an agent to believe magnitude claim C to degree p if and only if $\Pr(C) = p$ is the probability that results from applying the Born rule to this quantum state.

On this view, the content of a magnitude claim is governed by its place in a web of material inferences connecting it to other claims, and hence to perception and action. A quantum state offers advice on the content of a magnitude claim by indicating its place in this inferential web. It thereby marks a contextual element to the content even of claims about the properties of familiar objects like gross experimental apparatus and the moon. But by modeling the behavior of quantum states, quantum theory itself reassures us that only for claims about currently unfamiliar objects (or familiar objects in unrealizable situations) does the consequent modification of content amount to anything.

I go on to develop and defend this inferentialist account of the content of magnitude claims as these arise in applications of quantum theory. This section shows how the content of a magnitude claim about a physical object comes to depend not only on what that object is but also on its physical environment. This dependence of content on context does not fit standard philosophical models of how the pragmatics of discourse determine what proposition a claim expresses. But the rival inferentialist account may seem unable to nail down a claim's content sufficiently to explain how this may be unambiguously communicated in public discourse. Section 4 responds to this challenge by means of examples that illustrate how acceptance of quantum theory modifies the content of certain magnitude claims about physical objects as a result of specific alterations in the inferences these support.

It is natural to assume that in offering advice about the cognitive significance of a magnitude claim C about physical system s a quantum state is assigned *to s*. But to get advice about physical system s one sometimes applies a quantum model by assigning a quantum state to a *distinct* target system t.

In quantum field theory a quantum state is assigned to one or more quantum fields. As a target of the application, a quantum field is a physical system and so an *assumable* of quantum field theory. But the application does not yield advice about any quantum field magnitude. Quantum fields are not beables of quantum field theory. A model of quantum field theory includes mathematical objects called quantum field operators: a quantum field then consists in an assignment of one such operator at each point or region in space-time in accord with the theory's field equations. A quantum field operator is not used to represent any physical magnitude in an application of the model.

In an application of quantum field theory a quantum state assigned to a quantum field offers advice about some physical system *other than* the quantum field to which it is assigned. Section 5 considers how such a model may be applied to offer advice about the cognitive significance of claims—either about particles or about "classical" fields, depending on the physical environment in which it is applied.

I shall consider a number of magnitude claims to show how a quantum state $\hat{\rho}$ may be taken to gauge their content. In this section I consider only cases where $\hat{\rho}$ functions to offer advice concerning the content of magnitude claims about the physical system s to which it is assigned, and so plays no role in securing a referent for 's'. This is particularly obvious in my first example, where s is the moon!

In the 1980s, David Mermin wrote a pair of elegant little papers (Mermin [1981], [1985]). He had already answered the title question of one paper in the third sentence of the other: "We now know that the moon is demonstrably not there when nobody looks" ([1981], p. 397). Suppose, for the moment, that 'there' locates the moon within a "moon-sized" region R of space some 250,000 miles from the Earth with cross-section C in a plane at right-angles to a line joining the centers of Earth and moon. Choose Cartesian coordinates (x, y) in that plane and call D the diameter of C along the x-axis. Then the statement

C_1:   The x-component of the moon's position lies in D

is a canonical magnitude claim.

Taken literally, Mermin's answer implies that C_1 is false when nobody looks (at the moon). But I doubt he meant it literally: C_1 functioned as a metaphor for him, suggested by a rhetorical question of Einstein to which I shall return.[6] The demonstration Mermin appealed to concerns not the moon and its position, but the spin of a spin 1/2 particle (such as an electron or a silver atom). C_1 is a metaphor for:

C_2: The x-component of the particle's spin is r.

It is C_2 that Mermin designed his demonstration directly to refute (for any x-axis and any real value r) if no one measures the x-component of the particle's spin, even though a measurement of its spin x-component always yields one of two real values (about $\pm 0.50 \times 10^{-34}$ in appropriate units).

C_1 and C_2 don't just ascribe different magnitudes: they ascribe them to very different systems. Mermin could just as well have appealed to a more complex demonstration to argue for the falsity of the statement

C_3: The x-component of the particle's position lies in d

(where d is any sufficiently small interval of real numbers) if the x-component of the particle's position is not measured. But while that demonstration also could be given "with an effort almost certainly less than, say, the Manhattan project" (Mermin [1981], p. 398), the moon so differs from an electron or atom that quantum theory itself gives us overwhelming reason to think a similar demonstration that C_1 is false unless the moon is "observed" will forever exceed the powers of human or any other physically instantiated agents.

The moon and an electron or silver atom differ not only in size but in interactions with their physical environment. Perhaps surprisingly, while an electron or atom is sometimes so isolated that its environment can be neglected, that is never true of the moon. Newtonian physics gave an excellent model of the moon by neglecting all effects of the sun's illumination and impacts by stray matter and taking the moon's environment to affect its state only through gravitational forces. In classical physics, the way to incorporate these effects into an even better model is as additional external forces on a system (the moon) that can affect how its state changes, but not what counts as its state at any moment—they affect its dynamics but not its kinematics. In quantum theory, taking account of a system's interactions can alter the nature of its momentary state, as well as how its state changes.

As we saw in Chapter 3, Schrödinger [1935] introduced the term 'entanglement' to make this point as follows:

When two systems, of which we know the states by their respective representatives, enter into temporary physical interaction due to known forces between them, and when after a time of mutual influence the systems separate again, then they can no longer be described in the

---

[6] "... during one walk Einstein suddenly stopped, turned to me and asked whether I really believed that the moon exists only when I look at it" (Pais [1979], p. 907).

same way as before, viz. by endowing each of them with a representative of its own. . . . By the interaction the two representatives have become entangled. (p. 555)

I have used the symbol '$\hat{\rho}$' to stand for a density operator representation of a quantum state. In this quote, the author of wave mechanics used the term 'representative' to refer to a wave function $\psi(\mathbf{r}_1, \ldots, \mathbf{r}_n)$—a (typically) complex-valued function of the positions of the n particles in a system. It is often more convenient to regard a wave function as a vector $|\psi\rangle$ in an abstract vector space. Schrödinger's point, then, is that almost any interaction between two quantum systems assigned vector quantum states will result in a joint state in which neither system may be assigned a vector state.

But in Chapter 5 we saw that a system may be assigned a different kind of quantum state. When a composite of two or more quantum systems is assigned vector state $|\psi\rangle$, each of its subsystems may still be assigned a state, since there is a distinct mathematical object—a density operator—that plays that role. Suppose a composite system is assigned vector state $|\psi\rangle$. Then, for each of its subsystems s, $|\psi\rangle$ defines a unique density operator $\hat{\rho}_s$ with the following property: when applied to $\hat{\rho}_s$, the Born rule gives the same probability distribution for every magnitude on s as it does when applied to $|\psi\rangle$.[7] Insofar as generating Born probabilities is the core function of a quantum state, this makes the density operator $\hat{\rho}_s$ the quantum state of s. The notation $\hat{\rho}$ subsumes vector states as a special case, since each vector state $|\psi\rangle$ uniquely corresponds to a density operator state $\hat{\rho} = |\psi\rangle\langle\psi|$. An electron or atom is sometimes sufficiently isolated from its environment to be assigned a vector state, at least for a while. The moon's constant interactions with sunlight and stray matter may be weak, but they ensure that the moon can only be assigned a density operator quantum state.

We are almost in a position to see why quantum theory permits an agent to claim C_1 whether or not anyone is looking, while forbidding any claim of the form C_2 or C_3 for an atomic or subatomic particle except while it is subjected to the right kind of measurement. Very roughly, the answer is that the moon's position is constantly being measured by its environmental interactions with sunlight and stray matter, while a measurement of the particle's spin or position occurs only under very specific circumstances which don't always obtain. But we need to explicate such talk of "looking" and "measurement" to re-express this answer in kosher quantum-theoretic terms. This requires a quantum model of measurement. Chapter 6 noted that attempts to model measurement quantum-theoretically that give the quantum state a descriptive role founder on the quantum measurement problem. So the quantum state in the model must be understood to function non-descriptively.

It is natural to model the moon's interaction with its external environment as a continual series of collisions with small particles.[8] Even if the moon and each of these particles had a vector state before the collision, that would not be so afterwards.

---

[7] See Appendix A, § A.6.
[8] The assumption that these include photons—particles of light—is in need of justification, as we shall see in section 5.

214 MEANING

Detailed models of this type use plausible assumptions to show that whatever its hypothetical initial quantum state, the moon would extremely rapidly assume a density operator state $\hat{\rho}$ of a particular form, and then stay in such a state.[9]

This is not a state in which Dirac would allow one to say the moon has a precise position. But at every moment t, $\hat{\rho}(t)$ will define a set of vector states $|\psi_x(t)\rangle$ ($x \in \mathbb{R}$) with several special features:

(1) It is stable—if $|\psi_x(t_1)\rangle$ is an element of the set $\hat{\rho}(t_1)$ defines at t_1 then $|\psi_x(t_2)\rangle$ is an element of the set $\hat{\rho}(t_2)$ defines at t_2;
(2) $|\psi_x(t)\rangle$ approximates a classical mechanical state in the following sense: the Born probability distributions it yields for x-components of position and momentum are each concentrated around precise values (x, p_x respectively) and are consistent with the corresponding single probability distribution derived from a *joint* probability distribution on a space of classical states for a system of precise but unknown position and momentum;
(3) the classical state with values x, p_x obeys classical mechanical equations of motion.

These features suggest the following pragmatic rule for assigning content to each statement C_{1x} that locates the moon's center of mass at position x: assign a high content at t to each statement C_{1x} if the $|\psi_x(t)\rangle$ have features (1)–(3). Applied to $\hat{\rho}(t)$, the Born rule generates a probability distribution over the statements C_{1x}. Given their high content, quantum theory now advises an agent applying the model to believe C_1 to the degree corresponding to its probability under this distribution. The quantum model does not itself specify this distribution since it makes no assumptions about a hypothetical initial quantum state. But anyone looking at the moon is warranted in assigning a distribution already strongly peaked on the observed values of position and momentum. The model then warrants overwhelming confidence that these values will continuously evolve in accordance with classical physics, whether or not anyone is looking at the moon. So an agent is certainly entitled to claim C_1, whether or not s/he is looking at the moon.

Whether an agent is similarly entitled to make a claim of the form C_2 or C_3 concerning a particle depends on how it interacts with its environment. For Mermin's demonstration to work, an adequate model of the particle's interaction with its environment cannot allow a significant change in its quantum spin state before it is measured. But that state does not define a set of vector states with the special features that would justify application of a pragmatic rule for assigning significant content to a statement of the form C_2. Bluntly put, an agent should regard C_2 as devoid of empirical content when the particle has such a quantum state. Even though a

---

[9] According to a quantum model of environmental decoherence, the density operator of its center-of-mass translational state would become approximately diagonal in a pointer basis of approximate position and momentum states $|\psi_x(t)\rangle$: see Chapter 5. The moon's large mass makes this an excellent approximation.

thoughtless application of the Born rule to its quantum state would associate a number between 0 and 1 with C_2, an agent should not base partial belief in C_2 on this number.

Measurement of a particle's spin requires an external interaction. This will generally change its state in a way that can be modeled quantum-theoretically. As we saw in Chapter 5, a suitable interaction will require reassignment of the spin aspect of the particle's quantum state so it defines a set of vector states with special features that justify application of a pragmatic rule that grants a statement of the form C_2 a high degree of empirical content. A legitimate application of the Born rule to the particle's quantum state now yields a non-zero probability for each of two incompatible statements of the form C_2. An agent who accepts quantum theory but does not know the result of the measurement should use these as an authoritative guide in forming a partial belief in each statement. If the probability of one statement is near 1, an agent may feel entitled to make that claim; if not, the agent should suspend judgment concerning a fact of which s/he is currently ignorant.

The experiments on fullerene molecules I described in section 2 of Chapter 9 nicely illustrate the role of the environment in giving content to claims of the form C_3. A fullerene molecule is a form of carbon in which a large number of carbon atoms bond together in the shape of a football—soccer for C_{60}, rugby for C_{70}. While a fullerene molecule is fairly large, with considerable internal structure, it seems reasonable to call it a particle since its diameter of around 1 nanometer makes it over ten thousand times smaller than any visible speck of dust. But if passed through a carefully aligned array of narrow slits, a beam of fullerenes can display behavior typical of a light or water wave that passes through a number of slits (so that the parts going through different slits either cancel or reinforce each other) by forming an interference pattern on a detection screen. Such behavior may be understood quantum-theoretically by assigning the same quantum wave function to each beam molecule and then using the Born rule to calculate the probability of statements of the form C_3 locating a fullerene in a small region of the screen, where the x-axis is at right angles both to the slits and to the beam axis.

This assumes such statements have a high degree of empirical content here. The assumption is justified since interaction with the screen is correctly modeled by modifying a fullerene's quantum state into a form suited for applying basically the same pragmatic rule that assigned high empirical content to the statements C_{1x} about the position of the moon. In one recent experiment (Juffmann, et al. [2009]), the positions of C_{60} molecules on the screen were indirectly observed after they had landed on it and adhered to the screen like a fly on flypaper. Their positions were imaged using a scanning tunneling electron microscope, thereby providing strong evidence for many claims of the form C_3 about fullerenes in which d is an interval of only a few nanometers

But a fullerene usually interacts in this way only with the screen. The beam passes through a dark, high vacuum in the apparatus so hardly any fullerenes interact with gas molecules or light, while the material in which the slits are cut just constrains the fullerenes' vector quantum state to produce interference at the screen. So at no time before reaching the screen is a fullerene's quantum state of the right form to assign high empirical content to any statement of the form C_3, if d is an interval comparable to the separation between the slits. No statement that a fullerene passed through a particular slit has any empirical content.

Feynman ([1963], vol. 3, ch. 1, p. 9) said this about an electron as it passes through an analogous two-hole interference experiment:

> [I]f one has a piece of apparatus which is capable of determining whether the electrons go through hole 1 or hole 2, then one can say it goes through either hole 1 or hole 2. [Otherwise] one may not say that an electron goes through either hole 1 or hole 2. If one does say that, and starts to make any deductions from the statement, he will make errors in the analysis. This is the logical tightrope on which we must walk if we wish to describe nature successfully.

In the C_{60} experiment there was no piece of apparatus capable of determining which slit each fullerene goes through. But one cannot say which slit it goes through for a different, though related, reason: the environmental conditions for such a statement to be empirically significant are not met for these fullerenes. Such conditions are relevant to the possibility of determining which slit a fullerene goes through because if they were met, the environment itself could be so affected by its interaction with the fullerene as to incorporate a "record" that it went through one slit rather than the others. The presence of such a "record" in the environment would be a marker for the kind of environmental conditions required for a statement of the form C_3 to be empirically significant for a fullerene at the slits, whether or not any apparatus is capable of "reading" it. This is shown by another experiment in which a beam of C_{70} molecules was sent through a similar array of slits after first passing through a series of laser beams to heat the molecules to a high temperature.

A molecule that has absorbed energy from a laser may later radiate that energy as light. Just as one can track a firefly from the light it emits, the light emitted by a hot fullerene might be used to try to find out which slit it went through. But the slits are very closely spaced, so the emitted light would need to have a short enough wavelength to resolve the distance between them. A hotter fullerene emits more light and of a shorter wavelength than a cooler fullerene. Whether or not one sets up apparatus to collect any light to try to see through which slit a heated fullerene passes, the emitted light produces a "record" of its passage in the environment. As the beam is heated to a higher temperature, the interference pattern a cold beam would produce gradually disappears.

At first sight, emission of light into a vacuum may seem not to involve interaction with a fullerene's environment. But this is not true in a quantum-theoretic model of the fullerene's environment as the vacuum state of a quantized electromagnetic field. The effect of this electromagnetic environment is to decohere the fullerene's quantum state so that while each statement of the form C_{1x} about its position at the screen still has high empirical content, the Born probability density of that statement corresponds not to the low-temperature interference pattern, but to the pattern observed at a higher temperature. As the beam's temperature is increased, this approaches the "smoothed" shape one would expect to observe if each molecule passed through just one slit. At the same time, the fullerenes' increased interaction with their environment affects their quantum state so as to increase the empirical content of a claim of the form C_3 for a fullerene at the slits, where d is an interval comparable to the slit separation. One is entitled to claim that each sufficiently hot fullerene passes through just one slit, and to apply the Born rule to its quantum state to form credences about which slit the fullerene goes through.

I anticipate two objections to this account of the content of a claim of the form C_3 and other claims of the form s has $(M \in \Delta)$ in which s is an independently designated object such as the moon, an electron, an atom, or a molecule.

First objection: A claim has significant content if and only if it expresses a determinate proposition. While what content a claim expresses may depend on the context to which it relates (loosely, to the context in which it is made), context merely determines what proposition a claim expresses. Any variation of content with context can be represented by a function from context into proposition expressed. An adequate analysis of a claim's content must then supply an account of the content of each proposition in the range of that function in a referential semantics that provides its truth conditions: if the function is only partial, the claim has no content in a context in which it expresses no determinate proposition. So an adequate analysis of the content of a magnitude claim C will either assign it some specific content (varying from context to context) or no content at all (in other contexts). No analysis is adequate according to which what varies with context is not simply the specific content of the claim but also how much content it has.

Reply: One can give an account of the truth conditions of a claim of the form C: s has $(M \in \Delta)$ but this is trivial. For example: C is true if and only if the system to which 's' refers has a value for the magnitude to which 'M' refers that lies in the set of real numbers to which 'Δ' refers.[10] Once the (tensed) claim is indexed to a time, these truth conditions are independent of context, since the claim contains no explicit indexical elements. The problem with this referential approach is not that it is wrong but that it is too shallow to be helpful: it fails to illuminate the different ways a claim of the form C functions in different contexts. The claim functions within a web of inferences, and the extent of its content depends on the context provided by the presence of other claims in the web—in this case, an assumption about the quantum state of s is critical to determining the content of a claim of the form C about s.

Second objection: The proposed inferentialist alternative seeks to specify the empirical content of a claim of the form C by locating it in a web of material inferences—what other claims would entitle one to claim C, and to what other claims

---

[10] A committed extensionalist could drop talk of magnitudes in favor of an analysis in terms of the extension of a predicate that applies to an object if and only if the value of M on that object lies in set Δ.

218 MEANING

one would be committed by claiming C. But this would qualify as a serious analysis only if backed up by a complete specification of these inferences, and none has been offered. Indeed, it is doubtful that any such complete specification could be given, and even more doubtful that different agents could come to share the same web and so associate the same content with any particular claim of the form C.

Reply: To understand the function of a claim of the form C within quantum theory, it is not necessary to undertake the quixotic task of fully specifying a web of material inferences in which it is located. Claims of this form were used and understood well enough by scientists and others prior to acceptance of quantum theory. The task is merely to clarify the changes in use and understanding accompanying acceptance of quantum theory. We can rely on our previous understanding while characterizing these changes by showing how acceptance of quantum theory alters patterns of inference that grounded that understanding.

This second reply sets a task without accomplishing it. I begin that task in the next section by showing how acceptance of quantum theory affects the inferential function of analogous pairs of incompatible claims about three kinds of physical systems.

12.4 Some Conceptual Mutations

How does quantum theory displace key material inferences that contribute to the content of magnitude claims?[11]

These statements are readily recast as magnitude claims like C_2:

C_{red}: The red traffic light is on.
C_{green}: The green traffic light is on.
C_{one}: The computer memory bit stores one.
C_{zero}: The computer memory bit stores zero.

If a traffic signal and computer are operating normally, exactly one of each pair is true (assuming the traffic signal has no orange, and neglecting the brief period during which the lights and memory record are changing). A driver may defend his entitlement to claim C_{green} by appeal to his current visual experience or his memory of how the light looked a moment ago. Such inferences are fallible: someone may have secretly covered the illuminated red light with opaque material while strong reflected sunlight makes the green light appear to be on when it is off; or the driver's eyesight, memory, or cognitive functioning may have been rendered unreliable.

To conclusively establish the truth, either of C_{red} or of C_{green}, one would arrange close examination of the condition of the traffic lights by multiple observers with sense organs, cognitive skills, and measurement equipment subjected to rigorous testing. An examination could include direct visual and tactile inspection of the bulbs and filters, measurement of the intensity and spectral profile of the emitted light,

---

[11] "It is characteristic of modern science to produce deliberately mutant conceptual structures with which to challenge the world" (Sellars [1953], p. 337).

measurement of the bulbs' temperature, of the current flow through the bulbs and the rest of the circuit, and so on.

Suppose that such test results always provide overwhelming evidence for one of C_{red}, C_{green}, but a skeptic objects that this shows only that one or other of these claims is true whenever tests are performed, but that neither C_{red} nor C_{green} is true when no test is performed—in fact the traffic lights are red or green only when "someone looks". In response one can appeal to an account of how the signal works, according to which the tests performed have no effect on whether the lights are on or off.[12] Call a measurement of the lights' status noninvasive if it has no effect on their subsequent on/off status.

The skeptic may press his objection by questioning the evidence for this account. But then his skepticism becomes global in form as well as content. The account is embedded in general theories of how devices like light bulbs shine when an electrical current is passed through them, and how devices as diverse as human eyes and hands, thermometers, spectrophotometers, and ammeters function. Our confidence in those theories rests on much more than their ability to account for the operation of the traffic lights. To question those theories, the skeptic must indulge in a general inductive skepticism, either about the support relation between the evidence and these theories, or about that evidence itself. The former kind of "merely philosophical" skepticism may be set aside as irrelevant here. But a skeptic may question the account because he doubts whether observations or measurements on a system ever provide evidence as to how it is when unobserved: his doubt that the traffic lights are red or green "when no one looks" is simply an instance of this general kind of philosophical skepticism.

One can reply to such a skeptic who questions a claim C_{green} about an unmeasured traffic light by asking him what he takes to be the content of his doubt. We can at least begin to give a detailed account of evidence justifying an inference to the statement C_{green} that entitles one to make that claim: and we can embark on a detailed account of what is materially implied by C_{green} and so to what one is committed by claiming it. The latter might involve inferences, such as from only the green light's being on at time t_1 to only the red light's being on at t_2, but only the green light's being on at t_3, regardless of whether the lights are measured at or between any of those times. Observations confirming the inferences' conclusions lend support both to particular claims C_{red}, C_{green} at those times and to the general claim

G_T: Either C_{red} or C_{green} but not both

at any time the signal is operating. Indeed, on an inferentialist account of content, its place in such a smoothly functioning web of belief is what gives G_T its content. To simply replace G_T by the claim

G_T^*: G_T only when the lights are observed, otherwise neither C_{red} nor C_{green}

---

[12] While some tests could have a small effect on the current flow through the lights, the account could allow for this in determining the current flow in their absence. In any case, the effect is too small to alter their on/off status.

will cut so many inferential connections as to render the web useless. The alternative, of restoring all the inferential connections by making compensating modifications in the statements they connect, would simply produce a functionally equivalent web within which $G_T{}^*$ mimics the inferential role of G_T in the original web and so has essentially the same content. Neither option yields a genuine skeptical rival to the original account incorporating G_T.

Our unaided sense organs do not help us observe the status of a particular bit of static random access memory (SRAM) in a contemporary electronic computer. But there are many ways of measuring and recording whether it stores one or zero, and it is essential to the efficient functioning of the computer both that it always stores one or the other and that some of these measurements are noninvasive in the sense defined earlier. According to an inferentialist, each of C_{one} and C_{zero} gets its content from an inferential web connecting it to evidence for the claim and to what the claim commits one, and the general "shape" of the web is closely analogous to that which confers content on C_{red} and C_{green}. The conferred content warrants the exclusive disjunction

G_C: Either C_{one} or C_{zero} but not both

at any time the computer is operating. There is likely less reason to doubt that G_C is true when the bit status is unmeasured than to question whether the traffic light is red or green "when no one is looking".

While the operation of traffic signals and memory elements in digital computers can be understood quite well (at least in general terms[13]) without quantum physics, that is certainly not true of the analog to a single memory element in a quantum computer. The set of values available to a logical bit in a classical digital computer has two elements {0,1}. But the set of values available to a logical qubit in a quantum computer corresponds to the infinite set of elements of a two-dimensional complex vector space in which the vector quantum state of a system (such as an electron's spin) may be represented. However, a measurement of the contents of a qubit always gives one of two values {0,1} of a magnitude M.

A qubit must be realized physically in a quantum computer, just as a bit must be realized physically in a classical computer. One candidate for realizing a single qubit memory element is the focus of an experimental program designed to test what is called macro(scopic)-realism.[14] A key tenet of macro-realism (for a two-state system) is macro-objectivity:

(MO)   Any system which is always observed to be in one or the other of two macroscopically distinguishable states is in one of those two states at any time t, even if no measurement on that state is performed at time t.

---

[13] An SRAM element works because of the current/voltage profiles of its transistors. A detailed explanation of why they exhibit these profiles *would* require appeal to quantum theory.

[14] See Leggett [1998], [2002] for further details and proof of the claimed inconsistency.

The system in question is a kind of superconducting quantum interference device (SQUID). When operating, measurements on this device always find it in one of two states:[15] in one state a small current is flowing clockwise around a ring, while in the other state the same current is flowing anticlockwise around the ring. This small current is readily measurable: it is associated with the coordinated motion of a very large number of electrons (well over a billion). So these two states are plausibly considered macroscopically distinguishable—perhaps no less so than the one, zero states of a classical computer memory element.

We can define a magnitude M for the SQUID as taking value 0 if the current is circulating clockwise and 1 if it is circulating anticlockwise. For this SQUID to realize a qubit, there are times at which one must be able to assign it some specific vector quantum state. Applied willy-nilly to its instantaneous vector state, the Born rule yields a "probability" p for value 0 of M, and a "probability" $(1 - p)$ for 1. Much of the time, the Born rule applied to its vector state would yield a p-value intermediate between 0 and 1. But according to (MO), exactly one of these statements is true at any time the device is operating:

$C_{\circlearrowright}$: The current is circulating clockwise.
$C_{\circlearrowleft}$: The current is circulating anticlockwise.

Even if we assume that a measurement of M at t reveals the value M has at t, any apparent tension between this consequence of (MO) and Born probabilities outside $\{0,1\}$ for $C_{\circlearrowright}$, $C_{\circlearrowleft}$ is easily relieved by recognizing that the value of M may change in an unknown and even objectively random way between measurements. But if one makes a further assumption of noninvasive measurability, one can show that for the exclusive disjunction

G_Q: Either $C_{\circlearrowright}$ or $C_{\circlearrowleft}$ but not both

to be true at all times is inconsistent with certain consequences of the Born rule as applied to the vector states quantum theory prescribes for the SQUID at various times. Here that assumption is

(NIM) Consider a system which is always observed to be in one or the other of two macroscopically distinguishable states. No matter what quantum state that system is ascribed at t, there exists a noninvasive procedure for measuring which of its two macroscopically distinguishable states it is in at t; that is, a procedure that does not disturb the system's subsequent behavior, at least as far as concerns which of these states it is in.

This does not demonstrate the falsity of (MO) for two reasons. Since no practicable experiment has yet collected statistics to test the relevant consequences of the Born rule, this application may prove to be quantum theory's Achilles heel. But even if

---

[15] On rare occasions it is observed to be in some other macroscopically distinguishable state.

such statistics accord with quantum theory, one could consistently uphold (MO) by rejecting (NIM). Leggett ([1998], p. 20) argues that the dilemma as to which of these assumptions to jettison is a false one:

> Frankly I am not sure that this question is really very meaningful. The everyday language we use to describe the macroscopic world is based on a whole complex of implicit, mutually interlocking assumptions, so that once the complex as a whole is seen to fail it may not make much sense to ask which particular assumption is at fault. I am not sure, myself, that I could give a lot of meaning to [(MO)] under conditions where I had to admit that [(NIM)] fails.

As I understand him, Leggett does not claim that (MO) logically implies (NIM), but rather that by rejecting (NIM) one cuts key links in the inferential web that gives (MO) its content, with no clear way to patch up the web and imbue (MO) with any consequent new content.

Accepting quantum theory commits one to rejecting (NIM). (MO) does not logically imply (NIM), so one may try to retain (MO). But what could be the content of (MO) without (NIM)? How could one try to insert (MO) into the inferential web quantum theory weaves? Quantum theory here predicts probabilistic correlations between measured values of M at two times, provided no measurement occurs in the interim. A natural way to make (MO) relevant to these values is to assume that a careful measurement of M at t reveals which of $C_\circlearrowright$, $C_\circlearrowleft$ was true at t. But rejecting (NIM) blocks inferences from the direction of the current at t either to its direction at any later time or to the result of measuring M at any later time. So even if one does assume that a careful measurement of M at t entitles one to claim that a particular one of $C_\circlearrowright$, $C_\circlearrowleft$ was true at t, that claim commits one to nothing to which one is not already committed by applying quantum theory's probabilistic correlations between the measured values of M at t and at a later time. For one who accepts quantum theory as applied to this SQUID, there is simply no content to the claim that at any time one of either $C_\circlearrowright$ or $C_\circlearrowleft$ is true. By contrast with the traffic lights and (classical) computer memory bit, in this case it is the "skeptic" who insists on the truth of this exclusive disjunction in the face of quantum theory who has failed to give content to his/her claim.

This section has shown in some detail how acceptance of quantum theory affects the content of some claims of the form C by altering the inferential web that gives them content. I followed Leggett in making use of a binary distinction between times when a system is undisturbed and times when it is measured. That idealization employs the problematic term 'measurement'. In a quantum model not using that term, the SQUID's quantum state is briefly affected by some external device suited to record the result of this interaction as a value of M at some time t. It is the form of this quantum state that serves to indicate the content of claims about the value of M. In the SQUID qubit that content "crystallizes" then "redissolves" extremely rapidly during the brief external interaction. In contrast, quantum theory justifies the assumption that a classical computer memory element always reliably stores either a 0 or a 1 because

12.5 The Content of Denoting Terms

Einstein did not ask Pais whether he believed the moon is there when nobody looks, but whether he believed the moon exists only when he looks at it. Taking Einstein's reported question literally, he was worried that Pais's understanding of quantum theory would remove the empirical credentials of every claim about the moon by undermining the objectivity of the moon's existence.

In an inferentialist account it is claims or statements that serve as the primary vehicle of content. I have so far assumed that the variation of content with environmental context for a magnitude claim s has ($M \in \Delta$) afflicts only the property $M \in \Delta$ of a physical system s as it interacts with other quantum systems. But the forms of quantum theory currently considered fundamental—relativistic quantum field theories—raise questions about the content of the term 's' in a magnitude claim.

In discussing the meaning of a term like 'electron' or 'photon' one commonly distinguishes its denotation (the extension of a general term or the referent of a singular term) from its sense, intension, or stereotype. From this point of view, the varying inferential role in quantum theory of a magnitude claim C containing a term 's' may affect the sense, intension, or stereotype of 's' without affecting either its reference or the truth conditions of C. 'Electron' may continue to denote electrons, and 'photon' photons, even if acceptance of quantum field theory alters other aspects of these terms' meanings. But acceptance of quantum field theory threatens even the denotation of these terms, since the ontological status of "elementary particles" like electrons and photons in a relativistic quantum field theory is problematic.

This may seem surprising, since the relativistic interacting quantum field theories of the Standard Model provide our deepest current understanding of phenomena involving the detection of elementary particles including electrons and photons when beams of electrons or protons from a particle accelerator collide at high energies. But these theories do not yield this understanding by *describing* elementary particles. Rather, we improve our understanding by using models of quantum field theory to tell us when it is appropriate to describe physical phenomena in terms of elementary particles and then how we should expect them to behave.

To see how this comes about, start with the following assertion, which a recent book locates at the heart of quantum field theory:

Every particle and every wave in the Universe is simply an excitation of a quantum field that is defined over all space and time.   (Lancaster and Blundell [2014], p. 1)

The vector space used to represent the state of a quantum field is called Fock space. A unique vector $|0\rangle$ in this space is called the vacuum state: the Born rule implies probability 1 that a measurement of the energy of a system assigned this state will yield its minimum value and that a momentum measurement will yield the value zero. Since application of the Born rule to any other state in Fock space implies a positive probability that a measurement of energy will yield a higher value, all other states are called excited. The quote asserts a correspondence between a particle and an excited state, and a different correspondence between a wave and another excited state. But what kind of correspondence?

Perhaps the simplest relativistic quantum field theory (the Klein–Gordon theory) is often said to describe free (spinless) particles of mass m. A measurement of momentum on a system assigned state $|1_\mathbf{p}\rangle$ in this theory's Fock space will (with probability 1) yield value $\mathbf{p}$, while a measurement of energy will yield the value E characteristic of a classical relativistic particle of mass m with momentum $\mathbf{p}$.[16] A measurement of energy on a system assigned state $|2_\mathbf{p}, 1_{\mathbf{p}'}\rangle$ will yield the value E characteristic of three classical relativistic particles, of mass m and momenta $\mathbf{p}, \mathbf{p}, \mathbf{p}'$, respectively. This gives sense to the claim that a particle is an excitation of the Klein–Gordon quantum field.

Metaphorically, the excited state $|1_\mathbf{p}\rangle$ of the Klein–Gordon field contains a mass m particle of momentum $\mathbf{p}$, while the excited state $|2_\mathbf{p}, 1_{\mathbf{p}'}\rangle$ contains three such particles, with momenta $\mathbf{p}, \mathbf{p}, \mathbf{p}'$, respectively. Dirac would permit the more literal statement that the energy and momentum of the field in state $|1_\mathbf{p}\rangle$ is the same as that of a mass m particle of momentum $\mathbf{p}$, while the field's energy and momentum in state $|2_\mathbf{p}, 1_{\mathbf{p}'}\rangle$ is the same as that of three classical relativistic particles, of mass m and momenta $\mathbf{p}, \mathbf{p}, \mathbf{p}'$, respectively. If assignment of a quantum state were a way of describing the system to which it is assigned then to assign the Klein–Gordon field such states would be to describe that quantum field. One might contend that while this quantum field theory does not literally describe mass m particles, by saying what quantum state the field is in it does license one to speak in terms of mass m particles—as entities with no independent ontological status that may be said to emerge from the underlying field in such states. But no quantum theory says what state a system is in, as Chapter 5 explained (see § 5.3).

There are different excited states of the Klein–Gordon quantum field that correspond to waves: these are called *coherent states*, each written as $|\alpha\rangle$ for some complex number α.[17] A coherent state is an infinite superposition

$$|\alpha\rangle = \sum_n c_n |n_\mathbf{p}\rangle$$

---

[16] $E = \sqrt{\mathbf{p}^2 + m^2 c^4}$, where c is the speed of light. I follow a customary practice of ignoring mathematical niceties and writing precise momentum states as vectors in Fock space.

[17] The explicit form of a coherent state corresponding to momentum $\mathbf{p}$ is

$$|\alpha\rangle = \exp -\frac{|\alpha|^2}{2} \exp \alpha \hat{a}_\mathbf{p}^\dagger |0\rangle$$

where $\hat{a}_\mathbf{p}^\dagger |n_\mathbf{p}\rangle = \sqrt{n+1} |n_\mathbf{p} + 1\rangle$.

in which each Fock state $|n_\mathbf{p}\rangle$ for $n = 1, 2, \ldots$ is multiplied by a carefully chosen coefficient c_n. As the modulus $|\alpha|$ of α is increased it becomes more and more appropriate to speak of the emergence of a classical wave with momentum $\mathbf{p}$ from the Klein–Gordon field in state $|\alpha\rangle$. But even Dirac would not permit one to say a definite number of particles is present if the state of the field is a superposition of distinct Fock states $|n_\mathbf{p}\rangle$, but merely that the probability for a measurement of the number of particles with momentum $\mathbf{p}$ to record a value substantially different from $|\alpha|^2$ when the field is in state $|\alpha\rangle$ decreases as $\frac{1}{|\alpha|}$.

In fact neither particles nor waves emerge from properties of a quantum field in any state because to assign a quantum state to a quantum field is not a way of characterizing its properties. Quantum fields are assumables but not beables of quantum field theory. As I said in Chapter 4 and repeated in section 2 of this chapter, a quantum state is assigned to a system not to say what it is like but to prescribe what cognitive attitude it is appropriate to take toward relevant magnitude claims. In an application of quantum field theory these are not claims about the value of any quantum field magnitude. In serving as a source of sound advice to a physically situated agent on the content and credibility of magnitude claims about physical systems, a quantum state does not describe these physical systems, and need not be assigned to them.

In a quantum field-theoretic model, a quantum state is assigned to some purely schematic system such as a quantized electromagnetic field or an electron field. I call such quantum field systems schematic because while their physical existence is presupposed by an application of quantum field theory, that theory says nothing about their physical properties. In particular, a quantum field operator does not represent a physical quantum field magnitude. By assigning a quantum state to a quantum field system, one undertakes no commitment to the existence of any *specifiable* physical objects or magnitudes. So while one can say that quantum field theory is about quantum fields, accepting a quantum field theory does not mean believing the physical world contains a quantum field that theory describes.

Quantum field operators and quantum field states function within a mathematical model whose application is funneled through the Born rule, which assigns probabilities to magnitude claims of the form C: s has ($M \in \Delta$): call s the *subject* of C. The quantum state of a quantum field provides advice on when a statement C has enough empirical content to be an appropriate object of an epistemic state of partial belief: only then should an agent base credence on the Born probability of C. No statement of the form C whose subject system is a quantum field has any empirical content. But the quantum state of a quantum field sometimes assigns a high degree of empirical content to a claim of the form C about particles, such as the claim that a high-energy positron and electron with equal and opposite momenta will be converted into an oppositely charged muon pair with equal and opposite momenta. In a different environmental context it will be claims about "classical" field magnitudes that acquire a high degree of significance.[18]

To gauge whether claims about particles or about "classical" fields have a rich content in a particular environment one can use a quantum field-theoretic model of environmental interactions. Such a model is usually said to describe an interaction between a quantum field and its environment. But although environmental interactions are modeled by a mathematical expression coupling operator-valued quantum fields, this term does not in fact represent the value of any physical magnitude. As Chapter 5 explained, applying a quantum model of decoherence is a valuable way to gauge the significance of magnitude claims. But neither quantum fields nor interaction terms are physical magnitudes, and the model does not assign them values.

The quantum theory of light provides many examples of these two kinds of applications of quantum field theory. In quantum optics, some models of systems involving the quantum electromagnetic field license claims of the form C about photons, such as a claim that two photons have the same energy but opposite polarizations; and other models license claims of the form C about classical electromagnetic fields, such as a claim about the frequency of electromagnetic radiation emitted by a laser. These models of quantum field theory do not describe physical systems or their interactions. Only the magnitude claims associated with the models' applications describe physical systems—in this case photons or classical electromagnetic fields. Even when a model involves quantum fields that are said to interact, by itself it neither describes nor represents a physical interaction between physical systems. Though they form the basis for our deepest understanding of "elementary" particles and "fundamental" force fields, the interacting quantum field theories of the Standard Model in fact *describe* neither.

One can believe that some claims about electrons, electric fields, or photons sometimes have a high degree of empirical content without believing those claims. If no quantum field theory describes such things (or anything else that could constitute them) one may wonder how it can give us any reason to believe claims about them, including claims that they exist. Such reasons are provided by applications of the Born rule in circumstances that assign probability at or close to 1 to a significant claim of the form C about things like electrons, electric fields, or photons. But just as this sometimes gives us no reason to entertain a claim locating an electron or photon in some small region or roughly specifying the value of an electric field, there are circumstances in which we can have no reason to form a belief about whether electrons, photons, or electric fields are present. So while accepting a quantum field theory can justify one in believing that things like electrons, electric fields, and photons

---

[18] If the Higgs quantum field may be assigned a coherent state with a non-zero vacuum expectation value then perhaps this includes claims about the value of the Higgs field, passage through which is often said to give other elementary particles their masses.

exist, on the present view it also gives one reasons to deny that a specification of their features could constitute a complete and fundamental description of the world.

In explaining what I believe to be truly radical about quantum theory I have used Bell's term 'beable': quantum theory warrants our acceptance of its fundamental status in twenty-first-century physics, despite lacking any beables of its own. But by using quantum theory we are able better to describe and represent the physical world by magnitude claims and other statements in physical or ordinary language whose truth we may determine. These claims and statements are about beables, but quantum models introduce no new beables. Our descriptive situation has been improved in two ways. Quantum theory has proved a wonderfully successful guide as to which magnitude claims we should believe, and how strongly we should believe them. By doing so it has enabled the predictions and explanations that warrant its acceptance as a fundamental physical theory. But quantum theory has also helped us to appreciate that there are limits on the significance of magnitude claims, and to advise us on those limits. In this way it has warned against the overweening ambition completely to describe the world at a fundamental level.

Bell introduced his term 'beable' by explicit contrast with quantum theory's term 'observable':

> The concept of 'observable' lends itself to very precise *mathematics* when identified with 'self-adjoint operator'. But physically it is a rather woolly concept. It is not easy to identify precisely which physical processes are to be given the status of 'observations' and which are to be relegated to the limbo between one observation and another.... 'Observables' must be *made*, somehow, out of beables.   ([2004], p. 52 emphasis in original)

Since quantum theory lacks any beables of its own, it provides no resources from which to construct observables associated with self-adjoint operators. Bell had two reasons for finding this situation unacceptable: that there is a limbo, and that its precise extent is not easily identifiable. But quantum theory itself advises us on the limbo's extent by providing models of decoherence we can use to guide our application of quantum theory. Quantum theory can be precisely formulated with no term like 'observation', and there is nothing woolly about this advice. These models do not *precisely* specify the limbo's extent, but there can be no precise rule specifying how any physical theory is to be applied.

The limbo remains. Quantum theory advises us not to make descriptive claims in the absence of appropriate decoherence. It frustrates the ambition completely to describe the world at a fundamental level. But physics need not gratify that ambition, and the frustration may prove temporary. Quantum theory is a great advance on classical physics, but its success is no reason to think fundamental physical theories will always be quantum theories.

13

Fundamentality

The quantum revolution in physics occurred in the first third of the twentieth century. The period since then has been one of consolidation and expansion. The basic principles of quantum theory have been successfully applied to all kinds of physical systems at a vast range of scales of energy, time, length, and temperature. First applied at the atomic scale, quantum theory is now basic to our understanding of the behavior of molecules, the atomic nucleus, its constituent protons and neutrons, their underlying quark substructure, and (in the so-called Standard Model) all the other "elementary particles" associated with three of the four known fundamental interactions—electromagnetic, weak, and strong. This covers an enormous range of energies, from millionths of an electron volt to those characteristic of a single collision in the Large Hadron Collider (LHC) (more than 10^{13} electron volts).

Quantum theory has been successfully applied to large systems, including the 15-meter-long, 12,000-ton superconducting LHC magnets through each of which flows a current of over 10,000 amps at a temperature of $-271°C$ (less then $2°$ above absolute zero—colder than interstellar space). It has been used to understand processes at the center of the sun (at over 15 million $°C$) and also in the core of a neutron star (at a density of 10^{14} that of water). But while quantum theory has been successfully extended to apply to many relativistic phenomena, there is as yet no wholly satisfactory quantum treatment of phenomena in situations where the general theory of relativity predicts conditions where quantum effects are expected.

The unifying power manifested by the wide range of its successful applications clearly makes quantum theory fundamental to contemporary physics. But in this chapter I wish to step back to consider what general lessons we can learn from quantum theory about fundamental physics. Appeals to fundamental physics are quite common in contemporary philosophy.[1] I'll look at several of the roles people have allotted to fundamental physics in the next section. Physicists have also used some of the key terms used to express these roles when addressing a wider group than others pursuing the same branch of research—terms including 'building block', 'fundamental particle/field', 'law', and 'reduction'. Since that group includes some philosophers,

---

[1] This contrasts with periods in the twentieth century when most philosophers took it as their task to analyze basic concepts through a careful study of language, whether natural or symbolic.

philosophical appeals to fundamental physics have been influenced by what physicists say as well as by the work of an earlier generation of philosophers of science.[2]

13.1 Philosophy and Fundamental Physics

It may be helpful to begin with the thought that fundamental physics is concerned with what the natural, or empirical, world is like at the deepest level. Though vague and overly metaphorical, this thought already distinguishes fundamental physics from mathematics as well as most of natural as well as social science. But it cries out for clarification before we consider how the quantum revolution has changed, or should change, our view of fundamental physics.

We can begin by distinguishing between the goals of fundamental physics, the activity of pursuing those goals, and the significance of what has been achieved in the pursuit. Currently, the most lucrative prize that may be awarded to a scientist is not the Nobel Prize but the Breakthrough Prize. Prizes are now also awarded for Life Sciences and for Mathematics, but the Prize was founded first (in 2012) for transformative advances in Fundamental Physics. Presumably winners of that prize are deemed to have successfully acted in pursuit of the goals of fundamental physics. In contrast to the Nobel prize, most winners in its first three years were theoretical or mathematical physicists whose work has yet to receive experimental validation. Though directed toward that goal, their activities have so far resulted in little or no new knowledge of what the world is like at the deepest level.[3]

If we step back slightly from the unsettled cutting edge of contemporary research, we can find reliable knowledge of the world at a deep level and a broad consensus on the form such knowledge may be expected to take as we seek it at deeper levels. The relativistic quantum field theories of the Standard Model form the basis for much of our present knowledge of the deep structure of the world: The general theory of relativity currently provides our most reliable knowledge of phenomena associated with gravitation as well as the large-scale structure of space and time. We have no reliable knowledge of phenomena in the extreme conditions clearly in the domain of applicability of both quantum theory and general relativity, in the absence of a theoretically adequate and experimentally confirmed quantum theory of gravity. But there is wide (though not universal) agreement among researchers that further progress is to be sought by extending the general principles of quantum theory to apply to gravitational and other phenomena at energy scales much larger than those where the Standard Model has proved reliable. That puts quantum theory at the core of fundamental physics for the foreseeable future.

---

[2] These include not only logical empiricist heirs to the ideas of Rudolph Carnap and Hans Reichenbach (notably including Carl Hempel and Ernest Nagel), but also Karl Popper and those influenced by his work.

[3] The single exception is the special award in 2013 to leaders in the scientific endeavor that led to the discovery of the new Higgs-like particle.

The metaphor of levels ordered by depth may be cashed out in different ways. On perhaps the most literal understanding, it involves a relation of composition between entities, in which entities at one level are composed of those at a lower level. Fundamental physics would then be that branch of physics concerned with the physical entities that compose all others—the theory of atoms, if Democritus had been right.[4] The next section examines the claim that quantum theory concerns fundamental entities.

Another way to cash out the metaphor focuses on properties rather than (or as well as) objects that bear them. This view has philosophical antecedents in Descartes's view of extension as essential to matter and Locke's distinction between primary and secondary qualities. Fundamental physics may be taken to be concerned with fundamental physical properties of objects, whose possession determines all their other properties. Section 13.3 considers that way of taking it, with further discussion in 13.4 of a variant that considers structure more basic than objects and properties that display it. We'll see how far quantum theory contributes to physics's inventory of fundamental properties and structures.

The main topic of Section 13.4 is the view that what makes a branch of physics fundamental is the laws it takes to govern the behavior of physical systems, since these laws are ultimately responsible for any regularities in the world. An important issue here is how far the "laws" of classical physics reduce to those of quantum theory. But before addressing that issue we'll need to ask what could count as the laws or principles of quantum theory.

13.2 Entities

There are reasons why both physicists and philosophers have often taken it to be a goal of fundamental physics to describe the ultimate constituents of matter—the basic building blocks to which t'Hooft [1997] refers in his title. The ancient atomists' inspired guess that all ordinary matter is composed of atoms had been amply confirmed by the early twentieth century, just as the atom's component electrons, protons, and neutrons were being discovered and studied.

The successful decompositional strategy continued with confirmation of theories postulating quarks as constituents of protons, neutrons, and a host of other short-lived supposedly "elementary" particles. Interactions among them were successfully modeled by theories involving other particles (gauge bosons) associated to weak and strong forces as photons were associated to electromagnetic forces. The high energies necessary to test such theories probed matter at the increasingly short distance scales naturally associated with the size of nucleons and their constituent quarks. Such tests

---

[4] Even if the compositional structure had no "bottom" level, it could still support a distinction between more and less fundamental entities.

recently resulted in the successful discovery of the Higgs boson, which plays a key role in the Standard Model.

Philosophers have paid attention to these developments. They supplied one motivation for the micro-reductive hypothesis put forward long ago by Oppenheim and Putnam [1958] that assumed a compositional-level structure beginning with elementary particles and extending beyond physics to biology, psychology, and even sociology.

More recently, metaphysicians including van Inwagen [1990] and Merricks [1998] have assumed some lowest physical level of a compositional hierarchy in their investigation of what its occupants might come to compose, while Schaffer [2003] and others have contemplated the possibility that this hierarchy may extend downward without end.

Since the relativistic quantum field theories of the Standard Model are currently our best guide to micro-ontology, we should ask what these imply about the fundamental constituents of the physical world. But neither physicists nor philosophers have been able to agree on the answer. The quantum field theory of light (electromagnetism) is part of the Standard Model.[5] According to one famous physicist, Richard Feynman ([1985], p. 15):

I want to emphasize that light comes in this form: particles.

But according to another prominent physicist, Robert Wald ([1994], p. 2):

Quantum field theory is a theory of fields, not particles. Although in appropriate circumstances a particle interpretation of the theory may be available, the notion of "particles" plays no fundamental role, either in the formulation or interpretation of the theory.

Philosophers have offered arguments against particle ontologies for quantum field theory, but also closely analogous arguments against field ontologies. Doreen Fraser ([2008], pp. 841–2) argues against a particle interpretation:

Quantum field theory (QFT) is the basis of the branch of physics known as 'particle physics'. However the philosophical question of whether quantum field theories genuinely describe particles is not straightforward to answer. What is at stake is whether QFT, one of our current best physical theories, supports the inclusion of particles in our ontology.

[B]ecause systems which interact cannot be given a particle interpretation, QFT does not describe particles.

While David Baker ([2009], pp. 585–6) added:

The most popular extant proposal for fleshing out a field interpretation is problematic.... two of the most powerful arguments against particles are also arguments against such a field

---

[5] It is said to emerge from the electroweak field as a consequence of spontaneous symmetry breaking by the (quantum) Higgs field.

interpretation.... If the particle concept cannot be applied to QFT, it seems that the field concept must break down as well.[6]

In her recent book on quantum field theories and other advanced forms of quantum theory, Laura Ruetsche ([2011], p. 260) was able to conclude only that:

Is particle physics particle physics? The answer I've tried to support is: sometimes it might be.

It would be inappropriate to enter into the details of these arguments here. Instead I will indicate how applying to quantum field theories the view of quantum theory presented in earlier chapters of this book resolves the issue of a quantum field theory's ontology. What is a (relativistic) quantum field? Formally, it is an assignment of linear operators to points or regions of space-time.[7] As a mathematical object this "lives" in a mathematical model, not in the physical world. But if the model functioned representationally, it would be natural to assume that this object represented some physical magnitude. A *classical* field may be understood to play such a representational role: recall that Bell gave the classical electromagnetic field of Maxwell as an example of a local beable! But closer attention to the function of a *quantum* field in a model shows that it plays a very different role.

Assignment of a quantum field at a point or region of space-time is analogous to assignment of an operator (such as the Hamiltonian $\hat{H}$ or spin-component operator $\hat{S}_z$) to a system in ordinary (non-relativistic) quantum mechanics. A model of quantum mechanics may be applied to supply authoritative credences, not about the value of the operator, but about the associated magnitude (such as energy or spin-component). This magnitude is not part of the ontology of quantum mechanics itself. To provide authoritative advice on credences in magnitude claims, quantum mechanics must be supplied with relevant magnitudes—it does not introduce them as novel ontological posits. That was the sense in which they are assumables, not beables, of quantum theory.

Similarly, an application of quantum field theory will yield advice on credences in claims about magnitudes that theory does not introduce as novel ontological posits. What are these magnitudes? Consider the quantum field theory of electromagnetism (QEM). There are situations in which the quantum state of the quantized EM field will decohere in a basis of states associated with *classical* electromagnetic field values at space-time points. This will render significant rival claims about these values: it is then credences in *these* magnitude claims that an application of QEM will offer

---

[6] Baker [2009], pp. 585–6. As Ruetsche [2011] explains, the existence of inequivalent Hilbert space representations of the canonical commutation relations of quantum field theory operators poses parallel problems for field and particle ontologies. Unfortunately, the mathematics required to explain these problems lies beyond the scope of this book. Note that they arise even if particles are *not* required to be strictly localized or countable, and even if a particle may be merely ephemeral—the recently discovered Higgs particle endures for no more than 10^{-21} seconds before decaying into other particles.

[7] In standard Lagrangian quantum field theory these are operators on a Hilbert space of states: in algebraic QFT they are elements of a von Neumann or C* algebra.

advice about. There are also situations in which the quantum state of the quantized EM field will decohere in a photon-number basis. That will render significant contrary claims about the number (energies, momenta, polarizations) of photons present: it is then credences in *those* magnitude claims that an application of QEM will offer advice about.

Consider instead the theory QEF of the free quantized electron field: to call it free is to disregard interactions "between electrons and photons", as did the previous paragraph.[8] The electron field operator $\hat{\psi}$ is not self-adjoint, and in no application does QEF offer advice about credences in claims about a magnitude corresponding to $\hat{\psi}$ in the way that energy corresponds to $\hat{H}$ in non-relativistic quantum mechanics.[9] Given suitable decoherence of the quantum state of the electron field, QEF may be applied to offer advice about credences in rival magnitude claims about numbers (spins, etc.) of electrons and positrons and their currents. But neither QEM, QEF, nor any other quantum field theory offers advice on credences about *quantum* field values. Quantum fields are not a novel ontological posit of quantum field theories: they are assumables, not beables, of quantum field theory. We knew of some quantum fields (in particular those to which we apply QEM, QEF) through their classical manifestations. We have learned that there are others by building quantum models to account for experimental observations of a host of new particles at high energies: but the models succeed without describing the quantum fields themselves.

By accepting relativistic quantum field theories of the Standard Model one accepts that neither particles nor fields are ontologically fundamental. One accepts that there are situations where the most basic truths about a physical system are truths about particles, while claims about ("classical") fields lack significance. One accepts also that there are other situations in which the most basic truths about a physical system are truths about ("classical") fields, but claims about particles lack significance.

Physicists have good reasons to believe that the Standard Model is only a good approximation to some unknown theory, and that its predictions will fail at much higher energies than we will be able to create and investigate experimentally. But there is also a consensus that the search for such an unknown theory should proceed on the assumption that it will take the form of a quantum field theory (perhaps of strings rather than point particles). If that assumption were to prove justified, the resulting theory would offer the same kind of contextual ontology—in some contexts it would offer advice on credences concerning (hitherto unknown) particles (or strings), while in other contexts it would offer advice on credences concerning the values of (hitherto unknown) "classical" field magnitudes. But the unknown particles, strings, and fields

---

[8] Formally, the Lagrangian density in each case omits a non-linear coupling term involving operators for both fields at once.

[9] This is because the field is fermionic. Electrons are fermions, as are quarks. The quantized EM field operator is self-adjoint because it is bosonic. Photons are bosons, as are the quanta of the fields associated with the weak and strong interactions.

would still not be in the ontology of the quantum field theory itself: in that sense they would still be "classical" rather than quantum, even though no prior theory of classical physics posited their existence. A contextual ontology could not provide the ultimate building blocks of the physical world.

An analysis of a classical field suggests an alternative perspective on the ontology of a quantum field theory. According to this analysis, a classical field is an assignment of determinate properties (e.g. the value of electric field magnitude) to space-time points. So understood, it is space-time points that constitute the underlying ontology of the field, while field strength, though a beable, is not an entity but a property of space-time points.[10] In an application of a quantum field theory yielding a contextual assignment of credences in contrary magnitude claims about classical (or other "classical") field values, the only ontological commitment would then be to space-time points (or regions). Perhaps an application of QEM yielding a contextual assignment of credences in contrary magnitude claims about photons could also be understood to carry an ontological commitment only to space-time regions: talk of photon number, energy, momentum, polarization, etc. might then be analyzed in terms of properties of space-time regions (such as that occupied by an optical fiber between photon emission and detection events).

Viewed from this alternative perspective, a quantum field theory still describes no new entities but offers advice on properties of a space-time whose existence it presupposes. Many current approaches to a quantum theory of gravity assume that space-time itself will not be ontologically fundamental in a successful theory, but will emerge from an underlying structure in some limit. If a quantum theory has no novel ontology, it is not clear what this underlying structure could be. But suppose a quantum theory of gravity functioned as a source of advice on properties of a space-time whose existence is presupposed in its applications. Then the theory might also offer advice on which circumstances legitimize its application by underwriting that presupposition *without* describing any underlying structure. This suggests a novel understanding of emergence that would require no specifiable underlying physical structure for space-time to count as emergent.

13.3 Properties

The philosopher Willard Quine [1951] drew a distinction between the ontology of a theory and what he called its ideology. Influenced by logic, he thought of a theory as a set of sentences in some precisely characterized language. That characterization included a specification of its predicate symbols (e.g. F, G, R) and constants

---

[10] Vector and tensor fields would need to be understood in terms of properties of regions, or relations among neighboring points.

(e.g. a, b, c, l, m, n).[11] An interpretation of this language said to what entity each constant referred and to which entities the predicate symbols should be taken to apply.

In a simple geometric theory a, b, c may be interpreted as denoting three distinct points in a plane while l, m, n denote three distinct lines in that plane, where F is true of all and only points, G is true of all and only lines, and R is a relation that holds of points a, b just in case they are connected by a line. Here's the theory: $\Theta = \{Fa, Fb, Fc, Gl, Gm, Gn, Rab, Rac, Rbc\}$. This theory is true under the intended interpretation where a, b, c are the vertices of a triangle with sides l, m, n. These points and lines constitute the ontology of Θ. The ideology of Θ is whatever F, G, R are taken to apply to. One might think that F, G stand for the properties *being a point, being a line* while R stands for the relation *being connected to*, in which case these properties and this relation constitute the ideology of Θ.[12] One who thought this way would take the ideology of a theory to be the properties and relations required to make it true.

Since the 1970s, physicalism has been a hotly debated topic among philosophers of mind as well as philosophers of science. Many have tried to turn the vague slogan "everything is ultimately physical" into a precise yet substantive general thesis expressing a contemporary form of materialism. As Horgan ([1993], p. 565) notes:

Many philosophers were attracted by the thought that a broadly materialist metaphysics can eschew reductionism, and supervenience seemed to hold out the promise of being a non-reductive inter-level relation that could figure centrally in a non-reductive materialism.

Quine [1978] offered this argument for a nonreductive physicalism:

[N]othing happens in the world, not the flutter of an eyelid, not the flicker of a thought, without some redistribution of microphysical states. . . . [p]hysics can settle for no less. If the physicist suspected there was any event that did not consist in a redistribution of the elementary states allowed for by his physical theory, he would seek a way of supplementing his theory. Full coverage in this sense is the very business of physics, and only of physics. Anyone who will say, "Physics is all very well in its place"—and who will not?—is then already committed to a physicalism of at least the nonreductive, nontranslational sort stated above.

The suggestion is that a constitutive goal of fundamental physics is to find fundamental microphysical states, a complete specification of which would determine everything that happens in the world. That would require not merely a fundamental ontology but also a fundamental ideology sufficient to permit this complete specification of states. But there are two problems with this suggestion.

The quantum revolution has convinced most physicists that the future as well as the present of fundamental physics lies with quantum theories. In applying such

---

[11] Quine showed how to use predicate symbols to dispense with the constants, but I will ignore this to keep things simple.
[12] Skeptical of the existence of properties and relations, Quine thought otherwise: he would interpret F, G by giving their extension—the set of objects to which each applies; and R by giving *its* extension—the set of ordered pairs to which it applies (i.e. $\{\langle a, b\rangle, \langle b, a\rangle, \langle a, c\rangle, \langle c, a\rangle, \langle b, c\rangle, \langle c, b\rangle\}$).

theories we do assign quantum states to microphysical systems. But a quantum state assignment does not specify the physical condition of that system. The application of a quantum theory does presuppose that some non-vacuous magnitude claim about a physical system is true. But quantum theory does not say which, and it is by no means clear that the sum total of true magnitude claims provides the kind of full coverage that Quine demands of fundamental physics. If anything, the quantum revolution set back progress toward Quine's goal.

Nor do physicists seem overly concerned by their failure to meet Quine's demand. One architect of the Standard Model, Steven Weinberg ([2001], p. 11), advocates what he calls grand reductionism:

...the view that all of nature is the way it is (with certain qualifications about initial conditions and historical accidents) because of simple universal laws, to which all other scientific laws may in some sense be reduced.

He contrasts this with what he calls petty reductionism:

...the much less interesting doctrine that things behave the way they do because of the properties of their constituents.

and continues:

Petty reductionism is not worth a fierce defense. Sometimes things can be explained by studying their constituents—sometimes not. . . . In fact, petty reductionism in physics has probably run its course. Just as it doesn't make sense to talk about the hardness or temperature or intelligence of individual "elementary" particles, it is also not possible to give a precise meaning to statements about particles being composed of other particles.

Unlike Quine, metaphysicians influenced by David Lewis are comfortable speaking of properties and ready to grade them on a scale of less to more fundamental, according to how well they "cut nature at its joints", to use a now customary metaphor. Ted Sider ([2011], p. 29), for example, after applauding Quine's distinction between ontology and ideology, assumes that the perfectly joint-carving notions are those of physics, by which he presumably means fundamental physics (to exclude such notions as entropy, second-order phase transition, and Mott insulator). He traces this assumption back to "a certain 'knee-jerk realism'" (p. 18) according to which

the point of human inquiry—or a very large chunk of it anyway, a chunk that includes physics—is to *conform* itself to the world, rather than to *make* the world

because knee-jerk realism requires that the physical description of reality be objectively privileged ([2011], p. 20, emphasis in original).

The pragmatist approach to quantum theory I have taken in this book offers a third alternative here. The point of a quantum theory is neither to conform our thought to the world by describing or representing it the way it is nor to create or mold the world, but to tell us *what to make* of it. Less metaphorically, by applying quantum theory we are able more effectively and responsibly to entertain significant claims about the world and to form reasonable expectations as to which are true. Quantum theory helps us do this without providing any fundamental "joint-carving" notions of its own.

Quantum entanglement has been widely taken to have metaphysical consequences because it entails a kind of property holism. In my [2016] encyclopedia entry I defined this as

Physical Property Holism: There is some set of physical objects from a domain D subject only to type P processes, not all of whose qualitative intrinsic physical properties and relations supervene on qualitative intrinsic physical properties and relations in the supervenience basis of their basic physical parts (relative to D and P).

The entangled spin state

$$|o\rangle = \frac{1}{\sqrt{2}}(|\uparrow\rangle \otimes |\downarrow\rangle - |\downarrow\rangle \otimes |\uparrow\rangle)$$

of a pair of atoms formed by the dissociation of a diatomic molecule such as hydrogen provides a relevant example. Take the domain D to include pairs of atoms resulting from such dissociation, subject only to processes of a type that preserve the state $|o\rangle$, and the atoms to be the basic physical parts of such a pair. Dirac would allow one to say the pair has spin component 0 in any direction; but he would not allow one to assign any definite spin component to either of its atomic components. It is natural to include atomic spin components (as well as any other magnitudes, such as position) in the supervenience basis. But these do not determine the spin components of the pair: the same properties and relations in the supervenience basis are, for example, equally compatible with the state

$$|o'\rangle = \frac{1}{\sqrt{2}}(|\uparrow\rangle \otimes |\downarrow\rangle + |\downarrow\rangle \otimes |\uparrow\rangle),$$

which Dirac would allow one to assign spin component 0 in the z-direction but *no* spin-component in x- or y-directions. If the qualitative intrinsic properties and relations were just those Dirac allowed one to attribute, this would be an example of physical property holism. If the atoms were spatially separated, it also appears to be a counterexample to

Spatial Separability: The qualitative intrinsic physical properties of a compound system are supervenient on those of its spatially separated component systems together with the spatial relations among these component systems.

But Dirac was too permissive in allowing inferences directly from quantum state assignments to magnitude claims. As presented in Part I, the quantum state's role is merely to yield probabilities for significant magnitude claims. Even if an application of the Born rule would yield probability 1 for such a claim, this does not imply the magnitude claim is true, or even significant. The Born rule may be legitimately applied only to significant magnitude claims. In this example that would require an interaction with the atom pair that decoheres its total spin and/or its spin component in some particular direction.

There is no legitimate inference directly from a quantum state assignment to the truth of magnitude claims about a pair in state $|o\rangle$ that don't supervene on true magnitude claims about its constituent atoms. More generally, neither property holism nor spatial non-separability is a direct consequence of quantum entanglement. Nevertheless, there are circumstances in which environmental decoherence renders significant claims about values of magnitudes on compound systems but *not* on their component subsystems.[13] In such circumstances a hypothetical agent would be right to be certain about those claims the Born rule assigns probability 1—because they are true! So while property holism and spatial non-separability are not ubiquitous, acceptance of quantum theory does require that one acknowledge cases of these phenomena.

13.4 Laws

There is general agreement among physicists that what makes a branch of physics fundamental is its laws. The physicist Phillip Anderson is widely cited for his influential [1972] defense of the explanatory autonomy of most branches of physics as well as the rest of science:

[T]he more the elementary particle physicists tell us about the nature of the fundamental laws, the less relevance they seem to have to the very real problems of the rest of science, much less to those of society. ... The behavior of large and complex aggregates of elementary particles, it turns out, is not to be understood in terms of a simple extrapolation of the properties of a few particles. Instead, at each level of complexity entirely new properties appear, and the understanding of the new behaviors requires research which, I think, is as fundamental in its nature as any other. (p. 393)

But even he begins his essay by stressing that

The working of our minds and bodies, and of all the animate and inanimate matter of which we have any detailed knowledge, are assumed to be controlled by the same set of fundamental laws, which except under certain extreme conditions we feel we know pretty well. (ibid.)

Faced with such professional consensus, it is not surprising that most philosophers have come to accept that an important goal of fundamental physics is to seek out the fundamental laws that govern everything in the natural or empirical world, and that much progress has been made toward this goal. They have gone on to offer rival accounts of what it is to be a law and what makes a law fundamental, resulting in the kind of esoteric disputes to which few non-philosophers pay attention. But two

---

[13] In the experiment described in Chapters 9 and 12, the position of a C_{60} molecule's center of mass is rendered determinate by interaction with the screen to which it adheres, unlike the positions of its component atoms.

distinctive philosophical views of fundamental laws are important in assessing the significance of the quantum revolution.

Rather than propose an analysis, Tim Maudlin recommended that we take the notion of a law of nature as primitive but use it to understand a variety of related notions including counterfactuals, causation, physical possibility, and explanation. Taking his lead from physics, he highlights the centrality of laws of temporal evolution (LOTEs), importantly including FLOTEs—fundamental laws of temporal evolution.

> I am a physicalist about laws: the only objective primitive laws I believe in are the laws of physics. Speaking picturesquely, all God did was to fix the physical laws and the initial physical state of the universe, and the rest of the state of the universe has evolved (either deterministically or stochastically) from that. Once the total physical state of the universe and the laws of physics are fixed, every other fact, such as may be, supervenes. In particular, having set the laws of physics and the physical state, God did not add, and could not have added, any further laws of chemistry or biology or psychology or economics. ([2007], p. 158)

While also professing physicalism, David Papineau ([1993], p. 16, original emphasis) distinguished it from the weaker thesis of the causal closure (or completeness) of physics:

> I take it that physics, unlike the other special sciences, is *complete* in the sense that all physical events are determined, or have their chances determined, by prior *physical* events according to *physical* laws.

Maudlin and Papineau both assume here that a fundamental physical theory would take a certain form: it would contain or imply one or more FLOTEs sufficient (given the initial physical state of the universe) to determine (at least) the chance of all future physical events. There are several reasons why quantum theories do not take that form.

If quantum theory had FLOTEs, would they be deterministic or stochastic? Maudlin ([2007], p. 11) calls the Schrödinger equation the fundamental law of non-relativistic quantum mechanics, by analogy to Newton's second law of motion. But both are deterministic rather than stochastic,[14] even though we saw in Chapter 10 how quantum theory yields chances of events through applications of the Born rule. This suggests that the Born rule is also a fundamental law of quantum theory, though not a FLOTE; but that in conjunction with the Schrödinger equation it does prescribe stochastic development of an initial state.

The problem with that suggestion is that even when quantum theory is applied to yield chances through the Born rule, both the quantum state input to the rule and the chances it outputs must be understood as relative to the physical situation of a hypothetical agent. Even if there were a unique initial quantum state of the universe

---

[14] In fact one can describe recondite hypothetical circumstances in which Newton's second law does *not* prescribe the unique future development of an initial state. Ironically, if one takes the Schrödinger equation as its fundamental law then quantum mechanics appears to be more, not less, deterministic than classical mechanics!

relative to the physical situation of every hypothetical agent, such uniqueness would lapse as subsequent events made it necessary to assign different quantum states relative to their different situations. Moreover, some of the chances of an event are determined in part by events that are not prior to but space-like separated from it, even though these events are not among its causes. If quantum theory has laws, these are neither deterministic not straightforwardly stochastic.

Though it prescribes deterministic evolution of the quantum state of an isolated system, the Schrödinger equation would not be a FLOTE even if it held in all models of quantum theory, not just non-relativistic models. The quantum state does not represent the physical condition of the system to which it is assigned, and the ubiquity of entanglement means that strictly only the universe may be assigned a deterministically evolving pure quantum state—no other system is sufficiently isolated. But assigning a quantum state to the universe has no implications for the occurrence or chances of any events.

Adoption of the semantic approach to the structure of scientific theories at least downplays the significance of the category of scientific laws. Taking this approach, Bas van Fraassen [1989] and Ronald Giere [1999] have both argued that the notion of a law of nature is a historical anachronism and functions more as an element of scientific ideology than a useful tool for understanding the structure and function of scientific theories. According to Giere ([1999], p. 90):

> the whole notion of "laws of nature" is very likely an artefact of circumstances obtaining in the seventeenth century. To understand contemporary science we need not a proper analysis of the concept of a law of nature, but a way of understanding the practice of science that does not simply presuppose that such a concept plays any important role whatsoever.

Physicists like to see themselves as furthering the project Newton began by announcing his laws of mechanics and gravitation even without Newton's faith in God as exercising dominion over nature by decreeing the laws that govern its behavior. For van Fraassen, to call something a law of physics is to do no more than point to a subclass of models of a theory whose elements satisfy some mathematical equation. Giere agrees that laws do not govern the world, recommending instead that we think of principles that govern the activity of scientists as they construct mathematical models in which certain equations necessarily hold, but only by definition: it is a contingent matter whether a theoretical hypothesis applying such a model to the world is true or false.

The Schrödinger equation functions as a constraint on models of non-relativistic quantum mechanics. I have argued that it is not a FLOTE. But this does not settle the question of whether classical mechanics may be reduced to quantum mechanics, since that question may be understood in a way that does not presuppose that either theory states or posits laws. It has been claimed that just as Kepler's laws may be reduced to Newton's, Newton's second law of motion can itself be reduced to laws of quantum mechanics by deriving it from the Schrödinger equation. We saw in Chapter 9 how

FUNDAMENTALITY 241

Kepler's laws could be reductively explained by incorporating Keplerian models in Newtonian models. Perhaps Newtonian mechanics may itself be reduced to quantum mechanics by displaying a similar relation between models of these theories?

Those claiming reducibility often appeal to what is called Ehrenfest's theorem.[15] Here is an instance of that theorem:

$$d\langle\hat{\mathbf{p}}\rangle/dt = -\langle\widehat{\nabla V(\mathbf{r})}\rangle, \tag{13.1}$$

where $\langle\hat{A}\rangle$ stands for the expectation value $\langle\psi|\hat{A}|\psi\rangle$ of dynamical variable A represented by operator $\hat{A}$ in quantum state $|\psi\rangle$, $\mathbf{p}$ is the momentum of a particle, and $V(\mathbf{r})$ is the potential acting on it. The claim is that the proof of the theorem shows how to derive Newton's second law as a limiting case of the Schrödinger equation. We could evaluate this claim by defining models of Newtonian mechanics and quantum mechanics and asking whether each of the former can be incorporated into an appropriate model of the latter. But it is clear that this would not constitute a reduction without the identification of the value of a classical dynamical variable with its quantum expectation value.[16]

But there are two reasons why that identification is illegitimate. As has been frequently noted, a Newtonian particle always has a precisely defined momentum while a quantum particle does not.[17] A particle assigned a quantum state for which it has a relatively well-defined position (in the sense that it is very unlikely to be found far from there if its position is measured) will have a very ill-defined momentum, and vice versa. So in no state may both expectation values in equation (13.1) be reasonably identified with the corresponding precisely defined classical magnitudes.

The second reason is that a magnitude claim attributing a momentum in some small interval of values is significant only in a situation in which a system has experienced significant momentum decoherence, while a magnitude claim attributing a position in some small interval of values is significant only in a situation in which a system has experienced significant position decoherence. For a particle like an electron there is no situation that meets both conditions at once.[18] There are situations in which a particle has experienced no significant decoherence either in position or in momentum. Deep in interstellar space, a claim localizing an electron's position or momentum within small limits would lack significance, even though claims about its precise position and

---

[15] I use this just as a simple illustration of the problems facing a reductive strategy physicists have pursued in far more sophisticated ways, often appealing to models of decoherence whose quantum states they treat as beables.

[16] Even with this identification reduction would strictly require derivation of something slightly different, namely

$$d\langle\hat{\mathbf{p}}\rangle/dt = -\nabla\langle\widehat{V(\mathbf{r})}\rangle.$$

[17] Unless one abandons quantum mechanics in favor of Bohmian mechanics!

[18] It is interesting to compare electrons to the much larger and heavier fullerenes in the interference experiments described in Chapters 9 and 12. The state of a C_{60} molecule adhering to the silicon surface in the first experiment has experienced enough decoherence in a basis of approximate position *and* momentum for both intervals to be small at once, compared to its expectation values of position and momentum.

its precise momentum would both be significant in classical mechanics, according to which both magnitudes always have precise values.

There is an understanding of fundamental physics that meshes well with a view of theoretical reduction as involving incorporation of models of a reduced theory into those of a theory that reduces it. James Ladyman and Donald Ross ([2007], p. 44) understand it this way:

> [F]undamental physics for us denotes a set of mathematically specified structures without self-individuating objects, where any measurement taken anywhere in the universe is in part measurement of these structures. The elements of fundamental physics are not basic proper parts of all, or indeed of any, objects.... The primacy of fundamental physics as we intend it does not suggest ontological physicalism.

In common with other contemporary structural realists, they advocate a shift in focus from fundamental entities, properties, and laws to fundamental *structures*, specified mathematically. If a theory is associated with a set of mathematical models then these are indeed specified mathematically. But if the models contain only mathematical elements it is difficult to see how these could be measured, since measurement is a physical operation intended to provide information about the physical world.

A mathematical model might be held to specify a physical structure by representing it with that mathematical structure. Models of a Keplerian theory are naturally taken to represent planetary motions, as theoretical hypotheses stated in terms of them explicitly claim. Newtonian models may be understood similarly as representing the planets' masses and the gravitational forces acting on them. In this case, a structural realist may plausibly argue that Newton's theory is more fundamental because the *physical* structures specified by its mathematical models are more fundamental than those specified by Keplerian models: physical structures like those specified by Newtonian models occur on Earth and throughout the universe.

But quantum models should *not* be taken to specify novel physical structures, as I argued in Chapter 8. A quantum model is not applied by specifying a physical structure but by prescribing cognitive attitudes toward independently specified structures. Quantum theory does not itself posit any novel physical structures so it does not make a contribution to fundamental physics, understood in Ladyman and Ross's way. But this is not a good reason to deny quantum theory the status of fundamental physics. The appropriate conclusion is surely that quantum theory has revealed to us a new, non-representational way of contributing to fundamental physics. In the final section I begin to articulate the nature of this contribution.

13.5 How Quantum Theory is Fundamental

The status of quantum theory in contemporary physics is similar in many ways to that of Newtonian mechanics in classical physics. By focusing on these similarities we can see what made each fundamental to the physics of its day. Appreciating what they have in common will make it easier to highlight the essential differences that make quantum theory revolutionary for physics and revelatory for philosophy.

Newton entitled his masterwork *The Mathematical Principles of Natural Philosophy*: what he called 'natural philosophy' has come to be known as physics. Dirac called his classic *The Principles of Quantum Mechanics*. Though usage is not uniform, physicists tend to use the words 'principle' and 'law' slightly differently. Einstein's presentation of his special theory of relativity illustrates this usage.[19] He began ([1905], p. 891) by stating the following Special Principle of Relativity:

the same laws of electrodynamics and optics will be valid for all frames of reference for which the equations of mechanics hold good.

In a modern generalization, the laws of physics hold equally in every inertial frame of reference. In this usage, a principle is not a law but something more abstract. It functions as a constraint on laws, and so a guide to finding them. I noted Giere's skepticism about the value of the notion of a law in understanding the nature of scientific theorizing. This does not extend to his treatment of principles:

Principles, I suggest, should be understood as rules devised by humans to be used in building models to represent specific aspects of the natural world. Thus Newton's principles of mechanics are to be thought of as rules for the construction of models to represent mechanical systems, from comets to pendulums. They provide a *perspective* within which to understand mechanical motions.... What one learns about the world is not general truths about the relationship between mass, force and acceleration, but that the motions of a vast variety of real-world systems can be successfully represented by models constructed according to Newton's principles of motion. ([1999], pp. 94–5 emphasis in original)

Of course Newton himself talked of laws, and formulated what he called *laws* of motion and gravitation. The former may be thought of as general principles about forces and their effects on motion, while the latter concerned one specific force. Newton himself regarded his laws of motion as also providing a general framework for the treatment of non-gravitational forces obeying as yet unknown laws. They established principles of mechanics—principles for the construction of mathematical models for mechanical systems. The framework proved generalizable, to Lagrangian and then Hamiltonian mechanics. It was sufficiently flexible to permit the modifications needed to adapt it to the novel space-time structure of relativity.[20] These developments in classical mechanics after Newton are better understood as modifications in the kinds of models admitted by the evolving theory than as repudiation of Newton's laws of motion and their replacement by something radically different.

---

[19] A usage he extended to the general theory of relativity, which he took to be based on a *general* principle of relativity whose exact role in the theory remains controversial.
[20] Though adapting Newtonian mechanics to *general* relativity involved the radical change of conceiving gravity not as a force but in terms of space-time curvature.

An architect of the Standard Model, Steven Weinberg once said in an interview:

> Unification is where it's at. The whole aim of fundamental physics is to see more and more of the world's phenomena in terms of fewer and fewer and simpler and simpler principles.
> [http://www.pbs.org/wgbh/nova/elegant/view-weinberg.html]

In his two-volume work on quantum field theory Weinberg adopted the point of view that the theory is the way it is because it is the only way to reconcile the principles of quantum mechanics with those of special relativity. This meshes well with the idea that principles guide the construction of a theory's models. Though first successful in the non-relativistic domain, the principles of quantum theory lend themselves to extension to relativistic quantum field theory much in the same way that principles of Newtonian mechanics were readily extended to guide the construction of relativistic mechanical models. Both before and after the extension, in each case the principles permitted a tremendous unification of physics by the application of the resultant models to a wide range of previously unrelated phenomena. In each case this resulted in a welter of successful predictions and explanations as well as an increased ability to control a wide range of natural phenomena. I have mentioned several such applications of quantum theory in the introduction to this chapter and elsewhere.

Symmetry principles came to occupy an increasingly important place in twentieth-century physics. E. P. Wigner played a major role in highlighting their importance. He writes ([1967], p. 5) that "the significance and general validity of these principles were recognized, however, only by Einstein", and that Einstein's work on special relativity marks

> the reversal of a trend: until then, the principles of invariance were derived from the laws of motion.... It is now natural for us to derive the laws of nature and to test their validity by means of the laws of invariance, rather than to derive the laws of invariance from what we believe to be the laws of nature.

The importance of symmetry principles has prompted such remarks as this:

> Fundamental symmetry principles dictate the basic laws of physics, control the structure of matter, and define the fundamental forces of nature.
> [Lederman and Hill, http://www.sackett.net/Noether.pdf, p. 1]

Certainly a variety of symmetry principles were key to the construction of the Standard Model that many physicists would say gives our best current account of the fundamental laws of physics. But these are not laws of quantum theory: at *most* they are laws of particular quantum field theories, such as the quantum chromodynamics (QCD) often said to describe the strong interaction between quarks, mediated by exchange of gluons. One can derive field equations for interacting quark and gluon fields from the QCD Lagrangian density $\mathcal{L}_{QCD}$[21] by a standard procedure. These may be called equations of motion and analogized to an instance of Newton's second law of motion with a specified force. But they are really just constraints on the form of a quantum mathematical model, as becomes clear when one notes that they constrain field *operators*, not field *magnitudes*. These models may be applied to predict and explain physical phenomena just the way any quantum theoretic model may be applied—by guiding credences in significant magnitude claims. These will include claims about the mass of the proton and the energy spectrum of charmonium as well as a huge number of detection events in accelerators like the LHC: they will not include claims about the values of a quantized quark or gluon field.

[21] $\mathcal{L}_{QCD} = \widehat{\bar\psi}_i (i(\gamma^\mu D_\mu)_{ij} - m\delta_{ij})\widehat{\psi}_j - 1/4 \widehat{G}^a_{\mu\nu} \widehat{G}_a^{\mu\nu}$. Experts will know that this yields Euler–Lagrange equations for quark and gluon quantum field operators $(\widehat{\psi}_i, \widehat{A}^a_\mu)$ after applying Hamilton's principle requiring the associated *classical* action to be stationary under independent variation of the corresponding classical fields (ψ_i, A^a_μ), where $G^a_{\mu\nu}$ represents the gluon field strength defined by the gluon field potential A^a_μ.

14

Conclusion

Good fictional writing leaves it to the reader to draw the moral of the story. But quantum theory is not a work of fiction. Its unblemished record of successful applications in the physical world places it at the foundation of contemporary science. So I'll end by summarizing what I take to be the real philosophical lessons to be learnt from quantum theory's success.

Western philosophy and modern science share common roots in ancient Ionian Greece. This metaphor is often continued when one speaks of different sciences as branches, as of a single tree of inquiry whose trunk is philosophy; the branch point being marked by the development of new, more specialized techniques enabling questions to be more precisely formulated and answered. The science of psychology, for example, branched off only in the nineteenth century under the influence of the pragmatist philosopher Wiliam James.

Isaac Newton called his [1687] founding text of physics *The Mathematical Principles of Natural Philosophy*, thus marking both continuity with and departure from the *Principles of Philosophy* of René Descartes (himself often known as the founder of modern philosophy). But the term 'natural philosophy' may still be used to label inquiry into very general features of the world informed by the best contemporary science. So what are the implications of quantum theory for natural philosophy?

Quantum theory as presented in Part I has remarkably few implications for natural philosophy. The theory itself describes no new entities or magnitudes and so has little to say about the natural world. Novel mathematical entities do make an appearance in models of quantum theory, but it is not their function to represent anything in the physical world that could not be represented without them. The most important of these is the wave function, state vector, or other mathematical representative of a quantum state.

Many peculiar alleged implications of quantum theory follow only if one mistakenly endows a quantum state with an ontological significance I have argued that it lacks. One will conclude that a particle follows more than one path through an interferometer at once only if one wrongly assumes that each term in a superposition of its quantum state vector represents it as following a corresponding path. One will take Schrödinger's cat to be suspended between life and death only if one makes the contrary mistaken assumption that it is alive only when represented by a quantum state the Born rule would assign life probability 1, and dead only when represented by a quantum state the Born rule would assign death probability 1. But a quantum state represents none of these things: its role is restricted to offering advice on the significance and credibility of claims about such things as where particles will be manifested and when cats will be alive or dead.

Everettian (many worlds) interpretations make this mistake twice over. They take the universal quantum state as quantum theory's fundamental ontology and derive a secondary, branch-relative, ontology from the quantum state relative to each branch in an expansion of the universal wave function as a superposition in some basis. In a currently prominent version, this basis is determined (approximately) by the decoherence induced by evolution under the actual universal Hamiltonian, thereby avoiding the need to choose a preferred basis on non-physical grounds. In this way contemporary Everttians seek to recover our ontology of ordinary objects (including atoms and elementary particles, as well as tables, people, planets, and galaxies) as emerging dynamically with "our" world—merely one world in a constantly branching multiverse of similar branch-worlds.

Only if one took a quantum state to represent the behavior of a physical system to which it is assigned would one view its apparently discontinuous reassignment as representing a physical process to be described by new physics—perhaps even a radically new physics involving consciousness! But that view rests on a similar misunderstanding of the function of a quantum state assignment. Moreover, it is not obvious how such a physical process could occur in the space-time of relativity theory.

The conflict with relativity is even clearer in Bohmian mechanics, which assumes that a universal wave function plays a role in guiding the motion of particles. Even though it never collapses, if this wave function is suitably entangled the motion of distant particle B may be affected simultaneously by doing something to nearby particle A. This would be intelligible only in the context of a preferred notion of simultaneity that is not definable in the space-time structure of relativity theory. It would also be in explicit violation of Einstein's principle of local action.

Quantum theory as presented in Part I involves no instantaneous action at a distance. There is no essential conflict between a sharp formulation and fundamental relativity: non-relativistic quantum mechanics has been consistently generalized to the relativistic domain, just as was Newtonian mechanics.[1] It is certainly surprising that spatially separated experiments on systems assigned entangled quantum states reveal correlations that cannot be explained by a *factorizable* common cause. This is not something we encounter in daily life. But it does not mean that the consequences of events at one place propagate to other places faster than light: nor is it because the parts of these systems retain a connection so private and intimate as to defy theoretical description as well as experimental detection.

---

[1] The generalization did involve moving from a theory whose applications take for granted a particle ontology to a quantum field theory in some of whose applications particles emerge, while "classical" fields emerge in others.

248 CONCLUSION

Quantum theory does not teach us that our observations create the world they reveal. The moon is demonstrably there when nobody looks. The doctrine that the world is made up of objects whose existence is independent of human consciousness is in conflict neither with quantum mechanics nor with facts established by experiment: on the contrary, our evidence for (as well as every application of) quantum theory presupposes objects whose existence is independent of human consciousness.

It is not observation but a specific kind of environmental interaction that determines when a claim about the moon, an atom, or an elementary particle is significant enough to be usefully treated as either true or false. It is not a coincidence that such interactions occur when we make observations since an observation report must be a significant claim. But they occur whether or not we are making observations and would occur in a world without observers. We can model these (and other) interactions in quantum theory, and use these models as a check on the significance of claims about the moon, an atom, an elementary particle, or a "classical" field.

By assigning an entangled quantum state one does not represent any physical relation among its subsystems: entanglement is not a physical relation. Nevertheless, there are situations in which its interaction with its environment renders significant some claim about a compound physical system as a whole,[2] but no corresponding claims about its physical parts. Perhaps this amounts to a kind of physical holism. But entanglement alone entails no widespread physical holism or physical non-separability.

Quantum field theory introduces another class of novel mathematical entities, but these are not used to represent corresponding physical entities or magnitudes. Models of relativistic quantum field theories are applied in high-energy physics to predict and explain phenomena involving particles and fields. But such an application describes no *physical* quantum field, and neither particles nor "classical" fields are represented by elements of these models. As in other quantum theories, what results from a legitimate application of a relativistic quantum field theory in high-energy physics is advice on the significance and credibility of claims about physical systems. Some applications issue in advice concerning properties of particles including electrons, quarks, photons, and other bosons: other applications issue in advice about "classical" fields, including the electromagnetic field and the Higgs field. By taking this advice we are able to make significant and true claims about such things. But neither elementary particles nor fields (quantum or classical) are elements of the ontology of any quantum field theory.

Metaphysics and epistemology are two core fields of contemporary philosophy: metaphysics has to do with general features of reality, while epistemology is concerned with our knowledge of that reality. One age-old metaphysical question is whether

---

[2] Such as a C_{60} molecule, or a pair of hydrogen atoms formed by dissociation of a hydrogen molecule. Environmental decoherence may render significant a claim about the position of the C_{60} molecule though not about the positions of its component carbon atoms; or about the spin of the pair though not about the spins of its two hydrogen atoms.

reality is ultimately mental or physical. Another is the nature of consciousness and its relation to the (assumed) physical world. Quantum theory has no implications for either question since it is applied only to the physical world.

A third metaphysical issue concerns freedom of the will. Historically, many have believed that our claimed ability freely to choose what to do (or at least attempt) would be nothing but an illusion in a deterministic world in which everything that happens is the inevitable result of prior causes. Newtonian mechanics encouraged the belief that nature is indeed governed by such deterministic laws. After successfully applying Newtonian mechanics to the solar system, Laplace famously said:

> We ought to regard the present state of the universe as the effect of its antecedent state and as the cause of the state that is to follow. An intellect which at a certain moment would know all forces that set nature in motion, and all positions of all items of which nature is composed, if this intellect were also vast enough to submit these data to analysis, it would embrace in a single formula the movements of the greatest bodies of the universe and those of the tiniest atom; for such an intellect nothing would be uncertain and the future just like the past would be present before its eyes. [1820/1951]

The thought has occurred to many that quantum indeterminism might make room for genuine human freedom by releasing us from the iron grip of determinism. Others have countered that if our actions are mere random events, determined by nothing that went before, then neither are we genuinely free in an indeterministic quantum world.

But quantum theory itself describes the world neither as deterministic nor as indeterministic, since it does not describe the world at all. When correctly applied, quantum theory offers advice on how to set our degrees of belief in matters about which we are not currently in a position to be certain. It does not state laws relating those matters to earlier events, so it is neither a deterministic nor an indeterministic theory. It provides us with the best available advice about what to expect without describing any processes that might be thought to bring it about, with certainty or merely by chance.

Nevertheless we are able to use quantum theory to explain natural phenomena only because its legitimate applications also tell us what an event depends on, and at least some such dependencies count as causal. Quantum theory here helps one to see what should already have been apparent: rather than couching the free will issue in terms of global determinism or indeterminism, the important metaphysical question to ask is what kind of causal dependence of acts on prior conditions poses a threat to freedom?

Contemporary metaphysicians inquire into the nature of laws, causation, probability, composition, and ontology. In Chapter 13 I followed the proposal of other contemporary philosophers of science that we reject the category of laws of nature as unhelpful in our attempts to understand how physical theories work. Quantum theory clearly reinforces this lesson. What can it teach metaphysicians about these other topics?

In each case I think the lesson is methodological and deflationary. A metaphysician should not look to physical theories, and in particular to fundamental physical theories, in the attempt to find out what causation, probability, etc. is in the world: their models mention no such things. We introduce such terminology when applying their models to meet our needs as agents, limited by our physical situation. Having introduced it in that context, we can use talk of causation, probability, etc. to make true statements. But these statements are not about the physical world independent of our perspective as agents physically situated within it; and their truth is not determined by the state of the physical world, even when extended to include our physical states.

First, consider probability. A legitimate application of a quantum model issues in objectively true statements about the probability of certain magnitude claims. Their function is not to describe elements of physical reality ("beables") but to meet the needs of epistemically limited (since physically situated) agents by guiding their beliefs about those magnitude claims. The guidance is reliable insofar as it tracks statistical patterns of correlation between physical conditions described by claims backing the underlying quantum state assignment and magnitude claims of this same kind. But a probability assignment does not describe these patterns and its truth is not determined (though it is evidenced) by the relative frequencies they display: it can fulfill its function only by "floating free" of such actual frequencies.

David Lewis recognized quantum phenomena as providing the last holdout of objective chance against a subjectivist about probability. Applications of quantum theory teach us how probability may be objective even though it does not describe probabilities that arise from fundamental physics that attach to actual or possible events in virtue solely of their physical description. This reinforces the pragmatist lesson that objective truth need not be grounded in truths about fundamental physical magnitudes, and that the best way to understand a concept like probability is to ask not what in the world it represents but what function is served by claims about it.

Now consider causation. Quantum models say nothing about causal relations, and nor do probabilistic claims made in their legitimate applications. One who accepts quantum theory also accepts certain counterfactual connections between conditions backing assignment of a quantum state and such probabilistic claims. These are not causal counterfactuals because an event's probability is not itself an event. But suppose the probabilistic claim concerns the chance of an event that actually occurs. Then the counterfactual relation between the event's chance and some condition backing assignment of the relevant quantum state may certify that condition as a cause of the event. But this counterfactual connection will count as causal only if it is robust under hypothetical interventions countenanced as possible by one who accepts quantum theory.

This is because causal relations are not woven into the fabric of physical reality but visible in it only from the perspective of physically situated agents like us. They are objective for all that. But once again they are of significant interest only for physically situated agents who need to know how to achieve their ends by intervening in the world they inhabit. Quantum theory should remind us that causation, like probability, is not built into the world as described by physics: we think of the world in causal terms to satisfy our needs as physically situated agents. But in this case our physical situation imposes limits on what we can do as well as on what we can know, so those needs are practical rather than merely epistemic.

There is a surprisingly lively debate in contemporary analytic metaphysics about the extent of composition—about what objects are composed of what other objects. We ordinarily think a particular table is composed of (say) a flat wooden top attached to four wooden legs, and that each of these parts is in turn composed of particular atoms of different chemical elements. Perhaps encouraged by their take on the success of elementary particle physics, many metaphysicians assume that a final theory of fundamental physics will describe the ultimate parts of everything in the physical world (though some discuss the possibility that the hierarchy of physical composition may have no lowest level). The debate is then about what, if anything, such ultimate parts compose. Does an arbitrary collection of diverse ultimate parts compose a physical object, even when these are randomly scattered throughout the world? If not, what *is* required for a collection of ultimate parts to compose a physical object?

A simple set of axioms for parthood define what is sometimes called classical mereology—a formal theory of the *part of* relation. Applying these axioms to arbitrary collections of objects, so-called mereological universalists claim that *any* random collection of objects compose an(other) object (however uninteresting we may find this object). But mereological nihilists claim that the only real objects are those that have no other parts—what they call simples or mereological atoms (though their rivals could call them ultimate parts). How does quantum theory bear on this metaphysical debate?

Relativistic quantum field theories form the core of the so-called Standard Model of elementary particles. One might naively assume these theories describe the behavior of these particles—quarks, electrons (and other leptons), photons (and other gauge bosons), and the Higgs boson. Such particles are at least candidates for the title of mereological atoms—simple, non-compound things of which every object in the physical world is composed. But the naive assumption is incorrect.

The elementary particles of the Standard Model are not part of the ontology of its relativistic quantum field theories, and these theories do not describe their behavior. A quantum field theory has no ontology of its own, though application of its models does presuppose the bare existence of one or more quantum fields. A legitimate application of a model of a relativistic quantum field theory will issue advice about the significance and credibility of certain magnitude claims. In one environmental context, these may be claims about particles and their magnitudes (e.g. the numbers of photons present and their energies); in another environmental context, these may be claims about "classical" fields and their magnitudes (e.g. the frequency of the primary mode of a laser). Rather than saying what are the ultimate parts of the physical world, the Standard Model only gives us good advice on when to make significant claims about elementary particles (and how firmly to believe them), and when to make significant claims about "classical" fields (and how firmly to believe them). The advice is good because it is borne out by the frequency with which these claims turn out to be true.

Some metaphysicians may take this to warrant a charge of dereliction of duty against contemporary fundamental physics. The physical world must be composed of *something* at a fundamental level, and if our best theories do not tell us what that is then we must look for better theories that will. But it is not the duty of physics to seek out the ultimate building blocks of the world. Historically this has proved a successful strategy for advancing the primary goals of physics—predicting, controlling, and explaining physical phenomena. Quantum theory is not beholden to the demands of metaphysics: in the guise of the relativistic quantum field theories of the Standard Model, it teaches us that these goals may be sucessfully pursued by other means.

This brings me to the issue of scientific realism. There is a long-standing debate in the philosophy of science about the aim of scientific theorizing and the appropriate attitude to take toward a successful scientific theory. In this debate a scientific realist holds that science aims to give us, in its theories, a literally true story of what the world is like. On the contrary, an instrumentalist maintains that science aims merely to give us theories that are empirically adequate—they are effective instruments for truly describing all observable phenomena, enabling us correctly to predict results of observations we have not yet made and indeed may never make. The realist holds that the appropriate epistemic attitude to take toward a successful scientific theory is to believe that it is true: the instrumentalist, on the other hand, maintains that a scientist need only believe such a theory is empirically adequate.

Some may object to the view of quantum theory presented in this book as instrumentalist rather than realist. Consider, for example, the following passage from a recent book advocating an Everettian reading of quantum theory:

[T]he consensus view in philosophy is now that instrumentalism is incoherent as a way to make sense of science. Not all serious philosophers of science are *realists* in the sense that they think that our best scientific theories are approximately true, but there is a consensus that the only way to *understand a scientific theory* is to understand it as offering a description of the world.
(Wallace [2012], pp. 26–7 emphasis in original)

How might the mathematical models of a physical theory describe the world?

Ever since Newton, physicists have been in the business of constructing mathematical models of aspects of the world, 'models' in the sense that they are—or are intended to be—isomorphic to the features of the world being studied.   (Wallace [2012], p. 12)

So the description is given not by the model itself, but by a claim that a mathematical model is isomorphic to the features of the world being studied.

One way to apply a physical theory is to claim some such isomorphism. The most general claim would be that the theory is true insofar as it includes a model that is isomorphic to the structure of the whole world. That would be a very bold claim, for any theory. Typical applications would be less ambitious and more tentative: one might more plausibly claim that a particular model of the theory is approximately, or partially, isomorphic to some features of a small part of the world. But applications of quantum models claim no such isomorphism with any aspect of the world. That is what makes quantum theory a revolutionary theory. By applying its models we are able to improve the way we describe the world as well as what we believe about it without claiming any kind of isomorphism between model and world.

These descriptions and beliefs extend far beyond results of actual and potential observations to phenomena occurring anywhere in the universe at any time. As we have seen, quantum theory is routinely and legitimately applied to describe microscopic and submicroscopic phenomena on Earth and distant stars, and perhaps even to the evolution of the large-scale structure of the universe. But these descriptions come in the form of magnitude claims (or claims whose truth or falsity is determined by them): they do not include claims about quantum states or quantum fields. Quantum states are not useful fictions. Statements about them are often meaningful and true, but their function is not to describe novel physical entities or magnitudes but to guide us in forming beliefs about other physical entities and magnitudes that are not distinctive to quantum theory.

So the view of quantum theory presented here is not instrumentalist. Quantum theory is not just an instrument for generating predictions of observations made in specific experimental set-ups. A precise formulation of quantum theory may be given with no mention of observation or measurement. The notion of observation, or observability, has no special epistemic or semantic significance for quantum theory. Statements assigning quantum states and Born probabilities are perfectly meaningful and often true. But is this view compatible with scientific realism?

In addressing this question it will be helpful to recall something (quoted in Chapter 1) which Niels Bohr once said that is often taken as a decisive *rejection* of realism about quantum theory:

There is no quantum world. There is only an abstract quantum description. It is wrong to think that the task of physics is to find out how nature is. Physics concerns what we can say about nature. (Bohr, quoted in Petersen [1963, p. 12])

One can read three of these four sentences as compatible with the view of quantum theory presented in this book. Start with the first sentence. Did Bohr deny the existence of the physical world? I think not: how could the author of the Bohr model of the atom deny the existence of atoms? In conjunction with the second sentence, what Bohr can plausibly be read as denying is that the function of distinctively quantum language (specifically, talk of wave functions or quantum states) is not to describe the physical world but something else.

As description, quantum language aims at truth while still being capable of falsity. Indeed, an assignment of quantum state may be right or wrong. So what is the function of a quantum state? Though he does not say so here, there is no reason to suspect Bohr would reject the orthodox view that the main reason to assign a quantum state is to be able to use the Born rule to derive probabilities of various possible measurement outcomes.[3] His fourth sentence may be read as compatible with the pragmatist view of content developed in Chapter 12. We can say something significant about nature only when environmental conditions permit. By modeling environmentally induced decoherence, quantum theory can advise us on the significance of a particular magnitude claim in different environmental conditions: this is one of the things that physics concerns.

It is only in his third sentence that Bohr here said something incompatible with the view of quantum theory I have taken in this book. It is hard to deny that here he is rejecting a key realist tenet about the aim of science. The quantum revolution gives us no reason to reject that tenet. Among other things, quantum theory tells us what we can significantly say about nature and cautions us against saying more. But the whole point is then to say what we can about nature. That is a basic task of physics, an essential preliminary to pursuit of its goals of predicting, controlling, and explaining natural phenomena.

Einstein was a scientific realist, not out of deep philosophical conviction but for pragmatic reasons. He considered realism a program for doing science with a proven track record and no clear alternatives.

Physics is an attempt conceptually to grasp reality as it is thought independently of its being observed. In this sense one speaks of 'physical reality'.
(Einstein, in Schilpp, ed. [1949], p. 81)

but

the "real" in physics is to be taken as a type of program, to which we are, however, not forced to cling *a priori*. (Einstein, in Schilpp, ed. [1949], pp. 673–4)

As is well known, Einstein expressed dissatisfaction with quantum theory even though he was a driving force behind its development:

What does not satisfy me in [quantum] theory, from the standpoint of principle, is its attitude towards that which appears to me to be the programmatic aim of all physics: the complete description of any (individual) real situation (as it supposedly exists irrespective of any act of observation or substantiation). (Einstein, in Schilpp, ed. [1949], p. 667)

He sought a more fundamental theory that could provide such a complete description from which quantum theory's statistical predictions could be recovered as approximations in some appropriate limit.

---

[3] Elsewhere he makes it clear that he takes a measurement to be a physical process that occurs whether or not anyone is conscious of its outcome. He doesn't appeal to decoherence here. If he is read as denying that this process may be modeled in quantum theory then he could not accept the view of quantum theory presented in this book.

Einstein did not seriously consider the possibility that quantum theory *does* permit one completely to describe certain real situations that exist irrespective of any act of observation *without itself describing them*. In fact we can use the theory to advise us as to when these situations obtain, again without the theory itself describing them. We do not use quantum models as representations of physical reality. Instead, we use them as sources of authoritative advice on how better to deploy independently available descriptive and representational resources. By deploying them in this way we may indeed be able completely to describe many individual real situations. With quantum theory, this is how physics has continued to pursue what Einstein took to be its programmatic aim.

Bell pointed out one thing we cannot do with quantum theory:

[I]n classical mechanics any isolation of a system from the world as a whole involves approximation. But at least one can *envisage* an accurate theory of the universe, to which the restricted account is an approximation.   ([1987]; [2004], p. 125 emphasis in original)

Quantum theory cannot help us envisage a complete and accurate model of the universe. We could not use a model of quantum theory completely and accurately to represent the universe because any use of a quantum model calls upon representational resources from outside quantum theory (in the form of magnitude claims about physical systems): in particular, a universal quantum state cannot provide such a representation. One can apply a quantum model to yield advice about significant magnitude claims about a physical system only if that system interacts in the right way with other physical systems. No such application is possible if the target system is the entire physical universe, in all its complexity. But it does not constitute a criticism of quantum theory that one cannot envisage a complete and accurate representation of the universe, since no successful application of quantum theory to cosmology or anywhere else requires that one be able to do so.

In fact, I doubt that one *can* envisage a detailed and accurate representation of the universe by a model of classical mechanics. Obtaining and using a complete and accurate classical mathematical model of the universe would vastly exceed the combined observational and cognitive capacities of humanity or any other physically realizable community of agents, while only in use would such a mathematical model represent anything. Scientific realism should not consider it an (unattainable!) aim of physics to provide us with a complete and accurate representation of the entire universe. The goals of physics (as opposed to metaphysics) should be more modest, even for a realist.

In philosophy, realism is often associated with a correspondence theory of truth. Here the term 'theory' is used loosely simply to label the vague thought that the truth of a statement or belief consists in its bearing some appropriate relation or relations to reality: perhaps it corresponds to an actual state of affairs by virtue of the fact that each of its components refers to items present in that state, their properties, or relations among them. Some scientific realists maintain that the truth of a scientific theory is what explains its success, just as the truth of our everyday beliefs explains how we are able regularly to succeed in our endeavors. When portrayed as a scientific hypothesis (as in Putnam [1978]), this version of the correspondence theory takes reference to be some kind of causal relation between elements of our thought or language use and corresponding elements in the world.

It is plausible that perception sets up causal relations (or at least regular correlations) between cognitive states and corresponding structures in the world, even in animals and other organisms incapable of language use.[4] As Burge says:

Perceptual representation that objectively represents the physical world is phylogenetically and developmentally the most primitive type of representation.... Representation of the physical world begins early in the phylogenetic elaboration of life. ([2010], p. xi)

It is much less plausible that reference works that way in sophisticated uses of language or scientific theories. It may be true that cleanliness is next to godliness, but not because of causal relations underlying representational uses of the terms or concepts 'cleanliness', 'next to', and 'godliness'. In applications of quantum models we are dealing with highly sophisticated uses of scientific language and mathematics, typically to physical phenomena that are not accessible to our perceptual faculties. Doubtless ultimately piggybacking on perceptual representation, these uses permit convergent agreement on the truth of claims framed in sophisticated scientific language. In such discourse truth is not to be explained in terms of reference: if anything it is the other way around. Born probabilities provide a good example: a true Born probability assignment does not correspond to physical reality by virtue of causal relations between uses of the Born rule and physical entities or magnitudes.

In Chapter 8 I denied that the truth of a magnitude claim about the spin component of a silver atom is grounded in some prior physical correspondence relation between the claim's utterance and spinning silver atoms. Instead I advocated a deflationary treatment of the truth of this and other magnitude claims. Truth is not a property of such a claim, and so cannot be appealed to in support of scientific realism, now understood as a scientific hypothesis purporting to explain the success of science (in this case, quantum theory). To say that the sentence $\ulcorner A \in \Delta \urcorner$ is true is to say no more than that the value of A is restricted to Δ ("The silver atom has z-spin $+\hbar/2$" is true if and only if the silver atom has z-spin $+\hbar/2$). Since the content of that claim depends on the physical system's environment, its truth cannot be grounded in physical reference relations to any environment-independent properties of the system. Nor is it fruitful to seek further grounds of its truth in physical relations to (unknown but very complex) aspects of its environment. Best to say that what it is for a silver atom to have the

---

[4] The 2014 Nobel prize in physiology or medicine was awarded for the discovery of specific nerve cells in the brain of a rat with the cognitive function of providing an internal map of the rat's environment and a sense of place. It is hard to deny that causal relations with the environment determine the correspondence between cognitive state and environment responsible for the rat's ability to navigate through it.

property $z\text{-}spin\ +\hbar/2$ is simply for the claim "The silver atom has z-spin $+\hbar/2$" to be true, and to redirect inquiries to the justification for this claim.

Consider instead a magnitude claim about the momentum of a photon in an application of the quantum theory of the electromagnetic field. The truth of such a claim cannot be grounded in facts about a novel quantum field magnitude represented by a mathematical object in a quantum model. Nor can it be grounded in intrinsic properties of the photon, since the content of *any* claim about a photon depends on the environmental context. Here, too, it is fruitless to try to locate physical relations to that environment in support of a causal theory of reference for the term 'photon' as it appears in this claim. The claim gets its content from its inferential relations to other claims, and if such inferences justify belief in its truth they thereby justify one in believing a photon was successfully referred to by the claim. So reference is subordinate to truth rather than providing its ground.

Its treatment of probability, causation, explanation, content, and now truth and reference make it appropriate to call this view of quantum theory "pragmatist". The success of these treatments in making sense of quantum theory, in dissolving the measurement problem, the problem of non-locality, and the problem of the ontology of a quantum field theory, reveals the power of pragmatist ideas in the way that means most to a pragmatist: pragmatism works! Having argued that it is not instrumentalist, I am content to leave it to the reader to decide whether to classify the resulting view of quantum theory as realist as well as pragmatist.

APPENDIX A

Operators and the Born Rule

A.1 Vectors and Inner Products

The state vector is called a vector because it is an element of a vector space—a type of mathematical structure. To get an idea of what a vector space is, consider an example: the set of "arrows" (i.e. directed line segments of various lengths) emanating from a fixed point in ordinary three-dimensional Euclidean space forms a vector space in which the vectors are arrows (like $\vec{a}, \vec{b}, \vec{c}$ in Figure A.1).

The most important features of vector spaces are that addition of two vectors always produces another vector in the space, and multiplication of a vector by a number also always produces another vector in the space. To add two arrows in Figure A.1, "slide" one over until its tail coincides with the tip of the other and draw another arrow from the tail of the second to the tip of the first: multiplying an arrow by a real number r makes it r times as long while pointing in the same direction. Formally, a *vector space* over the real numbers is a set V together with operations of vector addition and multiplication by a scalar (i.e. a real number, not a vector) satisfying the following conditions:

For any vectors $\alpha, \beta, \gamma \in V$ and any real numbers r_1, r_2:

$$\alpha + \beta \in V$$
$$r_1 \alpha \in V$$
$$\alpha + \beta = \beta + \alpha$$
$$\alpha + (\beta + \gamma) = (\alpha + \beta) + \gamma$$
$$r_1 (r_2 \alpha) = (r_1 r_2) \alpha$$

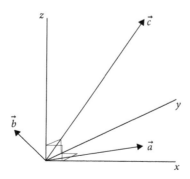

Figure A.1 The vector space $V(E^3)$

$$r_1(\alpha + \beta) = r_1\alpha + r_1\beta$$
$$(r_1 + r_2)\alpha = r_1\alpha + r_2\alpha$$
$$1\alpha = \alpha$$
$$(\exists \emptyset \in V)(\alpha + \emptyset = \alpha)$$
$$0\alpha = \emptyset$$

Vector spaces differ in their *dimension*. The space $V(E^3)$ of arrows from a fixed point in three-dimensional Euclidean space is three-dimensional since any vector in the space may be expressed as a linear sum of just three fixed *basis* vectors (e.g. arrows of unit length in the three mutually perpendicular directions defined by the x, y, z axes in Figure A.2). A *subspace* of a vector space is a set of vectors from that space that itself constitutes a vector space when taken together with the operations of scalar multiplication and vector addition inherited from the original space (and its $\emptyset$ vector). Here are some examples of subspaces of $V(E^3)$: the set $\mathcal{M}$ of arrows lying wholly in the x–y plane forms a two-dimensional subspace of $V(E^3)$; the set $\mathcal{N}$ of arrows that lie along the z-axis forms a one-dimensional subspace of $V(E^3)$; $V(E^3)$ itself forms its only three-dimensional subspace; and the set $\{\emptyset\}$ forms the null subspace $\varnothing$, its only zero-dimensional subspace.

Some vector spaces, including the kind used to represent the state of a system in quantum theory, have an *inner product* defined on them. Intuitively, it is the inner product that underlies talk of the length of a vector and of the angle between two vectors. An inner product is an operation that maps any ordered pair of vectors α, β onto a scalar written (α, β) and called the inner product of α with β. The inner product of two vectors $\vec{a}, \vec{b}$ in $V(E^3)$ is a real number equal to the product of the length $|\vec{a}|$ of arrow $\vec{a}$ with the perpendicular projection of $\vec{b}$ onto $\vec{a}$ (see Figure A.3). In this example the inner product is often called the scalar product and written $\vec{a}.\vec{b}$. If the angle between $\vec{a}$ and $\vec{b}$ is φ, then we have $\vec{a}.\vec{b} = |\vec{a}|\,|\vec{b}|\cos\varphi$, so this inner product is symmetric: $\vec{a}.\vec{b} = \vec{b}.\vec{a}$. Note that

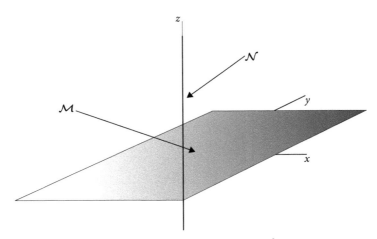

Figure A.2 Some subspaces of $V(E^3)$

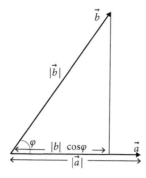

Figure A.3 The inner product in $V(E^3)$

$|\vec{a}|^2 = \vec{a}.\vec{a}$ and that if $\vec{a}, \vec{b}$ each have non-zero length then $\vec{a}.\vec{b} = 0$ if and only if $\vec{a}$ and $\vec{b}$ are perpendicular to one another.

The state vector in quantum theory is a vector in a particular type of vector space known as a *Hilbert space* (after the German mathematician David Hilbert). A Hilbert space is a vector space with an inner product, which is an abstraction from, and generalization in two respects of, the space $V(E^3)$. The generalizations are:

1. The numbers by which vectors may be multiplied may be complex as well as real. (Recall that complex numbers are those of the form $z = x + iy$, where x, y are real numbers and $i = \sqrt{-1}$).
2. A Hilbert space may be of infinite (or any finite integral) dimension.

These generalizations will not prevent us from using $V(E^3)$ as an illustrative example in what follows.

Formally, an *inner product* on a vector space V over the complex numbers $\mathbb{C}$ is a function $(,): V \times V \to \mathbb{C}$ such that for all $\alpha, \beta, \gamma \in V$ and all $c \in \mathbb{C}$

$$(\alpha, \beta + \gamma) = (\alpha, \beta) + (\alpha, \gamma)$$
$$(\alpha, c\beta) = c(\alpha, \beta)$$
$$(\alpha, \alpha) \geq 0$$
$$(\beta, \alpha) = (\alpha, \beta)^*$$

Here the symbol * stands for the operation of complex conjugation, defined by $z^* = x - iy$ if and only if $z = x + iy$. So this inner product is *not* always symmetric. But since (α, α) is always a non-negative real number we can define the *norm* $\|\alpha\|$ of a vector α by $\|\alpha\|^2 = (\alpha, \alpha)$. Two vectors α, β are said to be *orthogonal* iff $(\alpha, \beta) = 0$, and vector α is said to be *normalized* iff $\|\alpha\| = 1$. In the example $V(E^3)$ two vectors $\vec{a}, \vec{b}$ are orthogonal just in case the arrows are perpendicular to one another.

A.2 Operators and the Simple Born Rule

A system's dynamical variables are represented in a quantum model by linear operators of a certain kind acting on a vector space used to model it. An operator on a vector space is defined in terms of the vector that results from applying it to vectors in the space. Formally a *linear operator* $\hat{A}$ on a vector space V is a function from a subspace U of V into V that satisfies these conditions, for all $\alpha, \beta \in U$ and all scalars (i.e. real or complex numbers) c:

$$\hat{A}(\alpha + \beta) = \hat{A}\alpha + \hat{A}\beta$$
$$\hat{A}(c\alpha) = c\hat{A}\alpha$$

The previous section noted that infinite-dimensional vector spaces V are used in quantum theory. These admit linear operators whose domain of definition U excludes some elements of V: $U \neq V$. To treat these rigorously would involve mathematical digressions threatening to distract attention from the conceptual issues central to this book. So in this Appendix I will restrict attention to finite-dimensional Hilbert spaces, pointing out where this restricts the generality of its statements. I recommend Jordan [1996] as a good introduction to the mathematics of operators in infinite-dimensional Hilbert spaces.

Given the operator representing a particular dynamical variable and a state vector used to model the system to which that variable pertains, quantum theory prescribes probabilities for a measurement of that variable on the system in that state to yield particular values. I'll explain the prescription in this Appendix. But before proceeding I'll state one consequence which helps one to see in what sense a linear operator *represents* a dynamical variable.

If a state vector ψ is assigned, then the expectation value (i.e. the mean (average) value predicted for a measurement) of dynamical variable A on that system is equal to the inner product $(\psi, \hat{A}\psi)$, where $\hat{A}$ is the operator representing A. To make sure this will always be a real number, $\hat{A}$ must be a special kind of linear operator called a *self-adjoint* operator. An operator in a finite-dimensional space is self-adjoint if and only if, for any vectors α, β, $(\alpha, \hat{A}\beta) = (\hat{A}\alpha, \beta)$. Referring back to the definition of the inner product from section A.1, you can now see why the expectation value of any dynamical variable represented by a self-adjoint operator will always be a real number. We saw that the inner product of vectors φ, ψ is usually written as $\langle \varphi | \psi \rangle$. In this notation the requirement of self-adjointness becomes $\langle \alpha | \hat{A}\beta \rangle = \langle \hat{A}\alpha | \beta \rangle$, and the expectation value is written $\langle \psi | \hat{A} | \psi \rangle$.

The most perspicuous way to present quantum theory's mathematical prescription for probabilities of measurement results is by reference to a special class of self-adjoint linear operators called *projection operators*. The space $V(E^3)$ provides examples of such operators (see Figure A.4). Consider the linear operator $\hat{P}$ on this space that maps any vector onto a vector in the x–y plane in such a way that the new vector is the perpendicular projection of the original vector ("its shadow in the noonday sun"): this is a projection operator whose range is the two-dimensional subspace $\mathcal{M}$. Consider also the linear operator $\hat{Q}$ that maps any vector onto its perpendicular projection on the z-axis, with range $\mathcal{N}$. The identity operator $\hat{1}$ that maps any vector onto itself, and the null operator $\hat{0}$ that maps any vector onto the zero vector $\emptyset$ are also projection operators that project onto $V(E^3)$ and $\emptyset$ respectively.

Notice that repeated applications of these operators are redundant: $\hat{P}\hat{P}\vec{b} = \hat{P}\vec{b}$ and $\hat{Q}\hat{Q}\vec{b} = \hat{Q}\vec{b}$: this property is called idempotence. A *projection operator* on a Hilbert space is defined as an idempotent, self-adjoint linear operator on the space. By analogy with the definition of a subspace of a vector space, a *subspace of a Hilbert space* is any set of vectors from the space that itself forms a Hilbert space when taken together with the inner product, null vector, and operations of vector addition and multiplication by a scalar inherited from the original space. Each projection operator on a Hilbert space projects onto (i.e. has as its

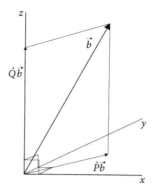

Figure A.4 Projection operators on $V(E^3)$

range) a subspace of the space, and each subspace uniquely defines a projection operator that projects onto it.

Now let A be any dynamical variable on physical system s and let Δ be any "nice" (Borel) subset of the set of real numbers. As we shall see, given the operator $\hat{A}$ representing A on the Hilbert space $\mathcal{H}_s$ of s, there is a natural way to associate with Δ a projection operator $\hat{P}^A(\Delta)$ on $\mathcal{H}_s$. Moreover, if Δ includes all values obtained in measurements of A on systems like s, then $\hat{P}^A(\Delta) = \hat{1}$. We can now state the basic quantum theoretical schema that yields predictions for the probabilities of outcomes of measurements of dynamical variable A on a system s:

If quantum system s is assigned state $|\psi\rangle$ then the probability that a measurement of A on s will yield a value in Δ is given by the inner product of $|\psi\rangle$ with $\hat{P}^A(\Delta)|\psi\rangle$.

$$\Pr{}_\psi(A \in \Delta) = \langle\psi|\hat{P}^A(\Delta)|\psi\rangle \quad \text{(Simple Born Rule)}$$

If Δ includes all values obtained in measurements of A on systems like s (so $\hat{P}^A(\Delta) = \hat{1}$), this gives $\Pr{}_\psi(A \in \Delta) = \langle\psi|\psi\rangle$. So to maintain the interpretation of this expression as a probability, the state ψ must be normalized: $\|\psi\| = 1$. Another special case occurs when Δ is a unit set $\{a\}$ in which case we may write

$$\Pr{}_\psi(A = a) = \langle\psi|\hat{P}^A(a)|\psi\rangle.$$

If $\hat{P}^A(a)|\psi\rangle = |\psi\rangle$, then (since $|\psi\rangle$ is normalized) $\Pr{}_{|\psi\rangle}(A = a) = 1$. One can also show that $\hat{P}^A(a)|\psi\rangle = |\psi\rangle$ if and only if $\hat{A}|\psi\rangle = a|\psi\rangle$.

An operator equation of the form

$$\hat{T}|\alpha\rangle = \lambda|\alpha\rangle$$

in which $\hat{T}$ is a linear operator, $|\alpha\rangle$ is a vector, and λ is a scalar is called an *eigenvalue equation*: a (non-null) vector for which this equation holds (for some value of λ) is called an *eigenvector* of $\hat{T}$, with *eigenvalue* λ. The set of eigenvectors corresponding to a particular eigenvalue always forms a subspace of the space. If this subspace is one-dimensional the eigenvalue is said to be *non-degenerate*: otherwise it is said to be *degenerate*. The importance of these notions in quantum theory is readily apparent from what has just been said. If the state vector $|\psi\rangle$ of a system is an eigenvector of $\hat{A}$ with eigenvalue a_i, then

$$\Pr{}_{|\psi\rangle}(A = a_i) = \langle\psi|\hat{P}^A(a_i)|\psi\rangle = \langle\psi|\psi\rangle = 1.$$

Hence if a quantum system s is in an eigenstate of $\hat{A}$ (i.e. if it is assigned a state vector that is an eigenvector of $\hat{A}$) then a measurement of A is certain (probability 1) to give outcome a_i, the corresponding eigenvalue. Moreover, if b is not an eigenvalue of $\hat{A}$, then for *all* states $|\psi\rangle$, $\Pr{}_{|\psi\rangle}(A = b) = 0$. This gives another sense in which operator $\hat{A}$ may be said to represent dynamical variable A: only if a_i is an eigenvalue of $\hat{A}$ can there be a non-zero probability of obtaining the result a_i in a measurement of A. In that sense the eigenvalues of $\hat{A}$ are the possible values of A and their discreteness reflects the quantization of A.

For an n-dimensional space $\mathcal{H}^n$ the quantum probabilistic algorithm simplifies further. In this case each dynamical variable A is represented by a self-adjoint operator $\hat{A}$ whose eigenvectors *span* $\mathcal{H}^n$: an arbitrary vector in $\mathcal{H}^n$ may be written as a linear sum of eigenvectors of $\hat{A}$. $\hat{A}$ is then said to have a *complete* set of eigenvectors. It can be shown that there exists a subset $\{|\alpha\rangle_i\}$ of eigenvectors of $\hat{A}$ whose elements are pairwise orthogonal (i.e. $\langle\alpha_i|\alpha\rangle_j = 0$ if $i \neq j$) and normalized ($\|\alpha_i\| = 1$), and which also spans $\mathcal{H}^n$: such a set is said to constitute an *orthonormal* basis for the space. An arbitrary vector $|\beta\rangle \in \mathcal{H}^n$ may be expressed in the form $|\beta\rangle = \Sigma_{i=1}^{n} c_i |\alpha\rangle_i$, where the c_i are complex numbers and each $|\alpha\rangle_i$ is an eigenvector of $\hat{A}$.

The generation of probabilities for a non-degenerate eigenvalue of the operator representing a dynamical variable is particularly easy to picture (see Figure A.5). In this case the probability that a measurement of the variable will yield that eigenvalue is given by the square of the length of the projection of the state vector onto the axis of eigenvectors corresponding to that eigenvalue. Quantum probabilities for non-degenerate eigenvalues are then generated in accordance with a simple generalization of Pythagoras's theorem!

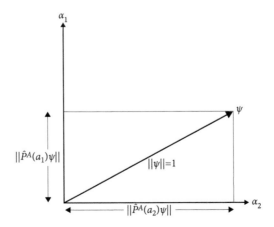

Figure A.5 Probabilities for non-degenerate eigenvalues

$$\Pr{}_{|\psi\rangle}(A = a_i) = \langle \psi | \hat{P}^A(a_i) | \psi \rangle$$
$$= \langle \psi | \hat{P}^A(a_i) \hat{P}^A(a_i) | \psi \rangle$$
$$= \langle \hat{P}^A(a_i) \psi | \hat{P}^A(a_i) | \psi \rangle$$
$$= \left\| \hat{P}^A(a_i) | \psi \rangle \right\|^2$$

The probability of getting a particular degenerate eigenvalue on measurement is scarcely more difficult to picture. It is obtained by projecting the state vector of the system onto the plane or higher-dimensional subspace formed by the eigenvectors with that eigenvalue.

A.3 More About Operators

I said that, for every dynamical variable A on s, there is a function that maps each "nice" subset Δ of real numbers into a corresponding projection operator $\hat{P}^A(\Delta)$ on the Hilbert space $\mathcal{H}_s$. Here's how that mapping arises.

In a finite-dimensional space $\mathcal{H}_s$, the eigenvectors of A span $\mathcal{H}_s$. This means that an arbitrary vector $|\beta\rangle$ may be written as

$$|\beta\rangle = \sum_i c_i |\alpha\rangle_i,$$

where

$$\hat{A}|\alpha\rangle_i = a_i |\alpha\rangle_i$$

and the $|\alpha\rangle_i$ are orthonormal. Hence

$$\hat{A}|\beta\rangle = \hat{A}\sum_i c_i |\alpha\rangle_i = \sum_i c_i \hat{A}|\alpha\rangle_i = \sum_i c_i a_i |\alpha\rangle_i.$$

Now let $\hat{P}_i$ project onto the subspace of vectors with eigenvalue a_i (this is one-dimensional if and only if a_i is non-degenerate). Since the $|\alpha\rangle_i$ are orthogonal we have

$$\hat{P}_j |\alpha\rangle_i = \emptyset \text{ (if } i \neq j\text{)}$$
$$= |\alpha\rangle_i \text{ (if } i = j\text{)}.$$

Consequently $(\Sigma_j a_j \hat{P}_j)|\alpha\rangle_i = a_i |\alpha\rangle_i$, and so

$$\hat{A}|\beta\rangle = \sum_i c_i a_i |\alpha\rangle_i$$
$$= \sum_i c_i (\Sigma_j a_j \hat{P}_j)|\alpha\rangle_i$$
$$= \sum_j a_j \hat{P}_j (\Sigma_i c_i |\alpha\rangle_i)$$
$$= (\sum_j a_j \hat{P}_j)|\beta\rangle.$$

So we can write the operator equation

$$\hat{A} = \sum_j a_j \hat{P}_j$$

where $\hat{P}_i \hat{P}_j = \delta_{ij} \hat{P}_i$ (where δ_{ij} is defined to equal 1 if $i = j$ and 0 if $i \neq j$) and $\Sigma_i \hat{P}_i = \hat{1}$. This operator equation gives a unique *resolution* of a self-adjoint operator representing a dynamical variable into its *component* projection operators. So in a finite-dimensional space the mapping $\Delta \to \hat{P}^A(\Delta)$ corresponding to A can now be simply expressed in terms of these component projections

$$\hat{P}^A(\Delta) = \hat{P}_1 + \hat{P}_2 + \hat{P}_3 + \dots$$

if and only if the set of distinct eigenvalues of $\hat{A}$ that are elements of Δ is $\{a_1, a_2, a_3, \dots\}$.

A.4 Commutation, Compatibility, and the Compound Born Rule

The simple Born rule gives probabilities for the possible outcomes of a measurement of a single dynamical variable. In certain circumstances (but not in all) quantum theory also specifies probabilities for the outcomes of joint measurements of more than one dynamical variable—for example, for the simultaneous measurement of the position of an electron on the x-axis and on the y-axis. In order to explain when and how quantum theory yields joint probability distributions it is necessary to introduce some more terminology.

First we say what it is for two linear operators to commute. Given a pair of linear operators $\hat{A}, \hat{B}$, their *commutator* $[\hat{A}, \hat{B}]$ is also a linear operator, which is defined by its action on a vector $|\gamma\rangle$ by

$$[\hat{A}, \hat{B}]|\gamma\rangle = \hat{A}\hat{B}|\gamma\rangle - \hat{B}\hat{A}|\gamma\rangle.$$

If $[\hat{A}, \hat{B}]|\gamma\rangle = \emptyset$ for all vectors $|\gamma\rangle$ then $\hat{A}, \hat{B}$ are said to *commute*.[1]

If $\mathcal{H}$ is finite-dimensional the eigenvectors of each of a pair of self-adjoint operators span $\mathcal{H}$: in that case these operators commute if and only if all pairs of projection operators from their respective resolutions commute—that is, if $\hat{A} = \Sigma_i a_i \hat{P}_i$ and $\hat{B} = \Sigma_j b_j \hat{Q}_j$ then

$$[\hat{A}, \hat{B}] = \emptyset \text{ if and only if } [\hat{P}_i, \hat{Q}_j] = 0 \text{ for all } i, j.$$

Operators in such a pair share a common set of eigenvectors. Two dynamical variables represented by a pair of commuting operators are called *compatible*: a set of pairwise compatible dynamical variables is called (jointly) compatible.

One can now state a generalization of the simple Born rule stated earlier. Suppose that $|\psi\rangle$ represents the state assigned to s and $A, B, \dots$ are compatible dynamical variables on s: then the probability that a joint measurement will yield a value for A in Γ, B in Δ, ... is given by the inner product of $|\psi\rangle$ with $\hat{P}^A(\Gamma)\hat{P}^B(\Delta)\dots|\psi\rangle$. Symbolically

$$\Pr{}_{|\psi\rangle}((A \in \Gamma) \& (B \in \Delta) \& \dots) = \langle\psi|\hat{P}^A(\Gamma)\hat{P}^B(\Delta)\dots|\psi\rangle \quad \text{(Compound Born Rule)}$$

Notice that when $A, B, \dots$ are compatible the order of the projection operators in this rule is immaterial.

A.5 Density Operators and Mixed States

According to the simple Born rule, the probability that the value of A on s lies in Δ on measurement if assigned state $|\psi\rangle$ is

---

[1] A more generous condition for commutation is required for an infinite-dimensional space since the operator $[\hat{A}, \hat{B}]$ may not be defined on all vectors $|\gamma\rangle$ in the space.

APPENDIX A: OPERATORS AND THE BORN RULE 267

$$\Pr{}_\psi(A \in \Delta) = \langle \psi | \hat{P}^A(\Delta) | \psi \rangle \qquad \text{(Simple Born Rule)}$$

Consider the quantity $\Sigma_i \langle \alpha_i | \hat{P}_\psi \hat{P}^A(\Delta) | \alpha_i \rangle$ where $\{\alpha_i\}$ forms an orthonormal basis for $\mathcal{H}_s$ and $\hat{P}_\psi$ projects onto the one-dimensional subspace spanned by $|\psi\rangle$.

$$\begin{aligned}\Sigma_i \langle \alpha_i | \hat{P}_\psi \hat{P}^A(\Delta) | \alpha_i \rangle &= \Sigma_i \langle \hat{P}_\psi \alpha_i | \hat{P}^A(\Delta) | \alpha_i \rangle \\ &= \Sigma_i \langle \alpha_i | \psi \rangle \langle \psi | \hat{P}^A(\Delta) | \alpha_i \rangle \\ &= \langle \psi | \hat{P}^A(\Delta) [\Sigma_i \langle \alpha_i | \psi \rangle | \alpha_i \rangle] \\ &= \langle \psi | \hat{P}^A(\Delta) | \psi \rangle.\end{aligned}$$

So $\Pr{}_\psi(A \in \Delta) = \Sigma_i \langle \alpha_i | \hat{P}_\psi \hat{P}^A(\Delta) | \alpha_i \rangle$ for an arbitrary orthonormal basis $\{\alpha_i\}$. This permits the introduction of a new notation that reflects this independence of basis. Define the *trace* of a linear operator $\hat{Q}$ by

$$Tr(\hat{Q}) \equiv \Sigma_i \langle \alpha_i | \hat{Q} | \alpha_i \rangle.$$

The simple Born rule may now be rewritten as

$$\Pr{}_\psi(A \in \Delta) = Tr \hat{P}_\psi \hat{P}^A(\Delta)$$

and the compound probability rule becomes

$$\Pr{}_\psi((A \in \Gamma) \,\&\, (B \in \Delta) \,\&\ldots) = Tr\left(\hat{P}_\psi \hat{P}^A(\Gamma) \hat{P}^B(\Delta) \ldots\right).$$

So far we have simply re-expressed the probability rules in a new notation. But now we can generalize these rules. The generalization proceeds by allowing that in certain circumstances it may be appropriate to represent the state of a quantum system *not* by a state vector but by a *density operator* $\hat{W}$: a self-adjoint operator on the system's Hilbert space whose eigenvalues $\{w_i\}$ satisfy the conditions

$$0 \le w_i \le 1, \quad \Sigma_i w_i = 1.$$

If and only if the only eigenvalues w_i of $\hat{W}$ are 0,1, there exists a vector $|\psi\rangle$ such that $\hat{W} = \hat{P}_\psi$ (so $\hat{W}$ is a projection operator: $\hat{W}^2 = \hat{W}$). In that case, $\hat{W}$ simply provides an alternative representation of the state $|\psi\rangle$: such a state is called *pure*. But if $\hat{W}^2 \ne \hat{W}$ then there is no vector $|\psi\rangle$ such that $\hat{W} = \hat{P}_\psi$, and $\hat{W}$ is said to represent a *mixed* state of a quantum system. The generalization of the Born rule for a system represented by an arbitrary density operator $\hat{W}$ is then

$$\Pr{}_W(A \in \Delta) = Tr \hat{W} \hat{P}^A(\Delta). \qquad \text{(Generalized Born Rule)}$$

The analogous generalization of the joint probability rule is obvious.

A.6 Interacting Systems

We saw that when physical systems (individually modeled by the assignment of pure states) interact, quantum theory characteristically models what happens by assigning the compound quantum system an entangled pure state in a bigger space.

Given a pair of Hilbert spaces $\mathcal{H}_1, \mathcal{H}_2$, one can define another Hilbert space, their *tensor product* $\mathcal{H}_1 \otimes \mathcal{H}_2$. If $\{\varphi_i\}, \{\psi_j\}$ are orthonormal bases for $\mathcal{H}_1, \mathcal{H}_2$ respectively, then $\{\varphi_i \otimes \psi_j\}$

is an orthonormal basis for $\mathcal{H}_1 \otimes \mathcal{H}_2$. For arbitrary vectors $\varphi_1, \varphi_2 \in \mathcal{H}_1; \psi_1, \psi_2 \in \mathcal{H}_2$

$$\langle \varphi_1 \otimes \psi_1 | \varphi_2 \otimes \psi_2 \rangle = \langle \varphi_1 | \varphi_2 \rangle \langle \psi_1 | \psi_2 \rangle,$$

hence $\|\varphi \otimes \psi\| = \|\varphi\| \cdot \|\psi\|$.

What about linear operators on the space? Let $\hat{A}_1$ be a linear operator on $\mathcal{H}_1$ and $\hat{A}_2$ be a linear operator on $\mathcal{H}_2$: then the action of the operator $\hat{A}_1 \otimes \hat{A}_2$ on $\mathcal{H}_1 \otimes \mathcal{H}_2$ is defined by its action on a basis vector $\varphi \otimes \psi$, namely

$$(\hat{A}_1 \otimes \hat{A}_2)\varphi \otimes \psi = \hat{A}_1\varphi \otimes \hat{A}_2\psi.$$

Of course, not every linear operator on $\mathcal{H}_1 \otimes \mathcal{H}_2$ can be expressed in the form $\hat{A}_1 \otimes \hat{A}_2$, just as not every vector in $\mathcal{H}_1 \otimes \mathcal{H}_2$ can be expressed in the form $\varphi \otimes \psi$.

Can one assign a density operator state acting on the Hilbert spaces $\mathcal{H}_1, \mathcal{H}_2$ of each of two quantum systems s_1, s_2 separately even when the pair is assigned an entangled state? It turns out that for each of systems s_1, s_2 there is a unique such density operator that generates probabilities that conform to the Born rule for outcomes of a measurement of any dynamical variable on s_1 or s_2 alone. Suppose $\hat{A}_1$ is the self-adjoint operator on $\mathcal{H}_1$ that represents dynamical variable A on s_1. One can think of a measurement of A on s_1 as a measurement on the system composed of s_1 and s_2 that in no way involves s_2: this is represented by the operator $\hat{A}_1 \otimes \hat{1}_2$. Hence A is represented on $\mathcal{H}_1 \otimes \mathcal{H}_2$ by $\hat{A}_1 \otimes \hat{1}_2$. We have supposed that, prior to the interaction between s_1 and s_2, the quantum system composed of s_1, s_2 is assigned a vector state Ψ with $\Psi(0) = \varphi(0) \otimes \psi(0)$. Despite the internal interaction between s_1, s_2, the system they compose is free of external interactions and so remains isolated. So at any time t the state assigned to the s_1, s_2 composite system is $\Psi(t) = \hat{U}_t\Psi(0)$ for some unitary operator $\hat{U}_t$. From the Born rule we then have

$$\text{Pr}_{\Psi(t)}(A \in \Delta) = \langle \Psi(t)|\hat{P}^A(\Delta)|\Psi(t)\rangle$$

where $\hat{P}^A(\Delta)$ is a projection operator on $\mathcal{H}_1 \otimes \mathcal{H}_2$ arising from the decomposition of $\hat{A}_1 \otimes \hat{1}_2$ into its component projection operators. Suppose there is a density operator $\hat{W}_1$ on $\mathcal{H}_1$ that generates this same probability measure in accordance with the generalized Born rule

$$\text{Pr}_{W_1}(A \in \Delta) = Tr\hat{W}_1\hat{P}^{A_1}(\Delta)$$

as applied to $\hat{W}_1$ for a measurement of A on s_1: then

$$Tr\hat{W}_1\hat{P}^{A_1}(\Delta) = \langle \Psi(t)|\hat{P}^A(\Delta)|\Psi(t)\rangle$$
$$= Tr\hat{W}\hat{P}^A(\Delta), \text{ with } \hat{W} = \hat{P}_{\Psi(t)}.$$

It can be shown that this last equation holds for all self-adjoint $\hat{A}_1$, Δ (and *arbitrary* density operator $\hat{W}$) if and only if

$$\hat{W}_1 = \Sigma_i \langle \chi_i|\hat{W}|\chi_i\rangle$$

for an arbitrary orthonormal basis $\{\chi_i\}$ for $\mathcal{H}_2$. The right-hand side of the last equation is written $Tr_2\hat{W}$, so we have

$$Tr\hat{W}_1\hat{P}^{A_1}(\Delta) = Tr\hat{W}\hat{P}^A(\Delta), \text{ with } \hat{W}_1 = Tr_2\hat{W}$$

for any dynamical variable A on s_1 represented by $\hat{A}_1$ on $\mathcal{H}_1$ and $\hat{A}_1 \otimes \hat{1}_2$ on $\mathcal{H}_1 \otimes \mathcal{H}_2$. Similarly, for any dynamical variable B on s_2 represented by $\hat{B}_2$ on $\mathcal{H}_2$ and $\hat{1}_1 \otimes \hat{B}_2$ on $\mathcal{H}_1 \otimes \mathcal{H}_2$

$$Tr\hat{W}_2\hat{P}^{B_2}(\Delta) = Tr\hat{W}\hat{P}^B(\Delta), \text{ with } \hat{W}_2 = Tr_1\hat{W}.$$

$\hat{W}_1, \hat{W}_2$ are sometimes called the *reduced states* assigned to s_1, s_2 (with respect to the joint state $\Psi(t)$). So the density operators $\hat{W}_1, \hat{W}_2$ faithfully represent the probabilistic aspects of the states of systems s_1, s_2 even if these systems interact with one another. To that extent, this justifies ascription of $\hat{W}_1$ to s_1 and $\hat{W}_2$ to s_2 before, during, *and after* their interaction. Prior to $t = 0$, $\hat{W}_1, \hat{W}_2$ represent pure states, but after they have begun to interact the state of each has become mixed, and these states continue to be mixed even after the interaction is over.

APPENDIX B

Two Arguments Against Naive Realism

By *naive realism* I mean a view that maintains both that every dynamical variable pertaining to a system always has a determinate (real) value, and that this is the value that a competent measurement on that system would reveal. By applying these assumptions to particular quantum systems we can derive a contradiction. So naive realism is false. Here are two different refutations of naive realism as applied to photon polarization. Each uses differently motivated auxiliary assumptions.

Assume

Naive Realism (NR)   That in a state of a system of three photons (e.g. $|GHZ\rangle$), each photon has a completely definite polarization in the following sense. For any axis $\hat{a}$ in a plane, either the photon is polarized parallel to that axis or the photon is polarized perpendicular to that axis. Moreover, either the photon is right (clockwise) circularly polarized or it is left (anticlockwise) circularly polarized; and

Faithful Measurement (FM)   That a competently conducted measurement of its linear polarization with respect to any axis $\hat{a}$ will simply reveal whether it was polarized parallel or perpendicular to $\hat{a}$, and a competently conducted measurement of its circular polarization will simply reveal whether it was right or left circularly polarized.

B.1 First Argument

If a photon passes a polarizer with axis set to $\hat{b}$, whether or not it then passes a polarizer set to $\hat{a}$ is quite independent of whether it had passed a polarizer set to $\hat{a}$ prior to encountering the polarizer set to $\hat{b}$. More generally. since any measurement of one component of polarization disturbs (the results of measurements of) other components of polarization, only one such measurement can be made at once.

So for each photon of a hypothetical three-photon system one can measure a chosen component of its linear polarization, or one can measure its circular polarization. Suppose we pick an axis at 45° to the vertical, so the option is now between measuring linear polarization with respect to this axis or measuring circular polarization. We can make the choice between these options independently for each of the three photons, for a total of eight—but we'll only need to consider four of these, which we can represent as $(\diagup\diagup\diagup), (\bigcirc\diagup\bigcirc), (\bigcirc\bigcirc\diagup), (\diagup\bigcirc\bigcirc)$, where the measurement on the first photon is indicated by the first entry in a triple, and similarly for the second and third entries. Notice that, for each option as to what to measure on a given photon, this lists two alternatives for the measurements on the other pair of photons: each is said to provide a different context for that measurement. Moreover, these contexts cannot be combined, because you can't measure a photon's linear and its circular polarization both at once.

Nevertheless, assumptions (NR) and (FM) imply that the result of measuring the circular or 45°-linear polarization of one of these photons is independent of which (if either) of the two alternative measurement contexts is chosen.

Now consider the polarization state $|GHZ\rangle$. We noted in Chapter 3 that this may be expressed not only as

$$|GHZ\rangle = \frac{1}{\sqrt{2}}(|H\rangle|H\rangle|H\rangle + |V\rangle|V\rangle|V\rangle) \qquad (B.1)$$

but also as

$$|GHZ\rangle = \frac{1}{\sqrt{2}}(|R\rangle|L\rangle|45\rangle + |L\rangle|R\rangle|45\rangle + |L\rangle|L\rangle|135\rangle + |R\rangle|R\rangle|135\rangle) \qquad (B.2)$$

and again as the two distinct expressions that result from this by permuting the positions of the first, second, and third entries in all components of this superposition at once. Indeed, there is yet another way of expressing the very same polarization state, namely

$$|GHZ\rangle = \frac{1}{\sqrt{2}}(|45\rangle|45\rangle|45\rangle + |135\rangle|135\rangle|45\rangle + |135\rangle|45\rangle|135\rangle + |45\rangle|135\rangle|135\rangle). \qquad (B.3)$$

Let the number +1 represent the outcome parallel to the 45° axis, and −1 represent perpendicular. If we apply the Born rule to $|GHZ\rangle$ in the form (B.3), we see that the probability is 1 that a joint measurement of 45°-linear polarization will yield a triple of outcomes whose product is +1 and 0 that it will yield a triple of outcomes whose product is −1. Applying assumptions (NR) and (FM), a measurement of the 45°-linear polarization of each of these photons will simply reveal whether it was polarized parallel or perpendicular to that axis. This implies (with probability 1) that not only was each photon definitely polarized either parallel or perpendicular to the 45° axis, but also an odd number were polarized parallel.

Now consider (B.2), and let the number +1 represent the Right outcome of a circular polarization measurement, and −1 the Left outcome. Applying the Born rule to (B.2) we can conclude that the probability is 1 that the product of the outcomes in a joint measurement of 45°-linear polarization on one photon and circular polarization on the other two is −1. Applying (NR) and (FM) we can conclude not only that each photon had a determinate circular polarization as well as linear polarization with respect to the 45° axis, but also that if two of the photons had the same circular polarization the third was polarized perpendicular to that axis, while if they had opposite circular polarizations the third was polarized parallel to that axis. In that case −1 represents not just the product of the outcomes, but the product of the actual polarization values prior to measurement.

Putting those partial conclusions together, it follows that the product of the three products of polarizations associated with (B.2) and its permutations itself equals −1, so the final product of *that* with the polarization values associated with (B.3) is also −1. But notice that each of the six determinate individual polarization values ($\bigcirc_1, \diagup_1, \bigcirc_2, \diagup_2, \bigcirc_3, \diagup_3$) appears twice in this final product, so the product must also equal +1. We have arrived at a contradiction. By *reductio ad absurdum*, (NR) and (FM) can't both be true—naive realism about dynamical variables in quantum theory is false.

One could avoid this conclusion by allowing the polarization values of one photon to be instantaneously altered by a distant measurement on another photon immediately before the first is measured. So this argument also rests implicitly on Einstein's assumption of local action. But it differs from other such arguments because it only appeals to the Born rule when that rule prescribes extremal probabilities of 1 and 0, which EPR and others would equate with certainty. The next argument makes no appeal to the Born rule and does not depend on any assumptions concerning the quantum state of a three-photon system.

B.2 Second Argument

Consider Figure B.1. We have been taking $\bigcirc_3$, for example, to represent either the value of the third photon's circular polarization or the outcome of measuring its circular polarization: assumption (FM) renders this ambiguity harmless. Instead, we now take $\bigcirc_3$ to represent photon 3's circular polarization—a dynamical variable whose measurement may yield one of two values (±1): similarly, take $/_2$ to represent 2's 45°-linear polarization, and so on. It is a basic assumption of quantum theory that every dynamical variable on a system is represented in a quantum model by a unique *linear operator* acting on the vector space in which that system's states are represented (see Appendix A). $\bigcirc_3$, for example, is represented by operator $\hat{\bigcirc}_3$ and $/_2$ by $\hat{/}_2$. Following Dirac, it is now common to use the term 'observable' to refer indiscriminately either to a dynamical variable or to its associated linear operator, as in the caption to Figure B.1 (in which the convention of representing operators by "hats" has been suspended to preserve this ambiguity !).

The three-photon dynamical variable $\widehat{///}$, for example, is also represented by an operator $\widehat{///}$ that acts on the tensor product space in which a three-photon polarization state like $|GHZ\rangle$ is represented in a quantum model. Indeed this operator may be expressed as a tensor product of operators in the individual polarization spaces as

$$\widehat{///} = \hat{/} \otimes \hat{/} \otimes \hat{/}.$$

Similarly

$$\widehat{\bigcirc\bigcirc/} = \hat{\bigcirc} \otimes \hat{\bigcirc} \otimes \hat{/}.$$

So each of the ten labels in Figure B.1 may also be taken to refer to the operator uniquely corresponding to a polarization dynamical variable on a system of one, or of three, photons.

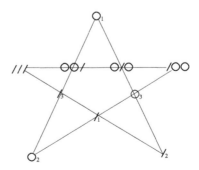

Figure B.1 Some related three-photon polarization observables

APPENDIX B: TWO ARGUMENTS AGAINST NAIVE REALISM 273

Not surprisingly, one way to measure ∕∕∕ is to conduct a joint measurement of ∕₁, ∕₂, ∕₃ and multiply the values of the three separate outcomes. This joint measurement is possible, since measurements of the polarizations of the three photons may be carried out by independent procedures that don't exclude one another (unlike measurements of the circular and linear polarization of a single photon). A set of dyamical variables that may be jointly measured is called *compatible*. This relation of compatibility is mirrored by a relation among their corresponding operators. If the result of applying two operators to a vector is always independent of the order in which these operators are applied, the operators are said to *commute*. A set of dynamical variables is compatible if and only if all pairs of their corresponding operators commute. Each quadruple of polarization variables that Figure B represents as lying on a tilted straight line is compatible.[1]

It turns out that all pairs of operators corresponding to the dynamical variables on the horizontal line *also* commute, which implies that

$$\{(\diagup\diagup\diagup), (\bigcirc\diagup\bigcirc), (\bigcirc\bigcirc\diagup), (\diagup\bigcirc\bigcirc)\}$$

is *also* a compatible set of three-photon polarization variables. Look again at Figure B.1. Each of the ten observables depicted there is a member of two different quadruples of compatible observables. Assuming (NR), each of these ten observables always has a determinate value: and assuming (FM), a measurement of any of them always reveals that determinate value. But the only values that measurement ever reveals are ±1. Unless God is extraordinarily malicious in preventing our measuring them when they take on other values, ±1 are the only values any of these observables has. The values of the observables lying along any tilted line in Figure B.1 always multiply to +1, since the one observable on the horizontal line is just the product of the other three. But the values of the four observables lying along the horizontal line *always* multiply to −1, no matter what the three-photon polarization state.[2] So once more, by *reductio ad absurdum*, (NR) and (FM) can't both be true—naive realism about dynamical variables in quantum theory is false.

This second argument was state-independent: it showed that in *no* quantum three-photon polarization state can each photon have both a determinate linear polarization and a determinate circular polarization. It made no direct appeal to Einstein's principle of local action. Rather, it assumed that the outcome of a measurement of a dynamical variable is independent of what other compatible variables are measured along with it, that is, of the measurement context. Naive realism implies that the result of a measurement is determined only by the possessed value of the measured variable, apparently automatically excluding such context-dependence.

However, if a measurement of a compatible variable could instantaneously affect that possessed value, it would thereby affect the outcome of its measurement. What Bell showed, and what the first argument uses, is that the second argument's assumption

---

[1] To make this precise one must think of $\bigcirc_3$, say, as a dynamical variable pertaining to all three photons, so its corresponding operator is actually $\hat{I} \otimes \hat{I} \otimes \hat{O}$, and so on, where $\hat{I}$ is an identity operator that does nothing to any vector on which it acts.

[2] The first argument showed this only for the state $|GHZ\rangle$. I have omitted the proof that the product of values of the observables on the horizontal line is always −1 since it involves knowledge of the explicit form of the operators $\bigcirc, \diagup$. For experts, these mirror those of the Pauli spin matrices and the result follows from their anticommutation relations (see Mermin [1993]).

of non-contextuality can be turned into a requirement of locality. For example, if the 45°-linear polarization of photon 1 has a value at some time that is revealed by measuring it, then that value cannot depend on whether it is measured along with $\diagup_2, \diagup_3$ or along with $\bigcirc_2, \bigcirc_3$ consistent with Einstein's principle of local action, since the measurements on photons 2, 3 may each be carried out at the same time but arbitrarily far away.

APPENDIX C

A Simple Model of Decoherence

First consider the general phenomenon of decoherence. Consider a measurement of observable B on a system in superposed state

$$|\psi\rangle = c_1 |\alpha_1\rangle + c_2 |\alpha_2\rangle \qquad \text{(C.1)}$$

where $\hat{A} |\alpha_i\rangle = a_i |\alpha_i\rangle \; (i = 1, 2)$

and $\hat{B} |\beta_i\rangle = b_i |\beta_i\rangle$.

The probability that this will give result b_1 is

$$\begin{aligned} Pr_\psi(B = b_1) \\ &= |\langle \beta_1 | \psi \rangle|^2 \\ &= |c_1 \langle \beta_1 | \alpha_1 \rangle + c_2 \langle \beta_1 | \alpha_2 \rangle|^2 \\ &= |c_1|^2 |\langle \beta_1 | \alpha_1 \rangle|^2 + |c_2|^2 |\langle \beta_1 | \alpha_2 \rangle|^2 + 2\,\text{Re}(c_1^* c_2 \langle \alpha_1 | \beta_1 \rangle \langle \beta_1 | \alpha_2 \rangle) \\ &= |c_1|^2 Pr_{\alpha_1}(B = b_1) + |c_2|^2 Pr_{\alpha_2}(B = b_2) + 2\,\text{Re}(c_1^* c_2 \langle \alpha_1 | \beta_1 \rangle \langle \beta_1 | \alpha_2 \rangle). \end{aligned} \qquad \text{(C.2)}$$

Notice that the first two terms on the right give the weighted sum of the probability of getting result b_1 in state $|\alpha_1\rangle$ and getting result b_2 in state $|\alpha_2\rangle$. If it weren't for the third "interference" term, we could have computed $Pr_\psi(B = b_1)$ by assuming that a system in state $|\psi\rangle$ is either in state $|\alpha_1\rangle$ (with probability $|c_1|^2$), or in state $|\alpha_2\rangle$ (with probability $|c_2|^2$). In certain circumstances, these interference terms all vanish, or at least become extremely small, compared to the non-interference terms. When that happens, terms like $|\alpha_1\rangle, |\alpha_2\rangle$ in a superposition like $|\psi\rangle$ are said to *decohere*. In that case, the state may be regarded as a mixture rather than a superposition of its components—at least for the purpose of computing probabilities of getting one result rather than another when measuring observable B on systems in that state.

Here is a simple model introduced by Zurek [1982] in which the joint evolution of the state of quantum systems α and ϵ induces decoherence in the state assigned to α. This can be used to model a simple physical system interacting with a compound physical system made up of a huge number of simple components. A two-valued dynamical variable P of system α (which may represent a spin component of a spin 1/2 particle, with value either up ($\Uparrow$) or down ($\Downarrow$)) interacts in the model with a system ϵ comprising a collection of N such systems. $|\Uparrow\rangle, (|\Downarrow\rangle)$ may represent spin up (down) states assigned to α. On the other hand, $|\uparrow\rangle_k, (|\downarrow\rangle_k)$ may represent spin up (down) states assigned to the the kth spin subsystem of ϵ.

If α, ϵ are assumed to be initially assigned pure, uncorrelated states

$$\psi_\alpha = (a |\Uparrow\rangle + b |\Downarrow\rangle), \qquad \text{(C.3)}$$

$$\psi_\epsilon = \prod_{k=1}^{N} (\alpha_k |\uparrow\rangle_k + \beta_k |\downarrow\rangle_k). \tag{C.4}$$

then the initial state

$$\Psi(0) = \psi_\alpha \otimes \psi_\epsilon \tag{C.5}$$

evolves linearly and unitarily in the model, becoming

$$\Psi(t) = \left(a|\Uparrow\rangle |\mathcal{E}_\Uparrow(t)\rangle + b|\Downarrow\rangle |\mathcal{E}_\Downarrow(t)\rangle\right) \tag{C.6}$$

at time t where

$$|\mathcal{E}_\Uparrow(t)\rangle = \prod_{k=1}^{N} (\alpha_k e^{ig_k t} |\uparrow\rangle_k + \beta_k e^{-ig_k t} |\downarrow\rangle_k) = |\mathcal{E}_\Downarrow(-t)\rangle. \tag{C.7}$$

The number g_k represents in the model the strength of the interaction between physical systems represented respectively by α and the kth subsystem of ϵ. For an arbitrarily selected large collection of N subsystems, these numbers are assumed to bear no systematic relations to one another. The corresponding mixed state of subsystem α is then

$$\hat{\rho}_\alpha(t) = |a|^2 |\Uparrow\rangle\langle\Uparrow| + |b|^2 |\Downarrow\rangle\langle\Downarrow| + ab^* r(t) |\Uparrow\rangle\langle\Downarrow| + a^*b r^*(t) |\Downarrow\rangle\langle\Uparrow|. \tag{C.8}$$

The coefficient $r(t) = \langle \mathcal{E}_\Uparrow(t) | \mathcal{E}_\Downarrow(t) \rangle$ appearing in the interference terms here is

$$r(t) = \prod_{k=1}^{N} \left[\cos 2g_k t + i \left(|\alpha_k|^2 - |\beta_k|^2\right) \sin 2g_k t\right]. \tag{C.9}$$

Cucchetti, Paz, and Zurek [2005] showed that $|r(t)|$ tends to decrease rapidly with increasing N and very quickly approaches zero with increasing t. More precisely, while $|r(t)|^2$ fluctuates, its average magnitude at any time is proportional to 2^{-N}, and, for fairly generic values of the g_k, it decreases faster than exponentially with time according to the rule $|r(t)|^2 \propto e^{-\Gamma^2 t^2}$, where Γ depends on the distribution of the g_k as well as the initial state of ϵ. This result is relatively insensitive to the initial state of ϵ, which need not be assumed to have the product form (C.4), though if the state initially assigned to ϵ is unchanged during the evolution $|r(t)| = 1$ so the state of α will suffer no decoherence. Since $r(t)$ is an almost periodic function of t for finite N, it will continue to return arbitrarily closely to 1 at various times: but if N equals the number of atoms in, say, a glass of water ($\sim 10^{26}$) then Zurek estimated that the corresponding "recurrence" time exceeds the age of the universe.

One thing this model illustrates is that when the quantum state of a system expressed as a superposition of component states associated with definite values of P becomes entangled in this way with that of another sufficiently complex system, its resulting quantum state may be regarded as a *stable* mixture rather than a superposition of these components for the purpose of computing probabilities of getting one result rather than another when measuring *any* observable Q on that system. I emphasize the word 'stable' to stress the fact that the coefficients $|a|^2$, $|b|^2$ appearing in the mixed state $\hat{\rho}_\alpha(t)$ do not change with time t, even though the very much smaller interference terms in that state continue to fluctuate slightly because of the time-dependence of $r(t)$.

Bibliography

Albert, D. [1992] *Quantum Mechanics and Experience* (Cambridge, MA: Harvard University Press).

Anderson, P. [1972] "More is Different", *Science* 177, pp. 393–6.

Anscombe, G. E. M. [1971] *Causality and Determination: an Inaugural Lecture* (Cambridge: Cambridge University Press).

Arntzenius, F. and Hitchcock, C. [2010] "Reichenbach's Common Cause Principle", *The Stanford Electronic Encyclopedia of Philosophy* (Fall 2010 Edition), Edward N. Zalta (ed.), URL = <http://plato.stanford.edu/archives/fall2010/entries/physics-Rpcc/>.

Aspect, A., Dalibard, J., and Roger, G. [1982] "Experimental Test of Bell's Inequalities Using Time-Varying Analyzers", *Physical Review Letters* 49, pp. 1804–7.

Aspect, A., Grangier, P., and Roger, G. [1981] "Experimental Realization of Einstein–Podolsky–Rosen–Bohm *Gedankenexperiment*: A New Violation of Bell's Inequalities", *Physical Review Letters* 49, pp. 91–4.

Bach, R. *et al.* [2013] "Controlled Double-Slit Electron Diffraction", *New Journal of Physics* 15, 033018.

Baker, D. [2009] "Against Field Interpretations of Quantum Field Theory", *British Journal for the Philosophy of Science* 60, pp. 585–609.

Barrett, J. [2014] "Everett's Relative-State Formulation of Quantum Mechanics", *The Stanford Encyclopedia of Philosophy* (Fall 2014 Edition), Edward N. Zalta (ed.), URL = <http://plato.stanford.edu/archives/fall2014/entries/qm-everett/>.

Bedingham, D. [2011] "Relativistic State Reduction Dynamics", *Foundations of Physics* 41, pp. 686–704.

Bedingham, D., Dürr, D., Ghirardi, G. C., Goldstein, S., and Zanghı, N. [2014] "Matter Density and Relativistic Models of Wave Function Collapse", *Journal of Statistical Physics* 154, pp. 623–31.

Beisbart, C. and Hartmann, S. (eds.) [2011] *Probabilities in Physics* (Oxford: Oxford University Press).

Bell, J. S. [1964] "On the Einstein–Podolsky–Rosen Paradox", *Physics* 1, pp. 195–200: reprinted in Bell [2004], pp. 14–21.

Bell, J. S. [1975] "The Theory of Local Beables", TH-2053-CERN, July 28: reprinted in Bell [2004], pp. 52–62.

Bell, J. S. [1981] "Bertlmann's Socks and the Nature of Reality", *Journal de Physique*, Colloque 2, suppl. au numero 3, Tome 42, C2 pp. 41–61: reprinted in Bell [2004], pp. 139–58.

Bell, J. S. [1987] "Are there Quantum Jumps?", in C. W. Kilmister (ed.) *Schrödinger: Centenary Celebration of a Polymath* (Cambridge: Cambridge University Press), pp. 41–52: reprinted in Bell [2004], pp. 201–12.

Bell, J. S. [1990] "La Nouvelle Cuisine", in A. Sarlemijn and P. Krose (eds.) *Between Science and Technology* (New York: Elsevier), pp. 97–115: reprinted in Bell [2004], pp. 232–48.

Bell, J. S. [2004] *Speakable and Unspeakable in Quantum Mechanics*, 2nd edition (Cambridge: Cambridge University Press).

Blatt, R. and Wineland, D. [2008] "Entangled States of Trapped Ions", *Nature* 453, pp. 1008–15.
Bohm, D. [1951] *Quantum Theory* (Englewood Cliffs, NJ: Prentice-Hall).
Bohm, D. [1952a] "A Suggested Interpretation of the Quantum Theory in Terms of 'Hidden Variables'. I", *Physical Review* 85, 166–79.
Bohm, D. [1952b] "A Suggested Interpretation of the Quantum Theory in Terms of 'Hidden Variables'. II", *Physical Review* 85, 180–93.
Born, M. [1926] "Zur Quantenmechanik der Stossvorgänge", *Zeitschrift für Physik* 37, pp. 863–7.
Born, M. and Einstein, A. [1971] *The Born–Einstein Letters*, translated by Irene Born (London: Macmillan).
Brandom, R. [1994] *Making It Explicit* (Cambridge, MA: Harvard University Press).
Brandom, R. [2000] *Articulating Reasons: An Introduction to Inferentialism* (Cambridge, MA: Harvard University Press).
Brukner, C. [2017] "On the Quantum Measurement Problem" in R. Bertlmann and A. Zeilinger (eds.) *Quantum [Un] Speakables II* (Berlin: Springer), pp. 95–117.
Brukner, C. and Zeilinger, A. [2003] "Information and Fundamental Elements of the Structure of Quantum Theory", in L. Castell and O. Ischebeck, eds. *Time, Quantum, Information* (Berlin: Springer) 325–54.
Brun, T. et al. [2002] "How Much State Assignments Can Differ", *Physical Review* A 65, 032315.
Burge, T. [2010] *The Origins of Objectivity* (Oxford: Clarendon Press).
Carnap, R. [1966] *An Introduction to the Philosophy of Science* (New York: Basic Books).
Cartwright, N. [1983] *How the Laws of Physics Lie* (Oxford: Oxford University Press).
Clauser, J. F., Horne, M. A., Shimony, A., and Holt, R. A. [1969] "Proposed Experiment to Test Local Hidden-Variable Theories", *Physical Review Letters* 23, pp. 880–4.
Cucchetti, F. M., Paz, J. P., and Zurek, W. H. [2005] "Decoherence from Spin Environments", *Physical Review* A 72, 052113.
de Broglie, L. [1927] "La mécanique undulatoire et la structure atomique de la matière et du rayonnement", *Le Journal de Physique et le Radium* 8, pp. 225–41.
de Finetti, B. [1968] "Probability: The Subjectivistic Approach", in R. Klibansky (ed.), *La Philosophie Contemporaine* (Florence: La Nuova Italia), pp. 45–53.
d'Espagnat, B. [1979] "The Quantum Theory and Reality", *Scientific American*, November, pp. 158–81.
d'Espagnat, B. [2005] "Consciousness and the Wigner's Friend Problem", *Foundations of Physics* 35, pp. 1943–66.
Deutsch, D. [1999] "Quantum Theory of Probability and Decisions", *Proceedings of the Royal Society of London* A455, pp. 3129–37.
Dick, P. K. [1968] *Do Androids Dream of Electric Sheep?* (New York: Doubleday).
Diosi, L. [1987] "A Universal Master Equation for the Gravitational Violation of Quantum Mechanics", *Physics Letters* A, pp. 377–81.
Dirac, P. A. M. [1930, 1958] *The Principles of Quantum Mechanics* (Cambridge: Cambridge University Press).
Dürr, D., Goldstein, D., and Zanghi, N. [2013] *Quantum Physics without Quantum Philosophy* (Berlin: Springer).
Einstein, A. [1905, 1965] "Über einen die Erzeugung und Verwandlung des Lichtes betreffenden heuristischen Gesichtspunkt", *Annalen der Physik* 17, pp. 132–48. English translation by D. Ter Haar, "On a Heuristic Point of View about the Creation and Conversion of Light", in Ter Haar, D. [1967] *The Old Quantum Theory* (Oxford: Pergamon Press), pp. 91–107.

Einstein, A. [1905] "Zur Elektrodynamik bewegter Körper". *Annalen der Physik* 17, pp. 891–921. English translation by G. B. Jeffery and W. Perrett [1923] in *The Principle of Relativity* (London: Methuen and Company, Ltd.).

Einstein, A. [1948] "Quanten-Mechanik und Wirklichkeit", *Dialectica* 2, pp. 320–4.

Einstein, A., Podolosky, B., and Rosen, N. [1935] "Can Quantum-Mechanical Description of Physical Reality be Considered Complete?" *Physical Review* 47, pp. 777–80.

Everrett, H. III [1957] " 'Relative State' Formulation of Quantum Mechanics", *Reviews of Modern Physics* 29, pp. 454–62.

Feynman, R. P. [1985] *QED: the Strange Theory of Light and Matter* (Princeton, NJ: Princeton University Press).

Feynman, R. P., Leighton, R. B., and Sands, M. L. [1963] *The Feynman Lectures on Physics* (Reading, MA: Addison-Wesley).

Fine, A. [1982] "Joint Distributions, Quantum Correlations, and Commuting Observables", *Journal of Mathematical Physics* 23, pp. 1306–10.

Frabboni, S. et al. [2008] "Nanofabrication and the Realization of Feynman's Two-Slit Experiment", *Applied Physics Letters* 93, 073108.

Fraser, D. [2008] "The Fate of 'Particles' in Quantum Field Theories with Interactions", *Studies in History and Philosophy of Modern Physics* 39, pp. 841–5.

Fuchs, C. [2016] "On Participatory Realism", posted at http://arxiv.org/abs/1601.04360, accessed July 15, 2016.

Fuchs, C., Mermin, N. D., and Schack, R. [2014] "An Introduction to QBism with an Application to the Locality of Quantum Mechanics", *American Journal of Physics* 82, pp. 749–54.

Geiger, H. and Marsden, E. [1913] "The Laws of Deflexion of α Particles through Large Angles", *Philosophical Magazine* 25, pp. 604–23.

Ghirardi, G. [2016] "Collapse Theories", *The Stanford Encyclopedia of Philosophy* (Spring 2016 Edition), Edward N. Zalta (ed.), URL = <http://plato.stanford.edu/archives/spr2016/ entries/qm-collapse/>.

Ghirardi, G., Rimini, A., and Weber, T. [1986] "Unified Dynamics for Microscopic and Macroscopic Systems", *Physical Review* D34, pp. 470–91.

Ghirardi, G., Pearle, P., and Rimini, A. [1990] "Markov Processes in Hilbert Space and Continuous Spontaneous Localization of Systems of Identical Particles", *Physical Review* A 42, pp. 78–89.

Giere, R. [1988] *Explaining Science* (Chicago, IL: University of Chicago Press).

Giere, R. [1999] *Science Without Laws* (Chicago, IL: University of Chicago Press).

Giustina, M. et al. [2015] "Significant-Loophole-Free Test of Bell's Theorem with Entangled Photons", *Physical Review Letters* 115, 250401.

Gleason, A. M. [1957] "Measures on the Closed Subspaces of a Hilbert Space", *Journal of Mathematics and Mechanics* 17, pp. 59–81.

Goldstein, S. [2013] "Bohmian Mechanics", *The Stanford Encyclopedia of Philosophy* (Spring 2013 Edition), Edward N. Zalta (ed.), URL = <http://plato.stanford.edu/archives/spr2013/ entries/qm-bohm/>.

Grangier, P., Roger, G., and Aspect, A. [1986] "Experimental Evidence for a Photon Anticorrelation Effect on a Beam Splitter", *Europhysics Letters* 1, pp. 173–9.

Greaves, H. [2007] "On the Everettian Epistemic Problem", *Studies In History and Philosophy of Modern Physics* 38, 120–52.

Greaves, H. and Myrvold, W. [2010] "Everett and Evidence", in S. Saunders *et al.* [2010], pp. 264–304.
Greenberger, D., Horne, M., and Zeilinger, A. [1989] "Going beyond Bell's Theorem", in M. Kafatos (ed.), *Bell's Theorem, Quantum Theory, and Conceptions of the Universe* (Dordrecht: Kluwer), pp. 69–72.
Hartle, J. and Hawking, S. [1983] "Wave Function of the Universe", *Physical Review* D 28, pp. 2960–75.
Hawking, S. [2011] "Why Are We Here?", talk at the Google European Zeitgeist Conference, Hertfordshire, England on May 17, 2011. https://www.youtube.com/watch?v=r4TO1iLZmcw accessed December 28, 2015.
Healey, R. [1989] *The Philosophy of Quantum Mechanics: an Interactive Interpretation* (Cambridge: Cambridge University Press).
Healey, R. [2016] "Holism and Nonseparability in Physics", *The Stanford Encyclopedia of Philosophy* (Spring 2016 Edition), Edward N. Zalta (ed.), URL = <http://plato.stanford.edu/archives/spr2015/entries/physics-holism/>.
Heisenberg, W. [1930] *The Physical Principles of the Quantum Theory* (Chicago, IL: University of Chicago Press).
Heisenberg, W. [1958] *Physics and Philosophy* (London: George Allen & Unwin).
Heisenberg, W. [1971] *Physics and Beyond* (New York: Harper & Row).
Hempel, C. G. [1965] *Aspects of Scientific Explanation* (New York: Free Press).
Hensen, B. *et al.* [2015] "Loophole-Free Bell Inequality Violation using Electron Spins separated by 1.3 Kilometres", *Nature* doi:10.1038/nature15759.
Herbst, T., Scheidl, T., Fink, M., Handsteiner, J., Wittmann, B., Ursin, R., and Zeilinger, A. [2014] "Teleportation of Entanglement over 143 km", *Proceedings of the National Academy of Sciences* 112(46), 14202–5.
Horgan, T. [1993] "From Supervenience to Superdupervenience: Meeting the Demands of a Material World", *Mind* 102, pp. 555–86.
Ismael, J. T. [2008] "Raid! Dissolving the Big, Bad Bug", *Nous* 42, pp. 292–307.
Jammer, M. [1974] *The Philosophy of Quantum Mechanics* (New York: John Wiley & Sons).
Jordan, P. [1934] "Quantenphysikalische Bemerkungen zur Biologie und Psychologie", *Erkenntnis* 4, pp. 215–52.
Jordan, T. F. [1996] *Linear Operators for Quantum Mechanics* (New York: Dover).
Juffmann, T. *et al.* [2009] "Wave and Particle in Molecular Interference Lithography", *Physical Review Letters* 103, 263601.
Kent, A. [2010] "One World versus Many", in Saunders *et al.* (eds) [2010], pp. 307–54.
Kuhn, T. S. [1962] *The Structure of Scientific Revolutions*, 3rd edition (Chicago: University of Chicago Press).
Ladyman, J. and Ross, D. [2007] *Every Thing Must Go* (Oxford: Oxford University Press).
Lancaster, T. and Blundell, S. J. [2014] *Quantum Field Theory for the Gifted Amateur* (Oxford: Oxford University Press).
Laplace, P. S. [1820/1951] *Philosophical Essay on Probabilities* (New York: Dover).
Leggett, A. J. [1998] "Macroscopic Realism: What is It and What do we Know about It from Experiment?", in R. Healey and G. Hellman (eds.) *Quantum Measurement: Beyond Paradox* (Minneapolis, MN: University of Minnesota Press), pp. 1–22.
Leggett, A. J. [2002] "Testing the Limits of Quantum Mechanics: Motivation, State of Play, Prospects", *Journal of Physics: Condensed Matter* 14, pp. R415–R451.

Leggett, A. J. [2005] "The Quantum Measurement Problem", *Science* 307 (11 February), pp. 871–2.
Lewis, D. K. [1986] *Philosophical Papers, Volume II* (Oxford: Oxford University Press).
Lewis, D. K. [1994] "Humean Supervenience Debugged", *Mind* 103, pp. 473–90.
Lewis, D. K. [2004] "Causation as Influence", in J. Collins, N. Hall, and L. A. Paul (eds.), *Causation and Counterfactuals* (Cambridge, MA: MIT Press), pp. 75–106.
Lockwood, M. [1989] *Mind, Brain and the Quantum* (Oxford: Blackwell).
Ma, X.-S. [2012] "Experimental Delayed-Choice Entanglement Swapping", *Nature Physics* 8, pp. 480–5.
Margenau, H. [1937] "Critical points in Modern Physical Theory", *Philosophy of Science* 4, pp. 337–70.
Maudlin, T. [1995] "Three Measurement Problems", *Topoi* 14, pp. 7–15.
Maudlin, T. [2007] *The Metaphysics within Physics* (Oxford: Oxford University Press).
Mermin, N. D. [1981] "Quantum Mysteries for Anyone", *Journal of Philosophy* LXXVIII pp. 397–408.
Mermin, N. D. [1985] "Is the Moon There When Nobody Looks?", *Physics Today* (April), pp. 38–47.
Mermin, N. D. [1993] "Hidden Variables and the Two Theorems of John Bell", *Reviews of Modern Physics* 65, 803–15.
Mermin, N. D. [2014] "QBism Puts the Scientist Back into Science", *Nature* 507, pp. 421–3.
Merricks, T. [1998] "Against the Doctrine of Microphysical Supervenience", *Mind* 103, pp. 59–71.
Monz, T., Schindler, P., Barreiro, J. T. et al. [2011] "14-Qubit Entanglement: Creation and Coherence", *Physical Review Letters* 106, 130506.
Nagel, T. [1986] *The View from Nowhere* (Oxford: Oxford University Press).
Needham, J. and Pagel, J. [1938] *Background to Modern Science* (New York: MacMillan).
Newton, I. [1687] *Philosophiae Naturalis Principia Mathematica* (London: Joseph Streater for the Royal Society).
Ollivier, H. et al. [2004] "Objective Properties from Subjective Quantum States: Environment as a Witness", *Physical Review Letters* 93, 220401.
Oppenheim, P. and Putnam, H. [1958] "Unity of Science as a Working Hypothesis", in H. Feigl, M. Scriven, and G. Maxwell (eds.), *Concepts, Theories and the Mind–Body Problem*. Minnesota Studies in the Philosophy of science, Vol. II (Minneapolis, MN: University of Minnesota Press), pp. 3–36.
Pais, A. [1979] "Einstein and the Quantum Theory", *Reviews of Modern Physics* 51, pp. 863–914.
Pan, J. W., Bouwmeester, D., Daniell, M., Weinfurter, H., and Zeilinger, A. [2000] "Experimental Test of Quantum Nonlocality in Three-Photon Greenberger–Horne–Zeilinger Entanglement", *Nature* 403, pp. 515–19.
Papineau, D. [1993] *Philosophical Naturalism* (Oxford: Blackwell).
Pearle, P. [1976] "Reduction of State Vector by a Nonlinear Schrödinger Equation", *Physical Review* D13, pp. 857–68.
Penrose, R. [2014] "On the Gravitization of Quantum Mechanics 1", *Foundations of Physics* 44, pp. 557–75.
Peres, A. [2000] "Delayed Choice for Entanglement Swapping", *Journal of Modern Optics* 47, pp. 139–43.
Peskin, M. E. and Schroder, D. V. [1995] *An Introduction to Quantum Field Theory* (Boulder, CO: Westview Press).

Petersen, A. [1963] "The Philosophy of Niels Bohr", *Bulletin of the Atomic Scientists* 19, pp. 8–14.
Pitowsky, I. [2006] "Quantum Mechanics as a Theory of Probability", in W. Demopoulos and I. Pitowsky (eds.), *Physical Theory and its Interpretation* (Dordrecht: Springer), pp. 213–40.
Popescu, S. and Rohrlich, D. [1994] "Quantum Nonlocality as an Axiom", *Foundations of Physics* 24, pp. 379–85.
Popper, K. R. [1967] "Quantum Mechanics without 'the Observer' ", in M. Bunge (ed.), *Quantum Theory and Reality* (New York: Springer), pp. 7–44.
Price, H. [1988] *Facts and the Function of Truth* (Oxford: Blackwell).
Price, H. [2011] *Naturalism without Mirrors* (Oxford: Oxford University Press).
Price, H. [2012] "Causation, Chance and the Rational Significance of Supernatural Evidence", *Philosophical Review* 121, pp. 483–538.
Price, H. [2013] *Expressivism, Pragmatism and Representationalism* (Cambridge: Cambridge University Press).
Putnam, H. [1968] "Is Logic Empirical?", in R. S. Cohen and M. W. Wartofsky (eds.) *Boston Studies in the Philosophy of Science V* (Dordrecht: Reidel, 216–241): reprinted as "The Logic of Quantum Mechanics", in H. Putnam [1974] *Philosophical Papers Volume 1*, pp. 174–97. (Cambridge: Cambridge University Press).
Quine, W. V. O. [1951] "Ontology and Ideology", *Philosophical Studies* 2, pp. 11–15.
Quine, W. V. O. [1978] "Other worldly" (Review of N. Goodman, *Ways of Worldmaking*), *New York Review of Books*, November 23 issue.
Ramsey, F. P. [1926] "Truth and Probability", in D. H. Mellor (ed. 1990), *Philosophical Papers*, (Cambridge: Cambridge University Press), pp. 52–109.
Reichenbach, H. [1956] *The Direction of Time* (Berkeley, CA: University of California Press).
Ruetsche, L. [2011] *Interpreting Quantum Theories* (Oxford: Oxford University Press).
Rutherford, E. [1911] "The Scattering of α and β Particles by Matter and the Structure of the Atom", *Philosophical Magazine* 21, pp. 669–88.
Saunders, S., Barrett, J., Kent, A., and Wallace, D. (eds.) [2010] *Many Worlds?* (Oxford: Oxford University Press).
Savage, L. J. [1972] *The Foundations of Statistics*, 2nd revised edition (New York: Dover).
Schaffer, J. [2003] "Is there a Fundamental Level?", *Nous* 37, pp. 498–517.
Schilpp, P. A. (ed.) [1949] *Albert Einstein: Philosopher-Scientist* (La Salle, IL: Open Court).
Schrödinger, E. [1935] "Discussion of Probability Relations Between Separated Systems", *Mathematical Proceedings of the Cambridge Philosophical Society* 31, pp. 555–63.
Schrödinger, E. [1936] "Probability Relations between Separated Systems", *Mathematical Proceedings of the Cambridge Philosophical Society* 32, pp. 446–52.
Sellars, W. [1953] "Inference and Meaning", *Mind* 62, pp. 313–38.
Sider, T. [2011] *Writing the Book of the World* (Oxford: Clarendon Press).
Sinha, U. et al. [2010]: "Ruling Out Multi-order Interference in Quantum Mechanics", *Science* 329, pp. 418–21.
Stern, W. and Gerlach, O. [1922] "Der experimentelle Nachweis der Richtungsquantelung im Magnetfeld". *Zeitschrift für Physik* 9, pp. 349–52.
t'Hooft, G. [1997] *In Search of the Ultimate Building Blocks* (Cambridge: Cambridge University Press).
Tittel, W., Brendel, J., Zbinden, H., and Gisin, N. [1998] "Violation of Bell Inequailties by Photons More Than 10 km Apart", *Physical Review Letters* 81, pp. 3563–6.

Tumulka, R. [2006] "A Relativistic Version of the Ghirardi–Rimini–Weber Model", *Journal of Statistical Physics* 125, pp. 821–40.
Tumulka, R. [2009] "The Point Processes of the GRW Theory of Wave Function Collapse", *Reviews in Mathematical Physics* 21, pp. 155–227.
Vaidman, L. [2014] "Many-Worlds Interpretation of Quantum Mechanics", *The Stanford Encyclopedia of Philosophy* (Spring 2016 Edition), Edward N. Zalta (ed.), URL = <http://plato.stanford.edu/archives/spr2016/entries/qm-manyworlds/>.
Valentini, A. and Westman, H. [2005] "Dynamical Origin of Quantum Probabilities", *Proceedings of the Royal Society* A 461, pp. 253–272.
van Fraassen, B. [1972] "A Formal Approach to the Philosophy of Science", in R. Colodny (ed.) *Paradigms and Paradoxes: Philosophical Challenges of the Quantum Domain* (Pittsburgh, PA: Pittsburgh University Press), pp. 303–66.
van Fraassen, B. [1973] "A Semantic Analysis of Quantum Logic", in C. Hooker (ed.) *Contemporary Research in the Foundations and Philosophy of Quantum Theory* (Dordrecht: Reidel), pp. 80–113.
van Fraassen, B. [1989] *Laws and Symmetry* (Oxford: Oxford University Press).
van Fraassen, B. [1991] *Quantum Mechanics: an Empiricist View* (Oxford: Clarendon Press).
van Fraassen, B. [2008] *Scientific Representation* (Oxford: Oxford University Press).
van Fraassen, B. [2014] "One or Two Gentle Remarks . . . ", *Philosophy of Science* 81, pp. 276–83.
van Inwagen, P. [1990] *Material Beings* (Ithaca, NY: Cornell University Press).
von Neumann, J. [1932, 1955] *Mathematische Grundlagen der Quantenmechanik* (Berlin: Julius Springer). First English edition *Mathematical Foundations of Quantum Mechaniics* (Princeton, NJ: Princeton University Press, 1955).
Wald, R. [1994] *Quantum Field Theory in Curved Spacetime and Black Hole Thermodynamics* (Chicago, IL: University of Chicago Press).
Wallace, D. [2012] *The Emergent Multiverse* (Oxford: Oxford University Press).
Weihs, G., Jennewein, T., Simon, C., Weinfurter, H., and Zeilinger, A. [1998] "Violation of Bell's Inequality under Strict Einstein Locality Conditions", *Physical Review Letters* 81, pp. 5039–43.
Weinberg, S. [2001] *Facing Up* (Cambridge, MA: Harvard University Press).
Wigner, E. [1967] *Symmetries and Reflections* (Bloomington, IN: Indiana University Press).
Woodward, J. [2003] *Making Things Happen* (Oxford: Oxford University Press).
Zeilinger, A., Weihs, G., Jennewein, T., and Aspelmeyer, M. [2005] "Happy Centenary, Photon", *Nature* 433, pp. 230–8.
Zurek, W. H. [1982] "Environment-Induced Superselection Rules", *Physical Review* D 26, pp. 1862–80.

Index

Note: Page references in bold faces indicate where the term is defined or at least most clearly explained.

accessibility 69–72, 88–9, 107–8, 171–2, 178, 182, 195–6, 207
action 114–15, 132, 169, 172, 202–3, 210, 249
  at a distance 8, 51, 53–5, 59, 65, 71–3, 165, 179
  instantaneous xii, 10, 12, 53–5, 88, 165
  spooky 2, 8, 53–6, 67, 101
agent 9–12, 70–4, 78, 85–9, 93–101, 110–15, 131–7, 146–8, 156, 168–79, 188, 207–18, 225, 238–40, 255
  situated 71, 89, 93, 134–5, 173–7, 181–5, 189, 194–9, 225, 250–1
  situation 76, 88, 93–4, 134–7, 178, 191, 208, 250
Albert, D. 97
Anderson, P. 238
Anscombe, G. E. M. 3
'apparatus' 85–6, 93
application 85, 122, 127, 130, 144–52, 161, 169–72, 214–15, 244
  of Born rule 59, 61, 70, 76–86, 89, 93, 97, 110, 131–5, 142, 147–8, 156–7, 160–4, 170, 176–9, 192, 205, 209, 215, 221–6, 237–9
  of quantum theory 1, 7, 10–11, 51, 74, 84–90, 94–7, 104–7, 117, 127–38, 142–3, 154–6, 161–3, 170–3, 179, 202, 206, 208–11, 225–8, 232–6, 244–57
argument (of complex number) 29–30, 49
Arndt, M. 157
Arntzenius, F. 55
Aspect, A. 45, 47, 51, 59, 64–5
Aspelmeyer, M. xi
assignment (of quantum state) 75–9, 86–9, 93–101, 110, 127–31, 135, 156–62, 170–9, 187, 191–7, 207–9, 224, 236–8, 247, 250–3, 267
assumable 127–8, 137, 161, 179, 201, 205, 207, 211, 225, 232–3

Bach, R. 27
backing condition 87, 159–60, 176–9, 207–9, 250
Baker, D. 231–2
Barrett, J. 111, 115
Barrett, J. A. 111
Bayes, T. 145
Bayesian 113, 145
  quantum 113, 145
  statistics 112, 131

basis
  categorical 178
  decoherence 11, 111
  orthonormal **264**
  pointer 86, 190–2, 214
  preferred 84–6, 111, 247
  supervenience 12, 237
  vector **42**, 260
beable 7–8, 11–12, 84, 127–8, 131, 137, 161, 175, 178–9, 182, 186, 200–11, 225–7, 232–4, 241, 250
  local 175, 179–**80**, 232
beam 15–18
beam splitter **18**
  polarizing 26, 77–8
Bedingham, D. 110
Beisbart, C. 175
belief, degree of 68–77, 88, 97, 131, 144–6, 168–9
Bell, J. S. ix–x, 4–5, 7, 36, 46, 53–7, 59–65, 69, 71–3, 84–6, 91–2, 94, 96, 101, 103, 106, 109, 165, 174–5, 178–81, 183, 186–7, 200–1, 227, 232, 255, 273
Bell argument 54, 56, 59–65, 72–3, 101, 165, 179–83
Bell inequality 59, 106, 200
Bell's correlations 55, 165
Bell's intuitive principle 56–9, 62–5, 72, **179**–82
Blatt, R. 52
Blundell, S. J. 223
Bohm, D. 5, 43, 45, 100, 103, 105–9, 111, 180
Bohmian mechanics 92, 103–9, 116–17, 241, 247
Bohr, N. 1, 8, 253–4
Born, M. 53, 143
Born
  probability 70–81, 88–9, 101, 105, 110, 113, 128–30, 138, 143, 147, 154–65, 171–2, 185, 190–3, 202–7, 213–16, 221, 225, 253
  rule 39, 59–61, 70–89, 93–8, 104–16, **126**, 130–8, 142, 147–8, 154–64, 170–80, 185–92, 204–17, 221–6, 237–9, 246–7, 254–69
    compound 267
    generalized 267
    simple **263**
    test of 147, 156, 162–3
  weight 112–15

branch (in Everett Interpretation) 6, 111–17, 247
Brandom, R. 11, 132, 201–2
Brink, D. xi
Brukner, C. 8
Brun, T. 195–6
Burge, T. 256

Carnap, R. 122–3, 148, 229
Cartwright, N. 154
cat 8, 10, 188, 246–7
canonical magnitude claim **80**, 83, 88, 128, 130–5, 162, 178, 196–7, 207–11
causal closure 239
causation xi, 10, 56, 66, 165, 172–83, 202, 239, 249–51, 257
cause, common (*see* common cause)
chance 3, 64, 71, 74, 82, 110, 121, 145–8, 165, 168–82, 204, 239–40, 249–50
   relativistic (*see* relativistic chance)
CHSH inequality 59–61, 63, 200
classical
   dynamical variables 130, 135, 205, 241
   electromagnetism 35, 106, 130, 226, 232
   ideas/lines of thought 37, 40, 50
   mechanics 66, 69, 114, 123, 128, 141–2, 155, 159, 214, 224, 239–43, 255
   physics 1, 3–4, 7, 9, 11, 17, 35, 86–7, 127–8, 130, 136, 139–45, 148, 153–4, 156, 163, 184, 196–9, 203–6, 212, 214, 227, 234, 242
   reality 117, 196–7
   system 81
   terms 12, 84, 93, 127, 179, 187, 201
   wave theory 44–5, 48–50, 225
"classical" 205, 211, 226, 233–**4**, 247–8, 251–2
Clauser, J. F. 61
collapse (*see* state collapse)
   effective 105
commutation 81, 193, 205, 232, **266**, 273
compatibility
   of dynamical variables 81, 85–6, **266**, 273
   of state assignments 66, 195
completeness assumption 37, 45–**6**, 54, 66–9, 90, 93, 100–1, 189, 254–5
complex number 29–30, 38, 40, 49, 82, 261
component
   of self-adjoint operator 265–6
   of statement 255
   of system 43, 78, 109, 230, 237–8, 248
   of vector 28–9, 33, 35ff.
composition 33, 230–1, 249–51
computer
   classical 11, 218, 220–2
   quantum 11, 50, 90, 220
configuration space (*see* space, configuration)
conditional factorizability 59–63, 72, 181

consciousness 1–3, 10, 12, 90, 93, 110, 185, 187, 189–90, 198, 247–9, 254
content xi, 10–11, **132**, 146, 184, 202
   conceptual 203–7
   empirical 84, 87, 106–7, 130, 133
   of denoting terms 202, 223–6
   of magnitude claim 210–22
   of quantum state assignment (QSA) 207–10
   of statement 6, 11–12, 137, 162, 184, 186–8, 192, 201–2
   of theory 6–7, 121–3, 125
   objective 10, 185, 193–6, 198–200
Copenhagen interpretation 37
correlations 49, 57–9, 64–5, 68, 75, 89, 91, 175, 177, 179, 181, 191–2, 222, 256
   Bell 55, 72, 165, 178–9
   EPR 47
   non-localized 10, 55, 67, 71–4, 138, 165, 174, 247
counterfactual 151–2, 160, 163, 172–9, 182, 239, 250
credence 86–8, 115, **131**, 133–6, 146–8, 156, 168–9, 173–6, 182, 190–2, 209, 217, 225, 232–**4**, 245
   conditional 173
Cucchetti, F. M. 276

de Broglie, L. 5, 103, 105–6
decision 114–16, 173–4
decoherence 10–11, 75, 79, 81–9, 96–9, 111, 133–5, 159, 161, 185, 190–8, 226–7, 233, 241, 247, 254, 275–6
   environmental 10, 158, 163, 187–8, 197–8, 208–10, 214, 238, 248, 254
de Finetti, B. 112, 144
dependence 153, 156, 186, 211
   causal 61, 65, 160, 173–6, 178, 181, 249
   counterfactual 160, 174, 177
   physical 74
   probabilistic 61–2, 74
Descartes, R. 203, 230, 246
descriptive function 69–70, 75, 93, 99–101, 156, 208, 213
d'Espagnat, B. xii, 1, 55, 57, 65, 184–7, 198
determinate 35, 41, 54, 96–7, 99, 161, 189–90, 217, 234, 238, 270–3
   outcome 86, 90–2, 97, 99
determinism 3, 107, 114, 175, 249
deterministic 29, 31, 92, 106–7, 111, 114, 145, 154–6, 172, 175, 177, 239–40, 249
Deutsch, D. 114
Dewey, J. 7
Dick, P. K. 186
dimension **42**, 50, 66, 76, 85, 104, 127, 152, 155, 167–8, 195, 205–6, 208, 220, 259–67
Diosi, L. 109–10

INDEX   287

Dirac, P. A. M. 15, 32, 36–41, 44, 46, 48, 51, 92, 96–8, 105, 110, 189, 204–5, 214, 224–5, 237, 243, 272
Dürr, D. 103–4, 106
dynamical variable **35–9**, 54, 67, 70, 73, 80–6, 91–3, 104–8, 126–35, 188–92, 197, 203–9, 241, 261–75

Einstein, A. ix, 22, 36, 40, 46, 51, 53–8, 62, 65–70, 80, 93, 100–1, 105–8, 110, 166, 185, 212, 223, 243–4, 247, 254–5, 272–4
Einstein's suggestion 54–5, 66–8, 70, 80, 101
empirical
  equivalence 108–10, 116
  inaccessibility 107–8
ensemble 67
entanglement 40–53, 64, 75–6, 86, 121, 195, 199, 212, 237–8, 240, 248
  swapping 59, 75, 177
entity 6, 7–10, 121, 127–30, 134–8, 152–4, 160, 176, 184–5, 205, 224, 230–5, 242, 246, 248, 253
epistemic
  access 112, 115, 169, 182, 207, 250
  function 69–70, 75, 93, 101, 104–5, 188
  situation (*see* situation, epistemic)
EPR (Einstein, Podolsky, Rosen)
  argument 45–6, 54, 67, 100–1, 180
Everett, H. III. 5, 92, 111–17, 247, 252–3, 256
Everettian quantum mechanics (EQM) 5, 92, 111–17, 247, 252
expectation value **60**, 163, 205–9, 226, 241, 262
explanation x–xi, 10, 12, 73, 123, 135, 138, 148–64, 176–9, 257
  and representation 8, 154
  of hydrogen spectrum 138, 164
  of Kepler's laws 148–51
  of Rutherford scattering 140–3, 148, 158–9
  of stability of matter 138, 163–4
  causal 55, 65, 152, 158–9, 202
  non-causal 152–3, 160, 202
  probabilistic 142–3, 148, 160, 162, 209
  statistical 141, 162

faithful measurement 80, 270
factorizability 180–1
  conditional 59–63, 72, 181
  quantum 72–3
Feynman, R. P. xiii, 15, 26–7, 216, 231
field 11, 53, 106–7, 124, 128, 137, 160, 174, 211, 226, 228, 231–2, 247–8, 251–2
  classical 232, 234
  electromagnetic 4, 17, 35, 50, 87, 94, 106, 128–30, 175, 204–5, 216, 226, 232, 257
  Higgs 2, 11, 130–1, 205, 226, 231
  Klein-Gordon 224–5

  quantum x, 11, 51, 106–8, 110, 122, 129–31, 137, 205, 211, 216, 223–6, 229, 232–4, 244–53, 257
  tensor 234
  vector 234
Fine, A. 200
Frabboni, S. 27
Fraser, D. 231
freedom 3, 60–2, 114, 249
Frege, G. 122
Fuchs, C. xi, 8, 145
fullerene 157–61, 215–17, 241
fundamental ix–x, 11, 59, 62, 85, 102, 110–12, 201, 226–7, 235–7, 247
  laws 11, 123, 230, 238–40, 244
  physics 11, 103–4, 121, 175, 202, 206, 228–30, 235–8, 242, 244, 250–2
  theory 8, 11–12, 36, 79, 84, 91, 103, 109, 111, 206, 223, 227, 239, 242–5, 250, 254
fundamental$_a$ 84–5
fundamentality 228–45
future, open 177

Galileo 2, 203
Geiger, H. 34, 44, 139–43, 148
Gerlach, O. 94
Ghirardi, G. C. xiii, 109
GHZ state **50**–2, 270–3
Gibbs, J. W. 67
Giere, R. 123, 127, 240, 243
Gisin, N. 48
Giustina, M. 59
Gleason, A. M. 155
Glick, D. xii
Goldstein, S. 103, 106
Grangier, P. 45, 51
Greaves, H. 112–13
Greenberger, D. 50

Haldane, J. B. S. 2
half-wave plate 24–6, 32, 35
Hartle, J. 94
Hartmann, S. 175
Hawking, S. ix, 94
Heisenberg, W. ix, 1–2, 27, 36, 141, 184, 204
Hempel, C. G. 149, 162–3, 229
Hensen, B. 59, 75
Herbst, T. 47
Hertz, H. R. 122
Higgs
  field (*see* field, Higgs)
  particle 106, 229, 232
Hilbert, D. 123–4, 261
Hilbert space (*see* space, Hilbert)
holism 237–8, 248
Holt, R. A. 61

Hooke, R. 139
Horgan, T. xii, 235
Horne, M. A. 50, 61

idealism 3, 103, 184
ideology 234–6, 240
ignorance principle 174
incompleteness 45–6
indeterministic 71, 92, 146–8, 174, 249
inertial frame 166–70, 243
inference 10, 162, 192, 203–8, 211, 218–19, 222, 237–8, 257
   material 132–3, 208–10, 217–18
   probability 209
   significance 209–10
inferential web 132, 162, 217–18
inferentialism 202–3
information x, 64, 73, 78–9, 83, 85, 91, 94, 109, 112, 144, 146–8, 156, 169, 176–7, 181, 192–9, 206–8, 242
   accessible 70–2, 74–6, 88, 136, 171, 189, 194–5
informational bridge 206–7
instrumentalism 8, 252–3, 257
interference 15–17, 21, 27, 51, 54, 77, 82–4, 96–7, 109, 158, 161–2, 215–17, 241, 275–6
   fringe 27, 157
   modeling 28–36, 122
   phenomena 18–21, 24–7, 48, 185
   single-particle 21, 48, 50
   two-particle 49–50
   two-slit 26–7, 108, 161, 216
interferometer
   balanced 19, 25, 29, 30–1, 48
   Franson 48–50
   Mach-Zehnder 18–19, 25–35, 41, 47–8, 51, 77, 96, 161, 246
   multiple-slit 157–8
   neutron 17–18, 20–1, 27, 36–7, 133
interpretation 4, 38, 44, 90, 92, 106, 109–11, 116–17, 123–9, 133, 138, 143–6, 199, 231–2, 235, 247, 263
   Copenhagen (see Copenhagen interpretation)
Interpretation 5–6, 9, 102–3, 108, 116
Interpreter 4–6, 9
intervention 165, 172–7
   counterfactual 182, 250
Ismael, J. T. xii, 169, 174
isomorphism 150, 252–3

James, W. 7, 132, 246
Jammer, M. 1
Jaynes, E. T. 112
Jones, J. P. xi
Jordan, P. 1, 3, 184
Jordan, T. F. 262

Juffmann, T. 157–8, 160–1, 215

Kant, I. 199
Kent, A. 111, 115
Kepler, J. 124–5, 148–53, 240–2
Kuhn, T. S. 9

Ladyman, J. 242
Lancaster, T. 223
Laplace, P. S. 107, 114, 249
law x–xi, 1, 3, 11, 37, 92–4, 123–6, 148–56, 162–3, 228–45, 249
Leggett, A. J. xi, 90, 220, 222
Lewis, D. K. 146–7, 165, 168–9, 172, 236, 250
light cone **167**–**82**, 195
limbo 84, 227
linearity 5, 29, 31, 37–9, 90, 92, 94, 97, 105, 109–11, 126, 189
linearly independent 42
local action, principle of **53**–**8**, 62, 64–5, 68, 74, 105, 110, 247, 272–4
local determinism 175
local causality 55, 59–65, 71–3, 165, 175, 179–82
locality 8, 106, 179–83 (see also non-locality)
Lockwood, M. 3
Lorentz invariance 107

Ma, X.-S. 177
macro-realism 220
magnitude 7, 9–11, 28, 35–41, 60, 73, 86, 98, 127–39, 147–8, 152–60, 166, 175–6, 182–91, 196–206, 211–13, 217–21, 225–6, 232–57, 276
   claim 80, 83, 87–9, 94–9, 126–37, 143, 147, 154–7, 161–4, 173–9, 185–7, 194–211, 217–18, 223–7, 232–41, 245, 250–7
many-outcomes theories 110–17
Margenau, H. 96
Marsden, E. 139–40, 143, 148
material inference (see inference, material)
matter 4, 7, 12, 15, 17–18, 24–8, 33–40, 90, 107, 138, 203, 212–13, 230, 238, 244
   stability of (see explanation of stability of matter)
Maudlin, T. 90–1, 175, 177, 239
Maxwell, J. C. 1, 17, 122, 175, 206, 232
meaning 124–5, 177, 188, 202–27, 236, 253
measure
   Hilbert space **128**, 154–5, 207–8
   probability (see probability measure)
measurement x, 6, 8, 12, 34–8, 43–6, 50–2, 54, 67–70, 76–87, 103–4, 109, 133, 158, 186–8, 213, 242, 253
   faithful (see faithful measurement)
   outcome 6, 10, 36, 39, 43–6, 49, 54, 60, 67, 73–80, 101, 107–12, 126, 131, 134, 185

problem 4, 90–100, 104, 109–10, 128, 187, 257
Mermin, N. D. xiii, 1–3, 8, 56, 58–9, 68, 113, 145, 184, 211–12, 214, 273
Merricks, T. 231
metaphysics ix–x, 8–9, 12, 117, 132, 148, 156, 177, 231, 235–7, 248–52, 255
mixed state (*see* state, mixed)
model 6–11, 15, 121–37, 243
  local 59
  common cause 58–9
  mathematical 6–8, 11, 15, 52, 154, 210, 225, 232, 242–3, 252
  Newtonian 123–5, 140–1, 149–53, 241–2, 255
  of decoherence 83–9, 97–9, 161, 188, 194–8, 209–10, 214–16, 227, 275–6
  of interference 28–38
  probabilistic 53, 59, 64, 139, 143, 154
  quantum (*see* quantum model)
  Standard (*see* Standard Model)
  state-space 123–7, 149–51
modulus
  of complex number 29, 49, 225
moon 2, 12, 55, 184, 210–17, 223, 248
multiparticle interference 48
multiverse 12, 111–14, 247
Myrvold, W. 112–13

Nagel, E. 229
Nagel, T. 188, 199
Newton, I. 1, 3, 7, 9, 11, 15–18, 21, 23, 27, 35, 55, 66, 114, 123–4, 126, 140, 148–53, 166, 169, 177, 181, 203–4, 206, 212, 239–47, 249, 252
Newton's rings 16, 27
noninvasive measurability 219–21
non-linear
  equation 5, 92, 109
  theories 92, 109–10
non-locality 40, 47, 110, 257
"non-locality" 53–74
non-localized correlations (*see* correlations, non-localized)
non-representational 7, 9, 202, 242
non-separability
  of quantum state 42–4, 50
  spatial 237–8, 248
norm
  of measure 144
  of vector **28**, 261
  social 132, 135, 156
normalized **28**, 261
null separated **167–8**

objectivity 10
  and observation 184–201
  as independent verifiability 196–8
  macro- 220

  of Born probabilities 88, 135, 147, 156, 207, 209, 250
  of causal relations 250
  of content 104, 185, 193–6, 198, 201
  of description 185–8, 191, 195–6
  of existence 223–7
  of magnitude claim 194–201
  of measurement outcome
  of properties
  of quantum state 88, 135, 156, 178, 194, 207
  scientific 112
  transcendental 199–201
observable 10, 37, 85–6, 175, 204–7, 227, 272
  preferred 197–8
observation x, 1, 12, 81, 84, 93, 103–4, 116, 136, 158, 219, 227, 248, 253–5
  and objectivity (*see* objectivity and observation)
Ollivier, H. 186, 196–7
ontology xi, 12, 51, 106, 111, 223–4, 231–6, 242, 246–51, 257
  contextual 233–4
operator 126–9, 133, 137, 142, 154–5, 189, 193, 205–7, 211, 225–6, 232–3, 241, 245, **259–68**
  density 76, 129, 195, 204, 213–14, 266–7
  Hamiltonian 126–8
  linear 126, 232, **262**
  projection (*see* projection operator)
  self-adjoint 37, 81, 128, 154, 197, 204–5, 209, 227, 233, **262**
  unitary 199, 268
Oppenheim, P. 231
orthogonal (of vectors) 28, 31, 33, 82, 91, 128, 189, 209, **261**, 264–5
outcome independence 63

Pais, A. 212, 223
Pan, J. W. 51
Papineau, D. 239
parameter independence 62
particle x, 2, 4–5, 11, 15–21, 25–7, 33, 35–7, 40–54, 57, 66–9, 81, 98, 101, 104–8, 111, 117, 122–3, 127–8, 137, 141, 152, 158–9, 205, 211–15, 223–6, 231–3, 241, 246–8, 275
  elementary 2, 12, 223, 226, 228, 230–1, 236, 238, 247–8, 251–2
  physics 231–2, 251
Paz, J. P. 276
Pearle, P. 109
Penrose, R. 109–10
Peres, A. 177
Peskin, M. E. 122
Petersen, A. 1, 253
phase 49, 192–5
phase space (*see* space, phase)

phenomenon x, 1, 7, 10–16, 21–3, 53–4, 80, 85, 102, 106–9, 117, 121–2, 131, 134–43, 147, 151–65, 173, 184, 198, 202, 209, 223, 228–9, 238, 244–5, 248–56, 275
  interference (*see* interference phenomena)
  probabilistic 138–9, 143, 147–8, 154–64, 209
philosophy ix, xi–xii, 8–9, 12, 102, 111, 122, 202–3, 228–30, 243, 246, 252, 255
  natural 243, 246
  of language 9, 117
  of physics 3, 102
  of science 8–9, 122, 252
  pragmatist 7, 202–3, 246
  quantum 53, 103–4
photon 2, 21–5, 34, 40–52, 59–60, 64, 67–8, 74, 80, 98, 106, 136–9, 148, 155, 161–2, 176–8, 194–5, 199–201, 213, 223, 226, 230, 233–4, 248, 251, 270–4
physicalism 11–12, 235, 239, 242
Pitowsky, I. 3
Podolsky, B. 45, 100
pointer
  basis **86**, 190–2, 214
  position **86**, 91, 98, 105, 126, 193
polarization 21–7, 34–5, 41–5, 54, 60–1, 67–82, 96, 136, 148, 155, 170–9, 194–5, 199–201, 208, 226, 233–4, 270–4
  axis 21–2, 32–3, 64
  circular 50–1, 270–3
  linear 32, 35, 47, 49–51, 98, 170, 179, 195, 270–3
  representation of 31–2, 50, 67
Popescu, S. 63
Popescu-Rohrlich game 63–4
Popper, K. R. 145–6, 229
possible world 6, 132
potential 241, 245
  electric 142, 159
  energy 163, 203–5
  gravitational 7
pragmatic 211–15, 254
pragmatism 257
pragmatist 7–10, 132–4, 138, 146–8, 173, 186, 201, 236, 250, 254, 257
  inferentialism 132, **202**–3
  philosophy (*see* philosophy, pragmatist)
prescription 68–72, 75, 78, 87, 96, 176, 262
prescriptive
  function 69–72, 75, 93–4, 99–101, 156, 208
  view of quantum states 69–72
Price, H. xi, 115, 132, 134, 173
Principal Principle 146–7, 165, 168
principles (of physics) 240–6
probabilistic theory 59, 62, 112–14, 147
probability xi, 3, 6, 10, 34–5, 39, 66, 79–89, 111, 143–8
  as chance 146–8
  as credence 131, 146, 209
  as degree of belief 131, 144–5, 209
  as frequency 141, 143–4, 209
  as propensity 145, 209
  Born (*see* Born probability)
  conditional 60–5, 73, 77
  distribution 66, 70, 72, 80–1, 155–60, 205–6, 213–14
  inference (*see* inference, probability)
  measure 69–70, 134, 187
  objective 71, 112, 147, 156, 175–6, 182, 209
  physical 66, 71
  pragmatist view 249–50, 256–7
  prior 112
  quantum 71
  relational 72, 88
  subjective 112–13, 145
  unconditional 62, 71, 74
product
  inner 28, 40, **260**–1
  scalar 28, 260
  tensor 32, 42–3, 50, 267, 272
projection
  operator 126, 154
propensity 74, 145–6, 174, 182
property 234–8
  correlational 41
  dispositional 178
  holism (*see* holism)
  objective 10, 186, 196–9, 201
Putnam, H. xi, 3, 231, 256

quantum
  computer (*see* computer, quantum)
  Darwinism 10, 185–6, 197–8
  factorizability 72–3
  field 211, 223–6, 232–3, 251, 257
  field theory 10–11, 51, 106–10, 122, 130–1, 205, 211, 223–34, 244–53
  mechanics x, 1, 15, 36–40, 55, 76, 84, 90–2, 106, 110, 116–17, 122, 126–30, 142, 159, 164–5, 174, 180–1, 199, 232–3, 239–44, 247
  model 9–10, 51–2, 85–9, 99, 126–38, 142, 147, 154–6, 160–3, 198, 202, 211–14, 222, 226–7, 233, 242, 250–7, 261, 272
  philosophy (*see* philosophy, quantum)
  physics 3, 103–4, 220
  state 4–12, 37–8, 45–6, 51, 54, 69–71, 135, 143, 160, 178, 186–9, 194–6, 204, 209, 225, 236, 253–5
    as information bridge 206–7
    assignment (QSA) 10, 73, 75–9, 86–9, 93–4, 99–101, 131, 135, 156, 162, 176–9, 187, 194, 207, 236–8, 247, 250

mixed (*see* state, mixed)
  two views of 65–8
  system 4–5, 12, 33, 38, 52, 85
  tomography 108
  quark 2, 130–1, 138, 205, 228–30, 233, 244–5, 248, 251
  qubit 208, 220–2
Quine, W. V. O. 234–6

Ramsey, F. P. 145
realism
  knee-jerk 236
  macro- 220
  naïve 185, 270–4
  scientific 252–6
reality criterion 46, 100–1
reduction 11, 228, 235–6, 241–2
reductionism 235–6
reference 132, 144, 211, 217, 223, 255–7
Reichenbach, H. 3, 55, 57, 65, 229
relational
  probability 71–2, 88
  state assignment 10, 88, 93, 194
relativistic
  chance 168–9
  invariance 107, 166–8, 170, 244
  space-time structure 165–9, 177, 243
relativity 1–3, 22, 56, 65, 88, 107–10, 122, 167–70, 181, 204, 228–9, 243–4, 247
representation xi, 2–10, 121–2, 127, 135–7, 151–6, 186, 201–3, 255–6
representational function 128–30, 133–4, 232
representationalist 132, 201
representative 40, 75–6, 91, 212–13, 246
revelation 9, 119, 121, 203
revolution ix, 9, 12, 15, 90, 103, 117, 121–2, 141, 202–6, 228–9, 235–6, 239, 243, 253–4
Rimini, A. 109
Roger, G. 45, 51
Rohrlich, D. 63
Rosen, N. 45, 100
Ross, D. 242
Ruetsche, L. 130, 232
Rutherford, E. 139–43, 148, 152–3, 155, 158–61
Rutherford scattering 142–3, 148, 152, 155, 158–61

Saunders, S. 111, 115
Savage, L. J. 114, 145
Scarani, V. xiii
Schack, R. 8, 145
Schaffer, J. 231
Schilpp, P. A. 3, 46, 53–4, 66–7, 100, 254
Schroder, D. V. 122
Schrödinger, E. ix, 40, 47, 50, 64, 75, 91, 138, 141, 160, 188, 212–13, 246

Schrödinger equation 4–5, 11, 39, 90–7, 104–5, 109, **126**–8, 142, 157–64, 239–41
scientific theory structure 122
  semantic approach 123–5
  syntactic approach 122–5
scientific theorizing 9, 202, 243, 252
Sellars, W. 132, 208, 218
semantic approach (*see* scientific theory structure)
separability
  of quantum state 43, 46
  spatial 237
shifty split 85, 187
Shimony, A. xi, 61
Sider, T. 236
significance
  cognitive 54, 211, 217
  empirical 11–12, 83–9, 96–9, 116, 130–7, 142–7, 155–63, 187–91, 198, 201–2, 207–10, 214–16, 226–7, 232–3, 237–8, 241–8, 251, 254–5
Simon, C. 98
Sinha, U. 162
situation
  agent (*see* agent situation)
  epistemic 88, 177, 207
  physical 71–2, 85–8, 93–4, 110, 134–7, 146, 170, 176–9, 191–4, 207, 239–40, 250–1
space
  configuration 66, 104
  Fock 223–4
  Hilbert 40, 42–3, 126–30, 133, 136, 147, 154–5, 205, 208, 232, **261**
  phase 7, 66–9, 197
  vector 28, 32, 35, 38, 40, 42–3, 50, 52, 76, 81, 85, 213, 220, 223, **259**–60
space-like separated 110, **167**–70, 174–82, 194, 240
space-time 88, 107, 111, 121, 129, 167–82, 232–4
  interval 167
spin 24–5, 42–7, 52–4, 83–107, 126–39, 205, 212–15, 220, 224, 232–8, 248, 256–7, 273–5
stability 7, 83, 90, 138, 142, 163
Standard Model 106, 223–33, 244, 251
state
  assignment (*see* quantum state assignment)
  coherent 224–6
  collapse 8, 92–104, 109–11, 156, 160, 185–91, 199, 247
  Fock 225
  mixed 76–83, 129, 164, 189, 195, 205, **266**–7
  polarization 24–7, 32–5, 41–50, 61, 67–8, 73–82, 98, 195, 200, 208, 271–3
  preparation 77–9, 87–8, 99, 113, 178, 208
  primary 69

state (*cont.*)
  pure **76**–8, 83, 86, 91, 157, 195, 208, 267–9
  secondary 68–9
  spin 24–5, 42–6, 52, 91, 99–100, 214, 237
  translational 32–3, 42–3, 48, 78, 96, 205, 214
  vector 75, 105, 213–15, 221, 268
statistical mechanics 66–70, 206
Stern, W. 94
Stern-Gerlach apparatus 95–100, 129, 133, 208
stochastic 106–8, 154–6, 177, 239–40
subjectivity 10, 88, 103, 114, 145–6, 158, 178–9, 186–8, 198, 250
subspace 76, 126–33, 147, 154, 195–6, 207, **260**–3
  measure (*see* measure, Hilbert space)
superposition 29, 37ff.
supervenience 12, 178, 235, 237
Suppes, P. 123
symmetry principles 244
syntactic approach (*see* scientific theory structure)
system
  abstract 208
  physical 11, 36–46, 52, 54, 69, 73, 80–8, 92–9, 109ff.
  schematic 225

target 10, 85–6, **127**, 159, 161, 211, 255
t'Hooft, G. 230
time-like separated **167**–8, 176
Tittel, W. 48
truth 132
  as correspondence 132, 209, 255
  conditions 6, 132, 217, 223
  deflationary view 132, 256
  value 97–9, 129–31, 133–6, **191**–3, 198
Tumulka, R. 110

uncertainty principle 27, 81

unification 141, 244
unitary evolution 31, 38, 91, 189

vagueness 84–5, 96
Vaidman, L. 111
Valentini, A. 108
van Fraassen, B. 4, 122–3, 126–7, 129, 133, 240
vector 28, 259–60
  component of (*see* component of vector)
  space (*see* space, vector)
verifiability 199
  independent 196–8
  intersubjective 186
von Neumann, J. 92, 94, 97–8, 105, 108–9, 126, 156, 160, 188–9, 205, 232

Wald, R. 231
Wallace, D. 111–12, 114–17, 252
wave
  function 40, 90–4, 97, 104–8, 111, 117, 129, 133–5, 142, 157–64, 180, 195, 204–6, 213–15, 246–7, 253
  effective **105**, 107
  theory 15–17, 21, 44–51
  transverse 21
Weber, T. 109
Weihs, G. 59
Weinberg, S. 236, 244
Westman, H. 108
Wigner, E. 1, 3, 10, 185–90, 193–4, 198–200
Wigner's friend 10, 185–200
Wineland, D. 52
Wittgenstein, L. 197
Woodward, J. 151
world line **168**, 177, 181

Zanghi, N. 103, 106
Zeilinger, A. xi, 8, 50–1, 161
Zurek, W. H. 83, 275–6